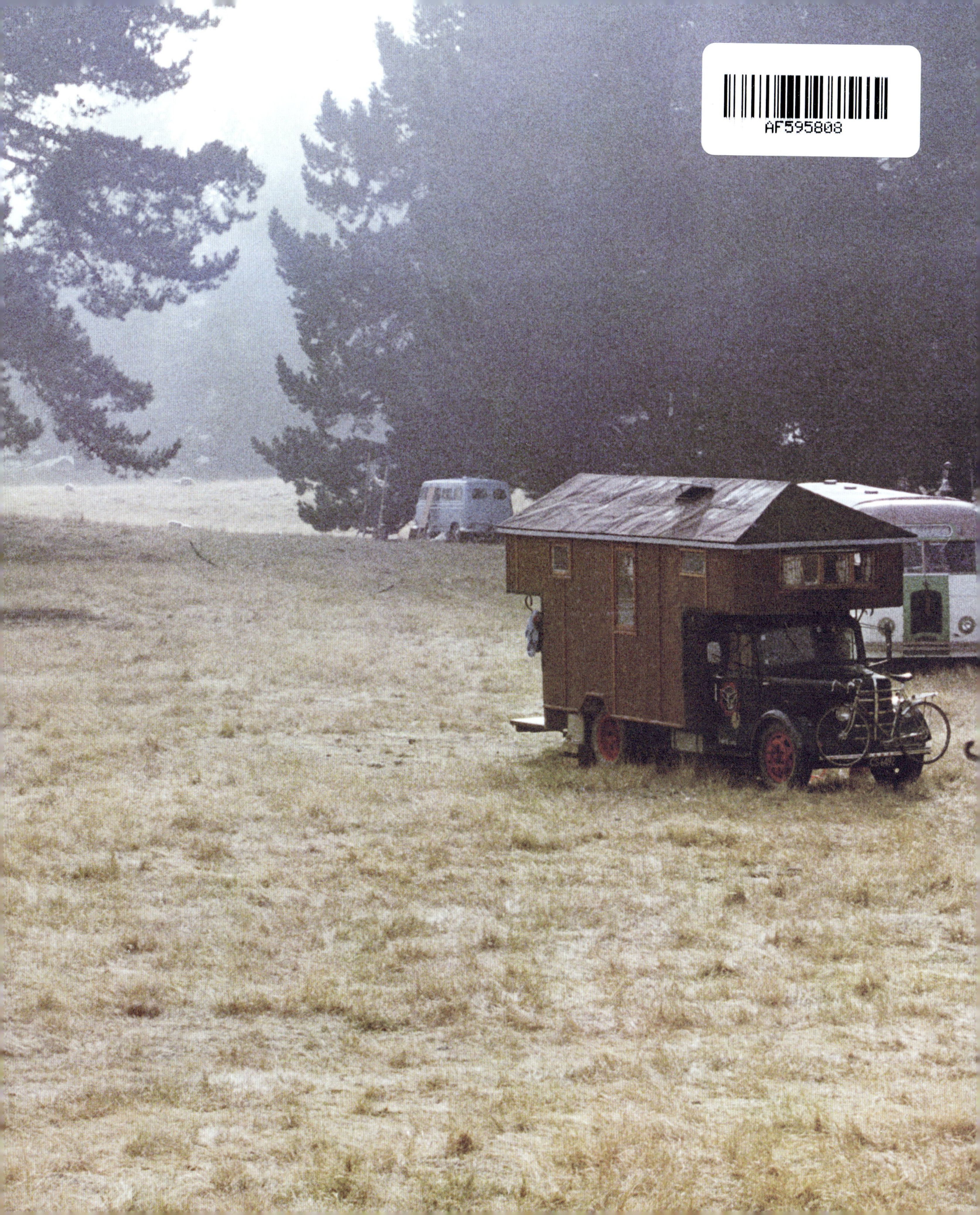

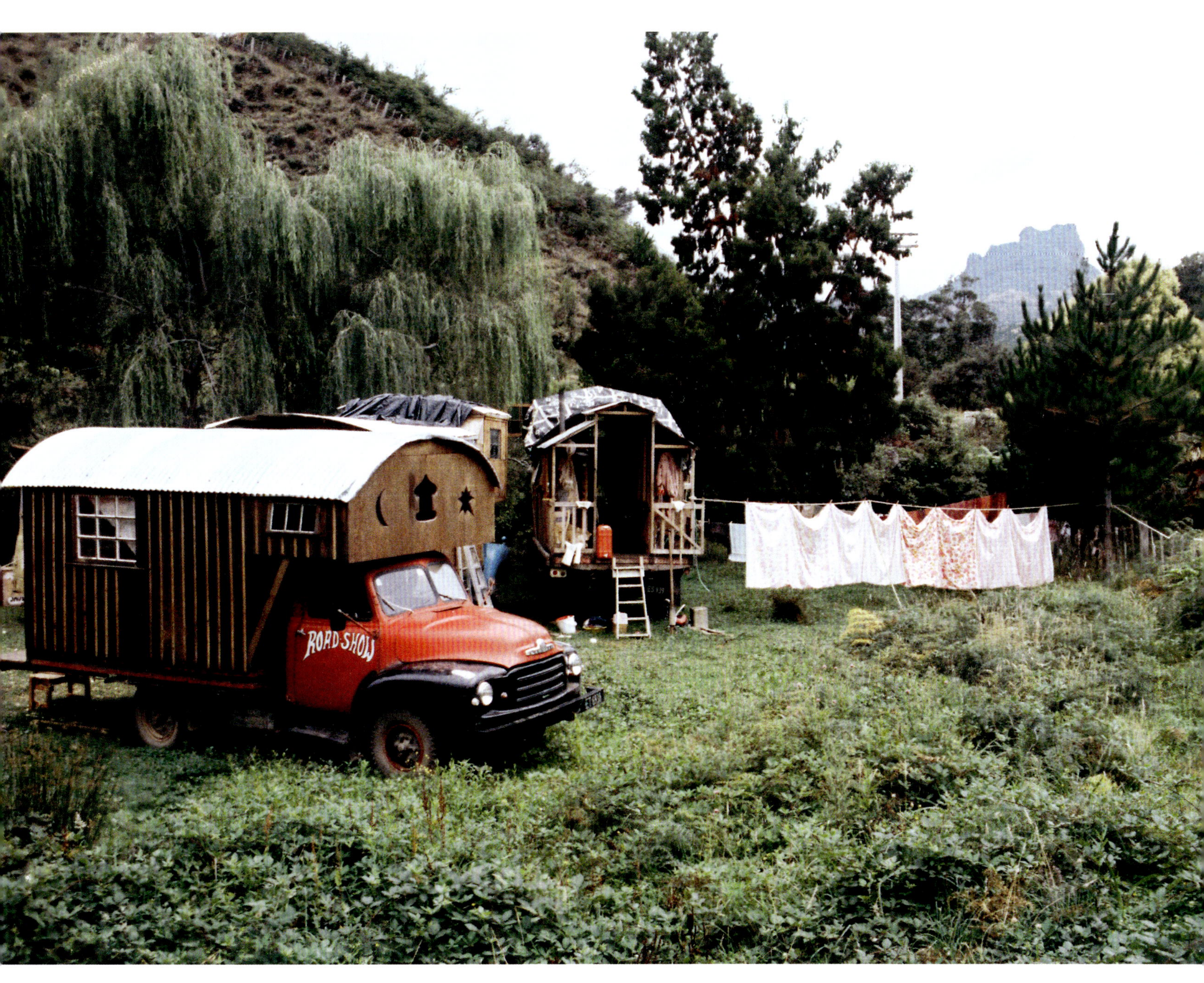

Road People of Aotearoa

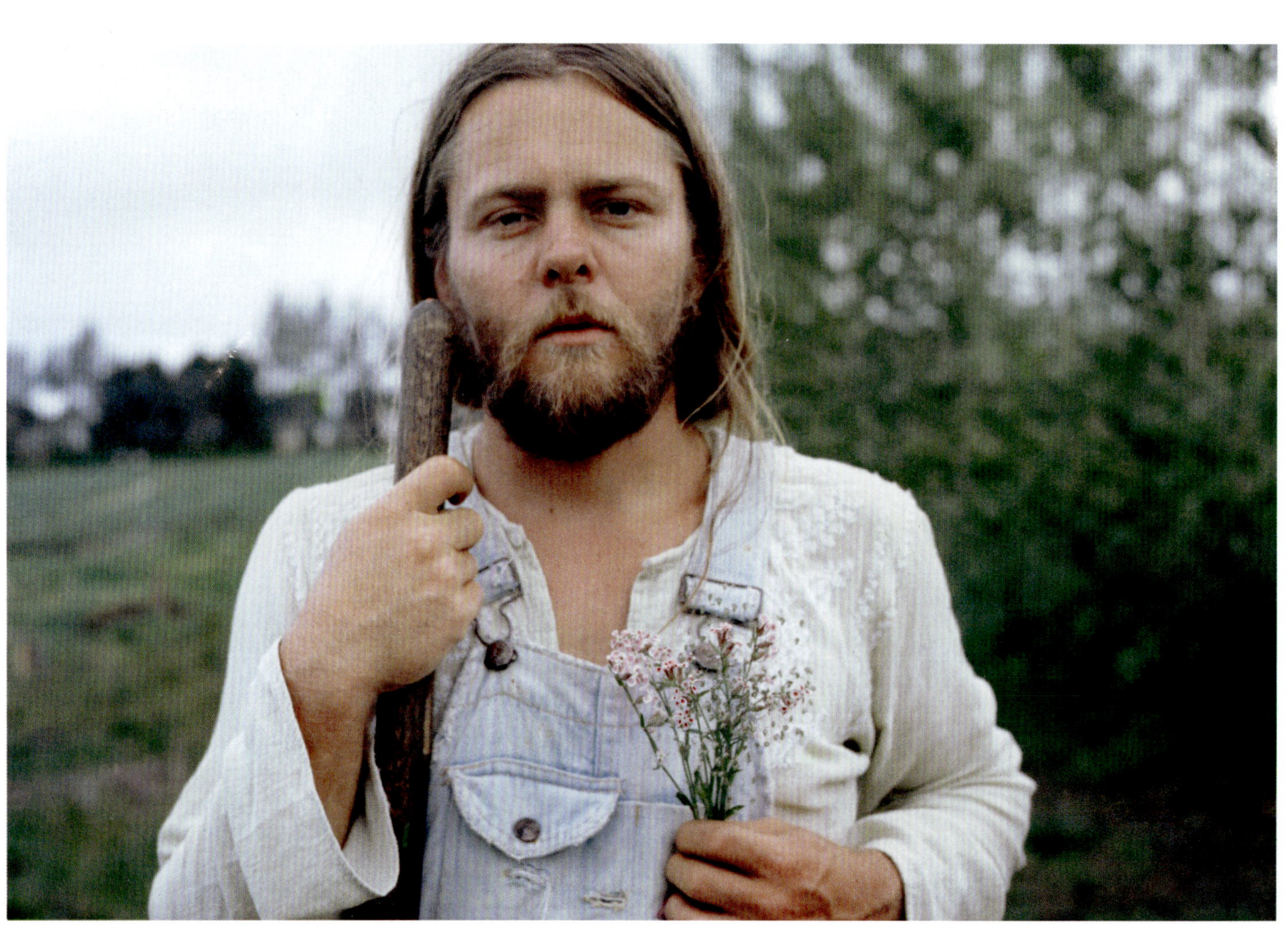

Road People of Aotearoa

House-truck journeys 1978-1984

Paul C Gilbert

with a foreword by Michael Colonna
and essays by Haru Sameshima & John B Turner

Rim Books Auckland 2021

House Truckers (a gentle revolution)

The first time I set eyes on a house truck was at the Nambassa Music, Crafts and Alternatives Festival in Golden Valley, north of Waihi, January 1978. What an amazing idea to put a bach on the back of a truck. I had read about these hippie mobile homes, built using plywood car packing crates, corrugated iron, second-hand windows and doors, wood ranges and Kiwi ingenuity, bringing together all of the ideals of our generation – self-sufficiency, recycling, community lifestyle and back-to-the-land be-here-nowness. But to see these in the flesh, so to speak, was amazing. These creations were gorgeous. And the people who owned these living works of art were eccentric and larger than life.

My kind of people!

Paul was there from the start, and he has recorded the history of this incredible phenomenon. This was a quiet revolution that came and went in its own time. In 1981, he travelled with the Acorn and Walnut Travelling Clown Troupe in his own converted bus, Iris, throughout the Waikato Region. We performed daily and usually stayed in camping grounds, next to community halls, or parked outside friends' houses. During the day we unpacked the show and put on two shows for the local community. Sometimes we visited small schools in remote villages with only four houses and a war monument, and sometimes a large show was performed at a museum or a marae. Such is the life of the itinerant.

For me the house truck and its owner are a fusion, a cooperative of form and function. There are two types of truckers. Some move very little, perhaps twice a year. I envy them their stillness and their ability to support themselves in faraway places. Others need to travel more often and make large journeys. They take in a lot more and eventually develop a wider comprehension of this remarkable country of ours, a sense of self and a sense of place.

Living in a house truck has not proved to be an efficient lifestyle in the new millennium, with high fuel prices and busy schedules. It is too cramped for children and pets. Many of the original houses leaked and have rotted away. The vintage trucks are collectors' items now, or are rusting somewhere in the backblocks. Most of the original members of The Road Show are still performing and creating, but some have moved into campervans, which are lighter, faster and less conspicuous. Others have bought land, built houses and put down roots. And still others now live on yachts or houseboats.

The dream lives on.

Michael Colonna (aka Walnut the Clown), 2015

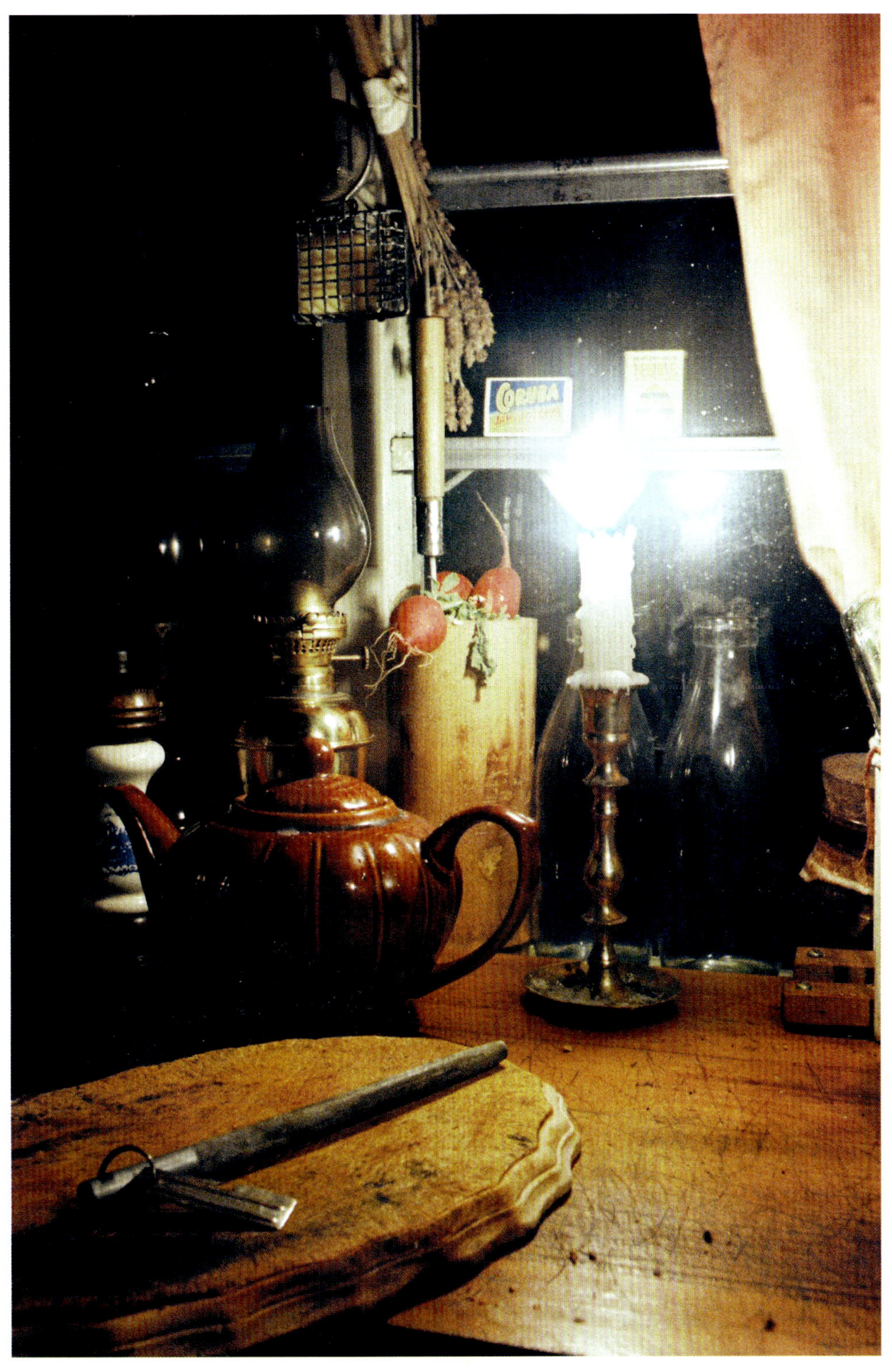

Interior of 'Iris', Paul Gilbert's house bus, n.d.

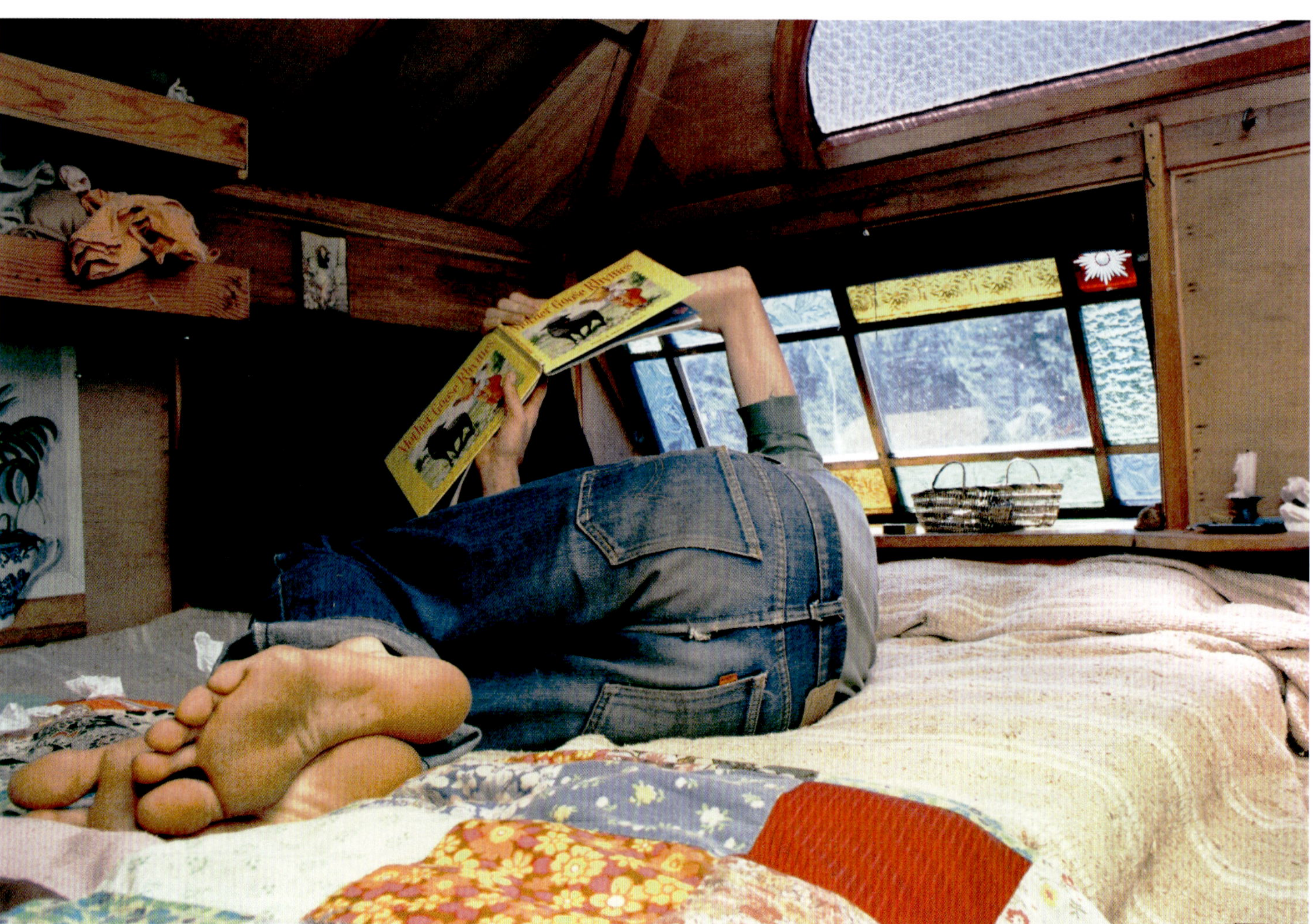

Ewan reading to Aya, House Truck & Mobile Shelter Convention, Aratiatia, near Taupo, 1984

Geographically, New Zealand is a distant and remote island nation, first inhabited by Polynesian voyagers as their southernmost Pacific outpost, and later as a 'discovery' by European explorers, whalers and colonisers. 'The Land of the Long White Cloud', Aotearoa is the location and identity of this photographic documentation, and its road people are the subject of this book.

As a port for American and French whalers and sealers and, eventually, as a new colony of Britain, the migration of northern hemisphere ideas and behaviour was mirrored enthusiastically in New Zealand, and interpreted with the tools and materials at hand – generally in ways not so very different from those of the northern hemisphere. This has not changed in anything but the speed with which those ideas have been transported and made available, although, in the period that this book covers, there was still a characteristic delay, perhaps as a result of New Zealand's inherent conservative outlook and its remoteness, and the Kiwi adaptive trait of making do with the materials to hand, expressed as a do-it-yourself, number 8 wire fix.

There has always been a sector of New Zealand society that has embraced mobility to access their lifestyle. In the early 1970s Kiwis were, in many ways, still mirroring the fashion and social trends that filtered down across the vast Pacific, albeit with that conservative time delay. The flower power of the hippie movement, along with anti-war protests, communes and music festivals, was about to migrate into Kiwi society. The social networkings of this period were reasonably scarce, with smallish subgroups forming connections and embracing the imported philosophies, informed by the generation of increasingly well travelled Kiwi youth and, of course, magazines and television. The Woodstock Festival was seen as a defining peak of this internationally, and various replications of this began to appear locally. The Nambassa Festivals were the quintessentially Kiwi response.

Along with the music, communes, dress codes, and the anti-war sentiment and the networking of like minds, there was the mobility of its tribal members. This could be seen in the colourful, long-haired hitchhikers, and the various vehicles that identified as being of that tribe. And of these, none was more confronting, unusual or more symbolic of this tribal identity than the New Zealand iteration of the 'rolling home' or house truck.

As the local version of what Ken Kesey's bus represented, the *Whole Earth Catalogs* touched on, and Jane Lidz's book *Rolling Homes* celebrated, my chance encounter with one of New Zealand's first house trucks prompted me to engage in a documentation of the road people of Aotearoa. This material has so far been aired in public only in the truncated format of an exhibition at Real Pictures Gallery in Auckland, in 1979. A book was always planned, but the gestation period has been prolonged and my preference has been for capturing photographic material for my other ongoing documentations.

Meanwhile, the subtribe of road people has continued, and to a degree, evolved. The commonalities are easily recognised: they are still a small tribe, and a passionate and colourful alternative addition to our social and physical landscape. This record is of the period of the late 1970s and early 1980s. The children of that era are now parents themselves and some have approached me for a record of the unusual childhood they experienced. This book is, in part, a response to that — and a record of a chapter in Aotearoa's subculture.

The photographs are all taken with the acceptance of my documentation as a family member of this subtribe. As such, it is a mixture of objective and subjective viewpoints that would not have been possible without the trust and generosity of those featured. I am deeply indebted to all who shared the music, the food, and the love along this journey.

Ko tahi tātau — We are one together

Paul C Gilbert, 2015

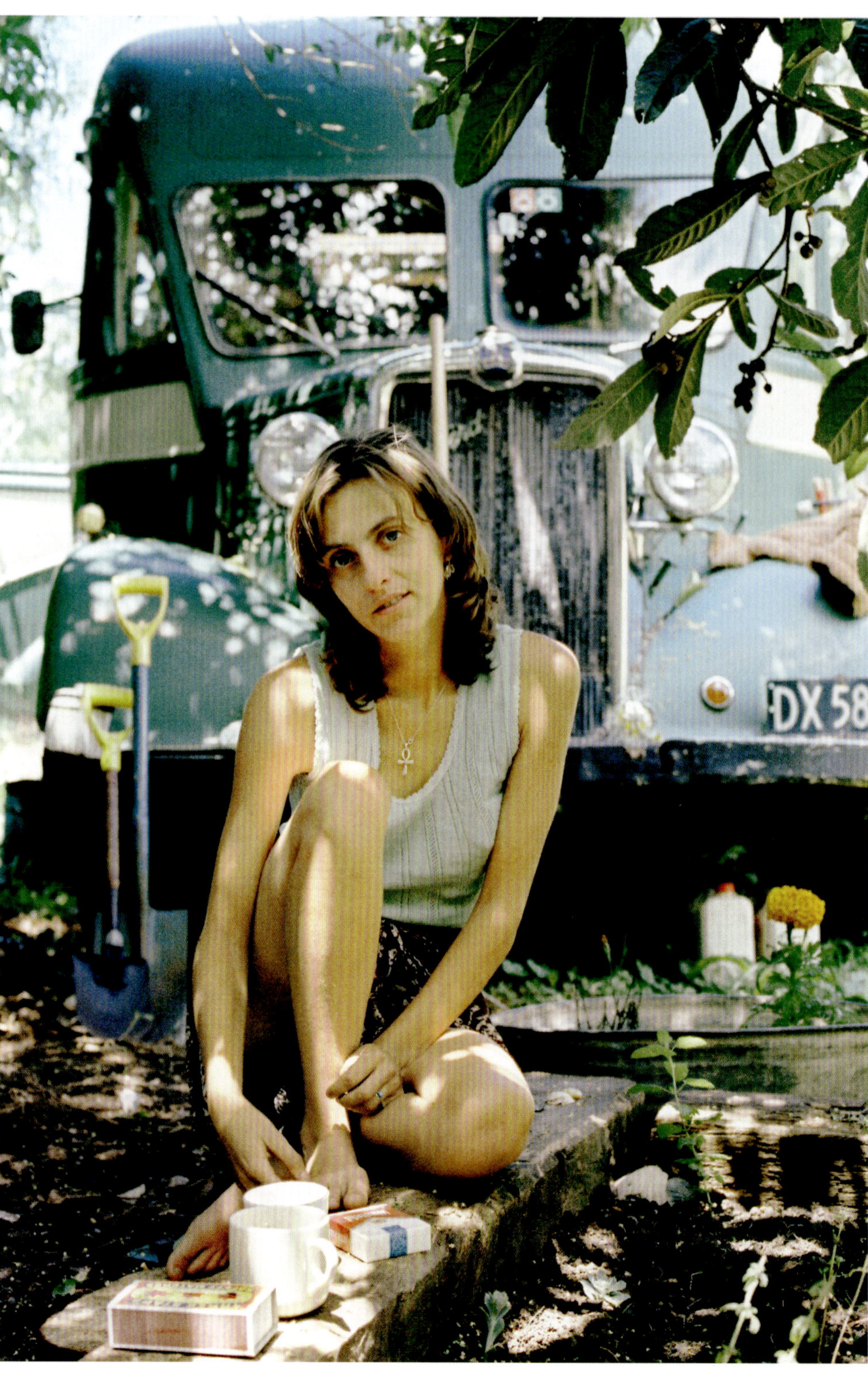

Sue Honeyman and 'Iris', Paul's bus, Shakespeare Road, Milford, Auckland, 1982

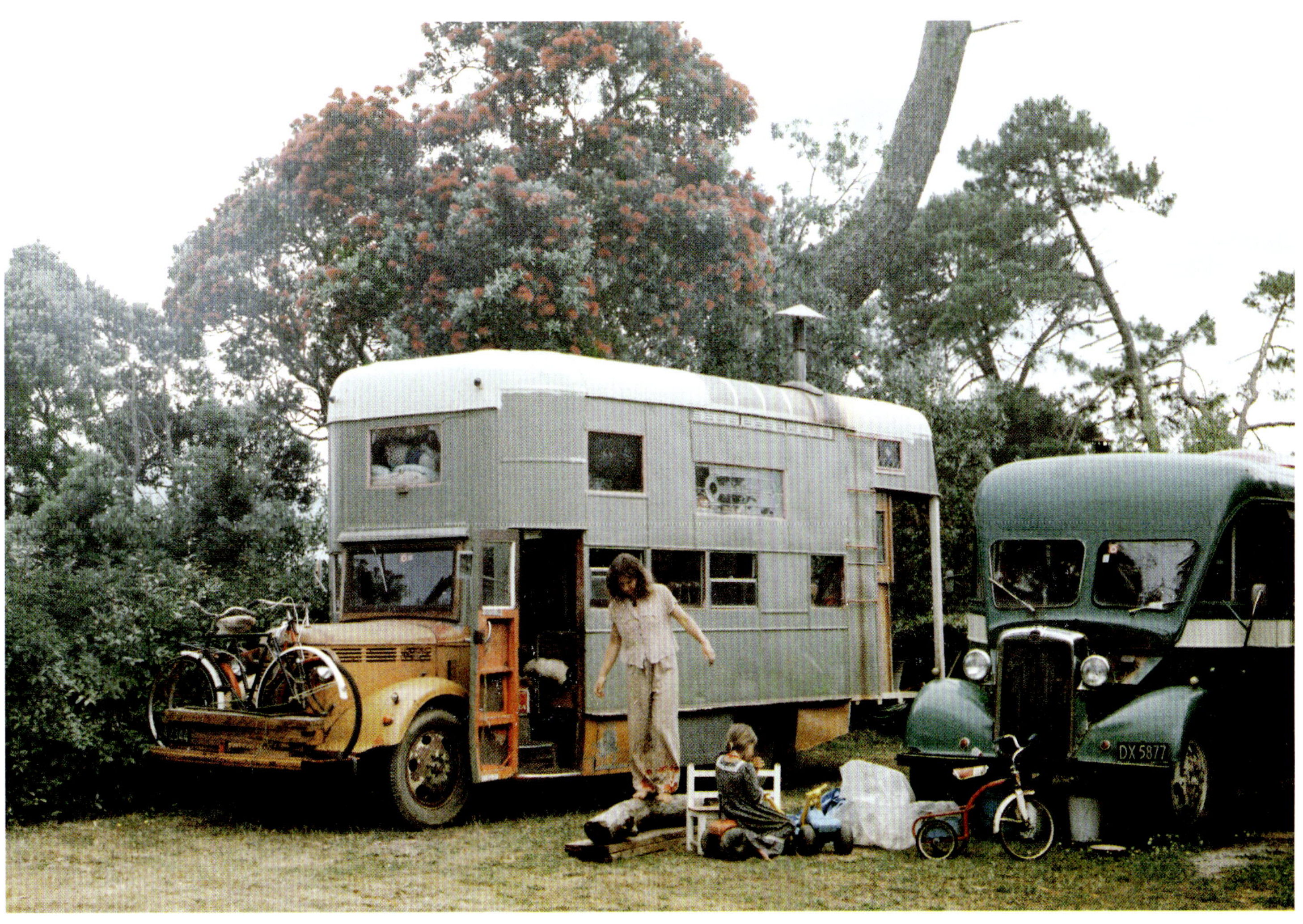

Mobile bus homes, Clifton Road, Takapuna, Auckland, 1981

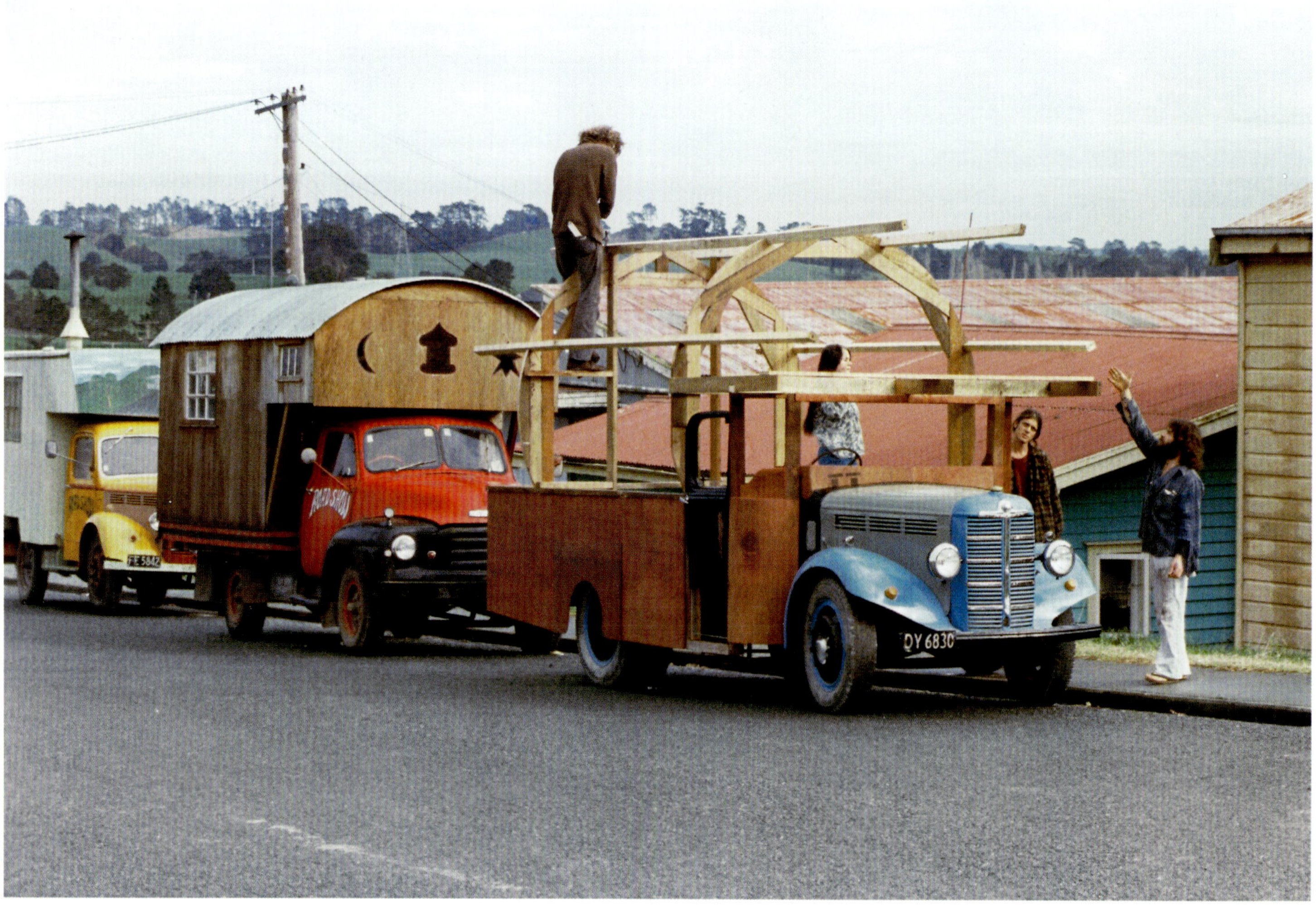

Rob's truck (yellow), Dale's truck (red) and Podge and Diane's truck (blue) being constructed, with Rob, Diane, John and Podge, Worker's Road, Wellsford, 1978

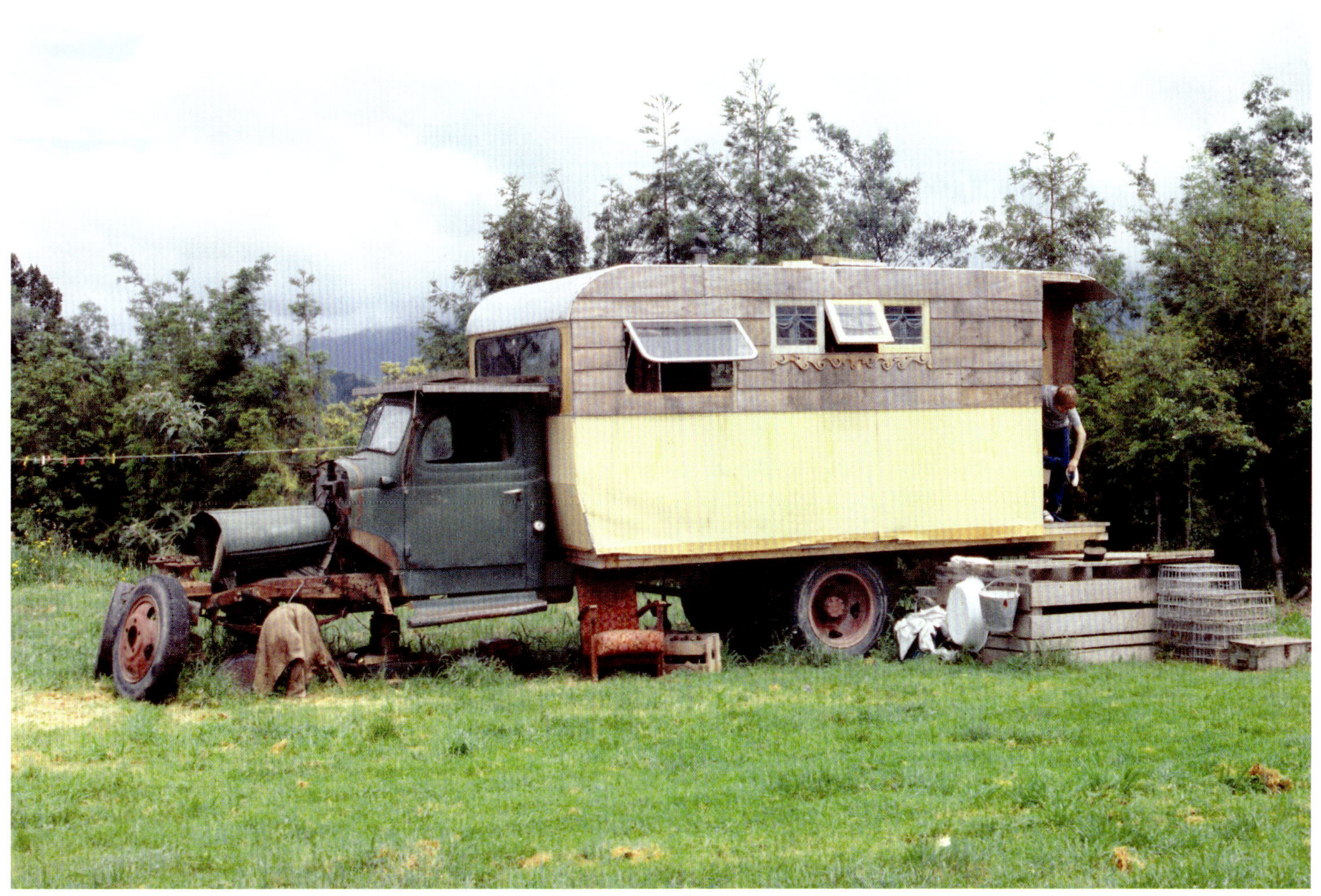

Robert and Claire's caravan, Katikati, Bay of Plenty, 1981

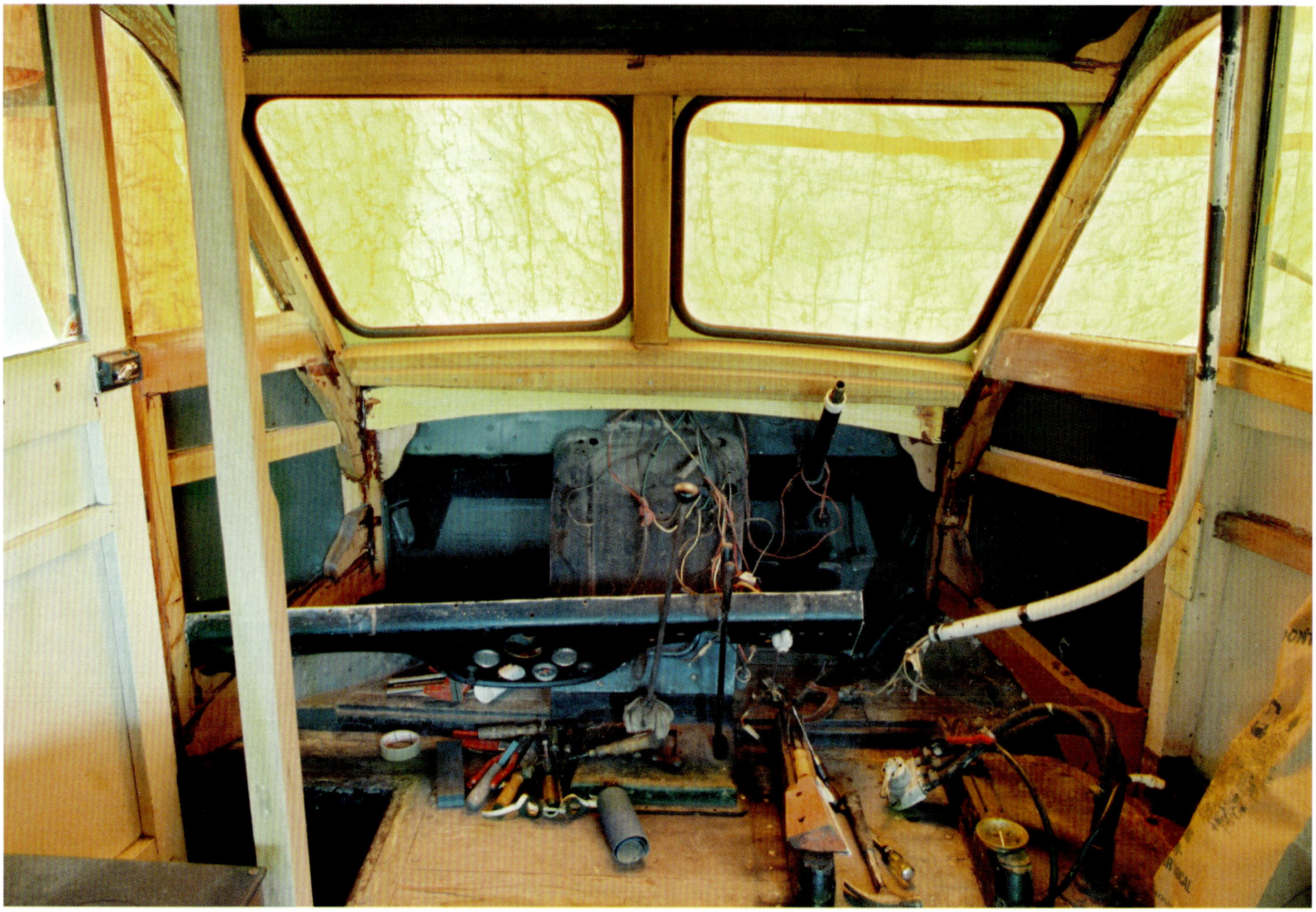

'Iris' being restored, c.1982

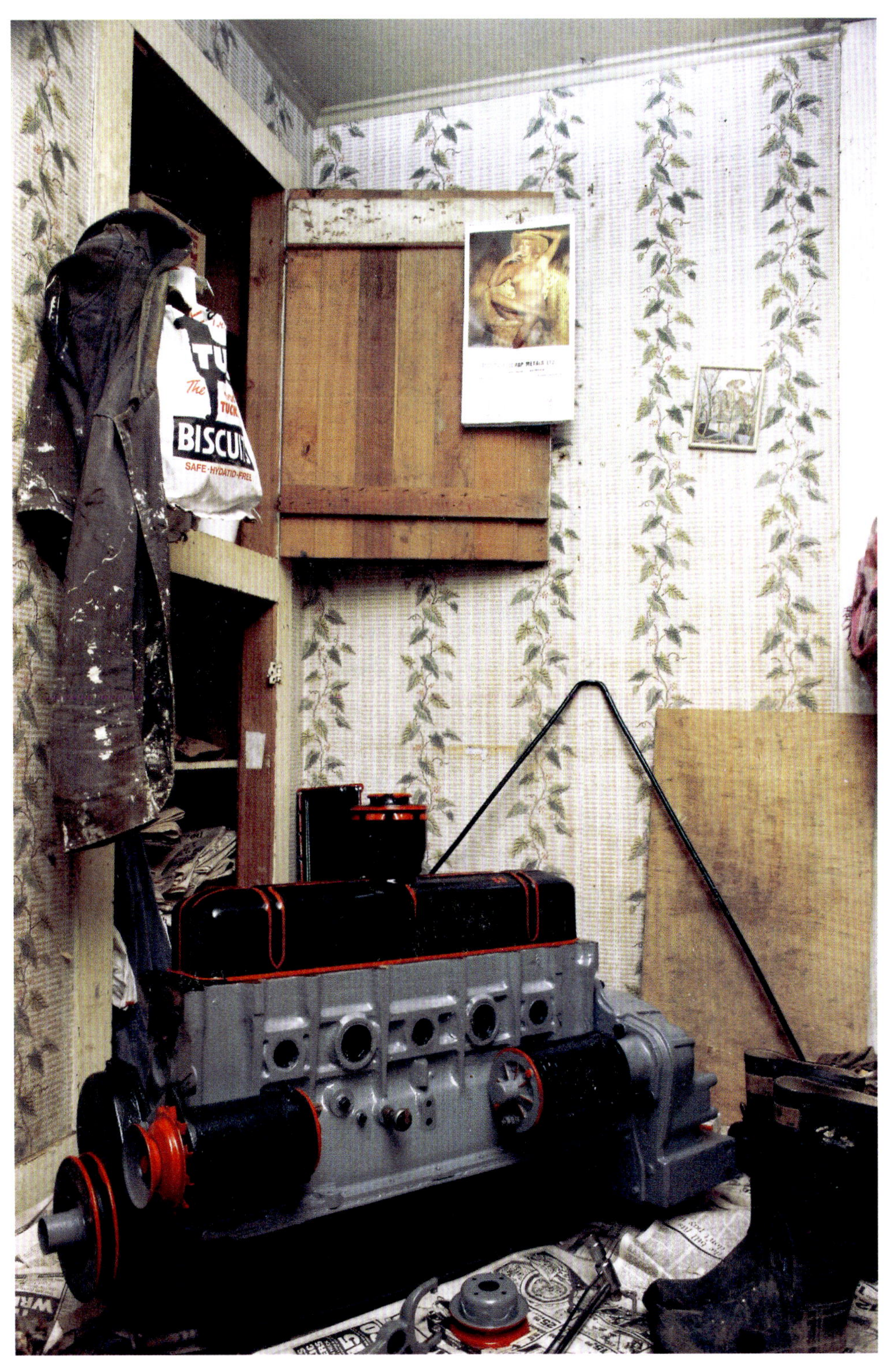

Unidentified engine being restored, c.1981

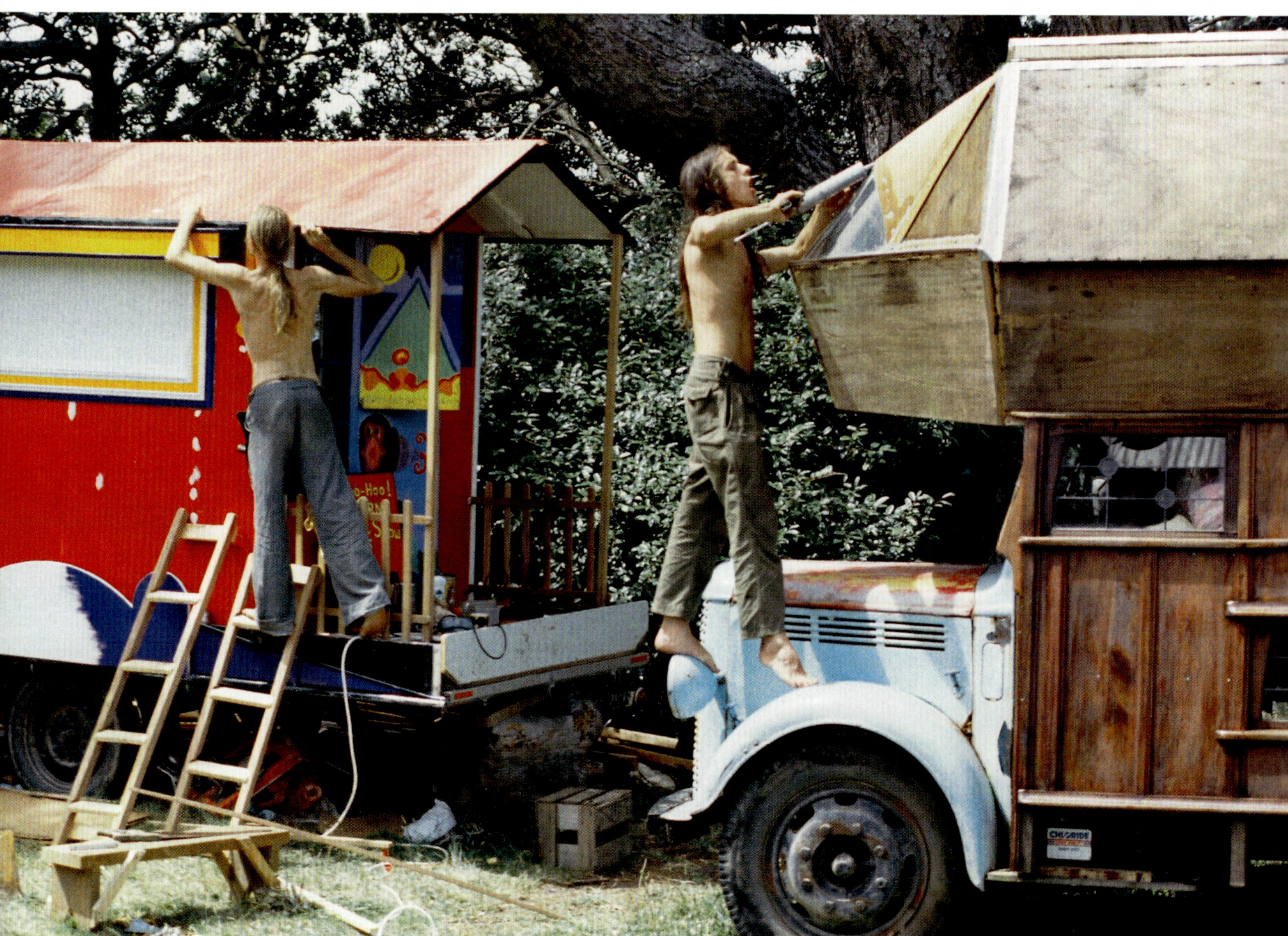

Getting ready for the Northland and Coromandel road show, Jonathon Acorn and John Tucker, Clifton Road, October 1978

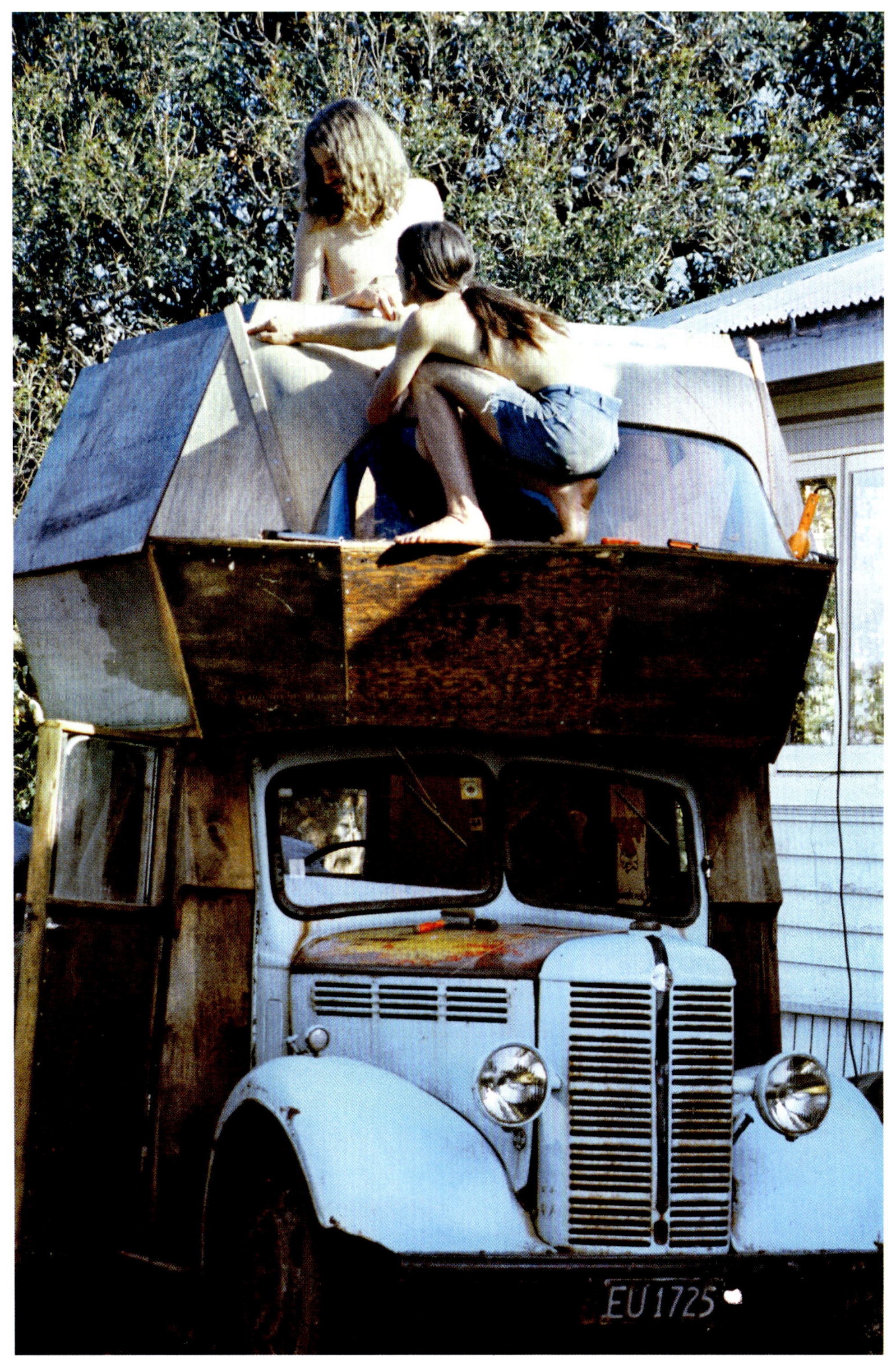

Constructing the stage-truck, Clifton Road, October 1978

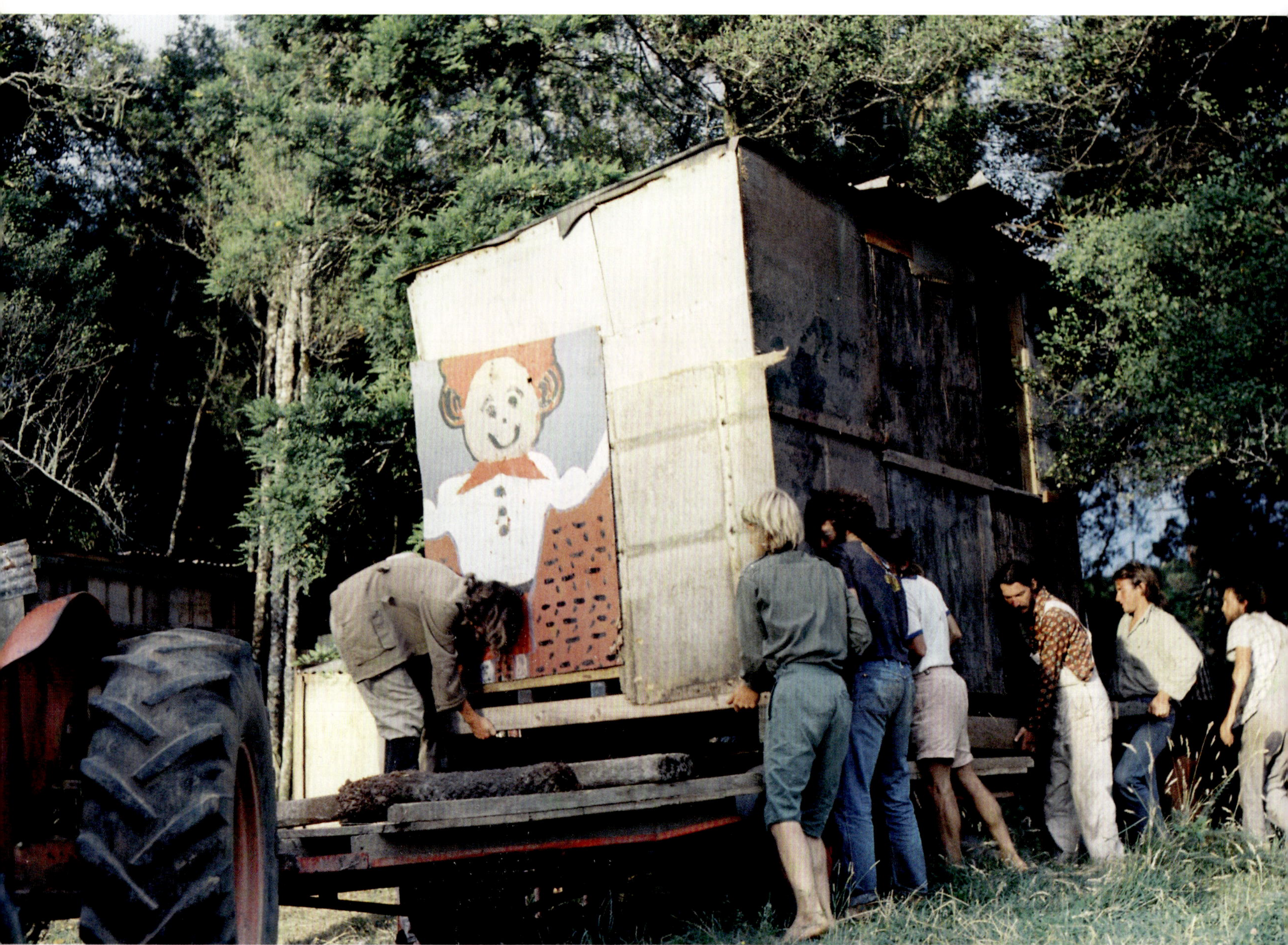

Unidentified shed being loaded onto a truck, c.1983

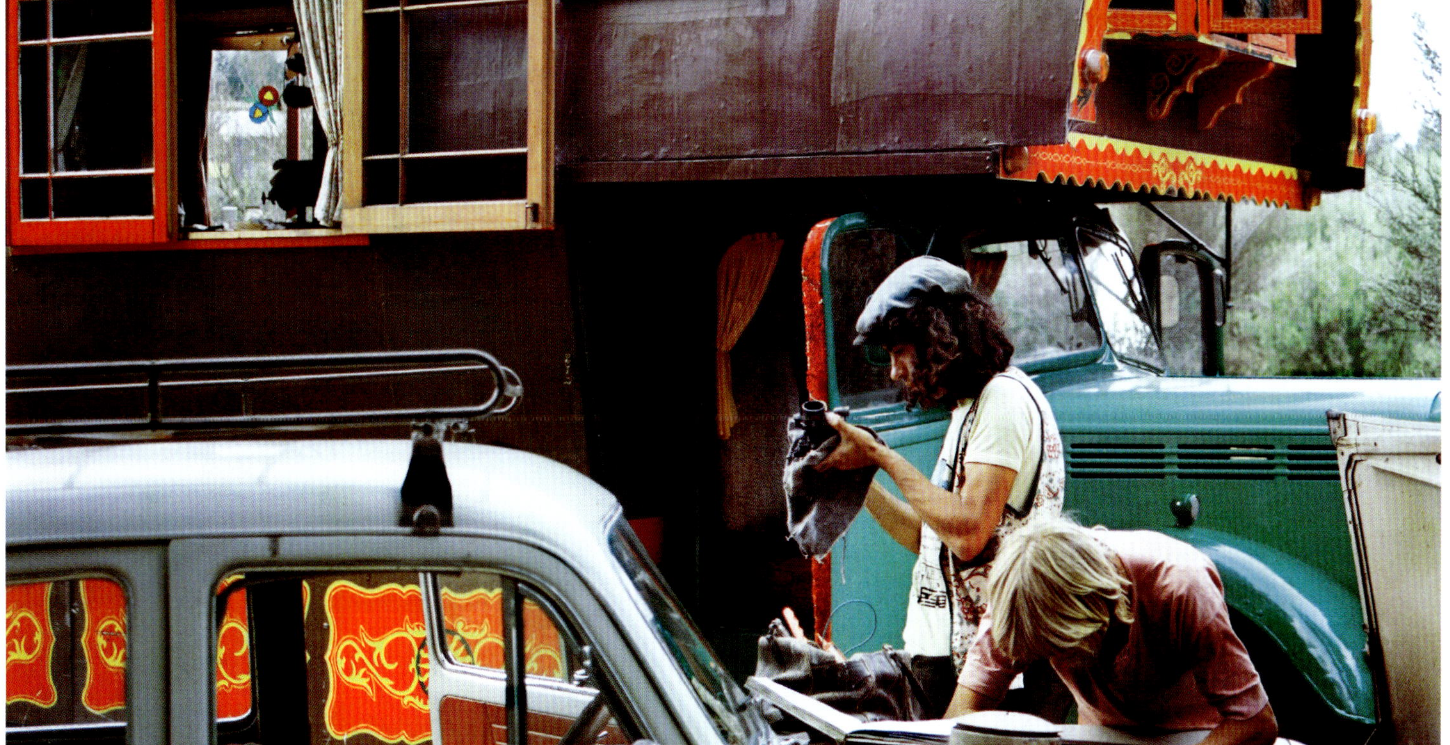

Dom and Stephan working on a Ford Prefect, with Dom and Jane's truck in background, Aratiatia, 1984

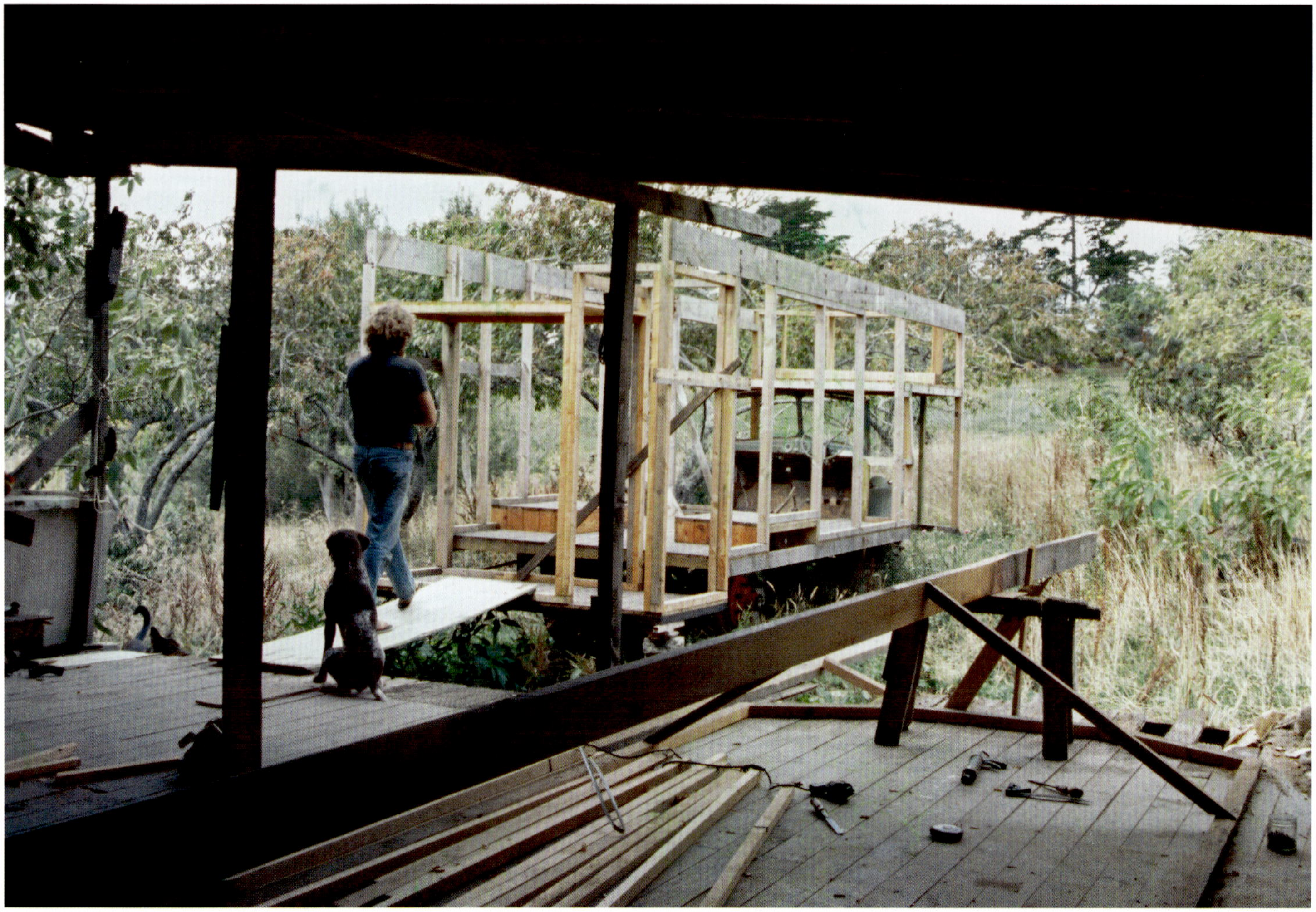

Harry checking out Chris's truck during construction, Taupaki, Rodney District, Auckland, c.1983

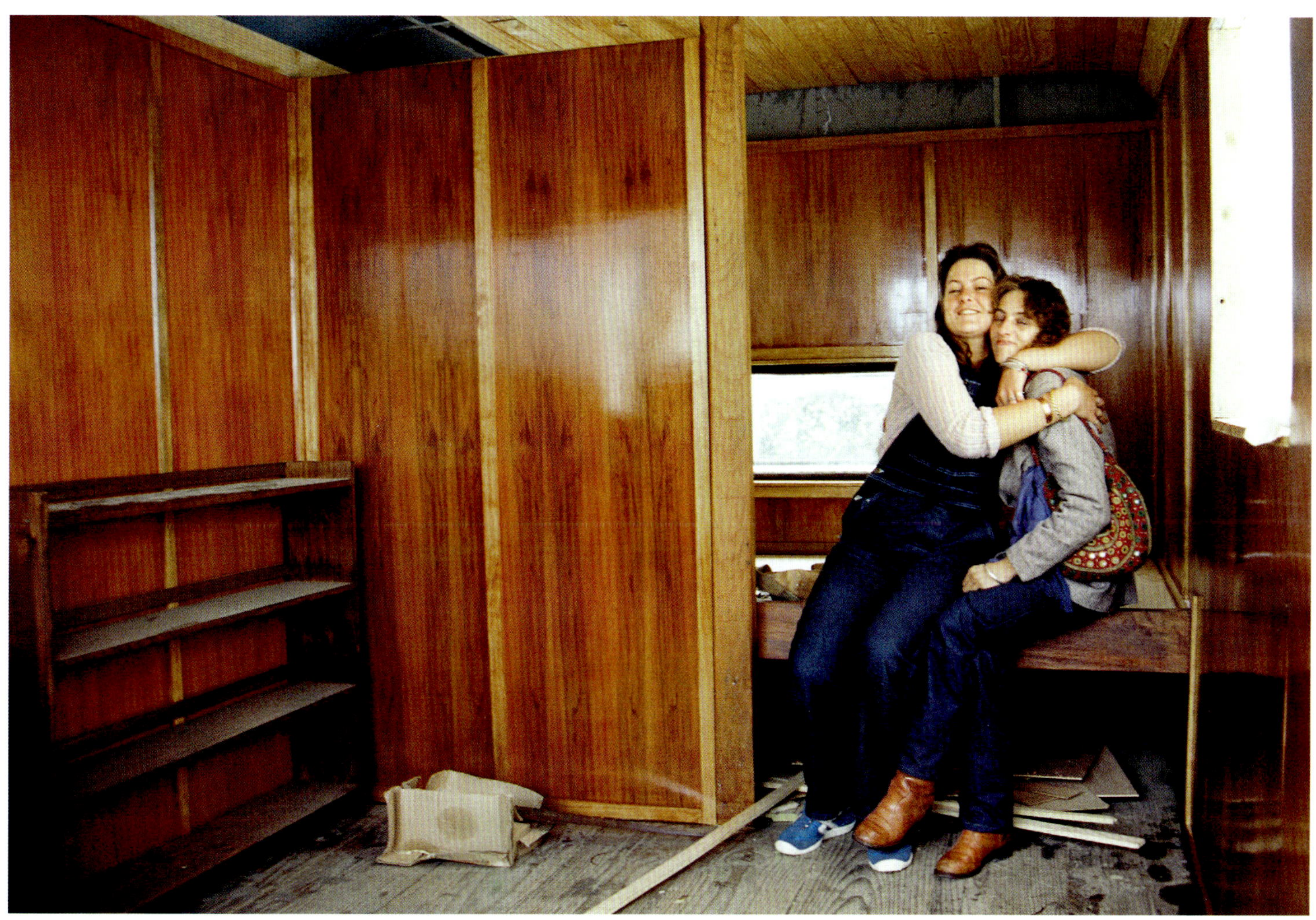

Rose and Sue, in Harry and Rose's truck under construction, Glenfield, Auckland, c.1983

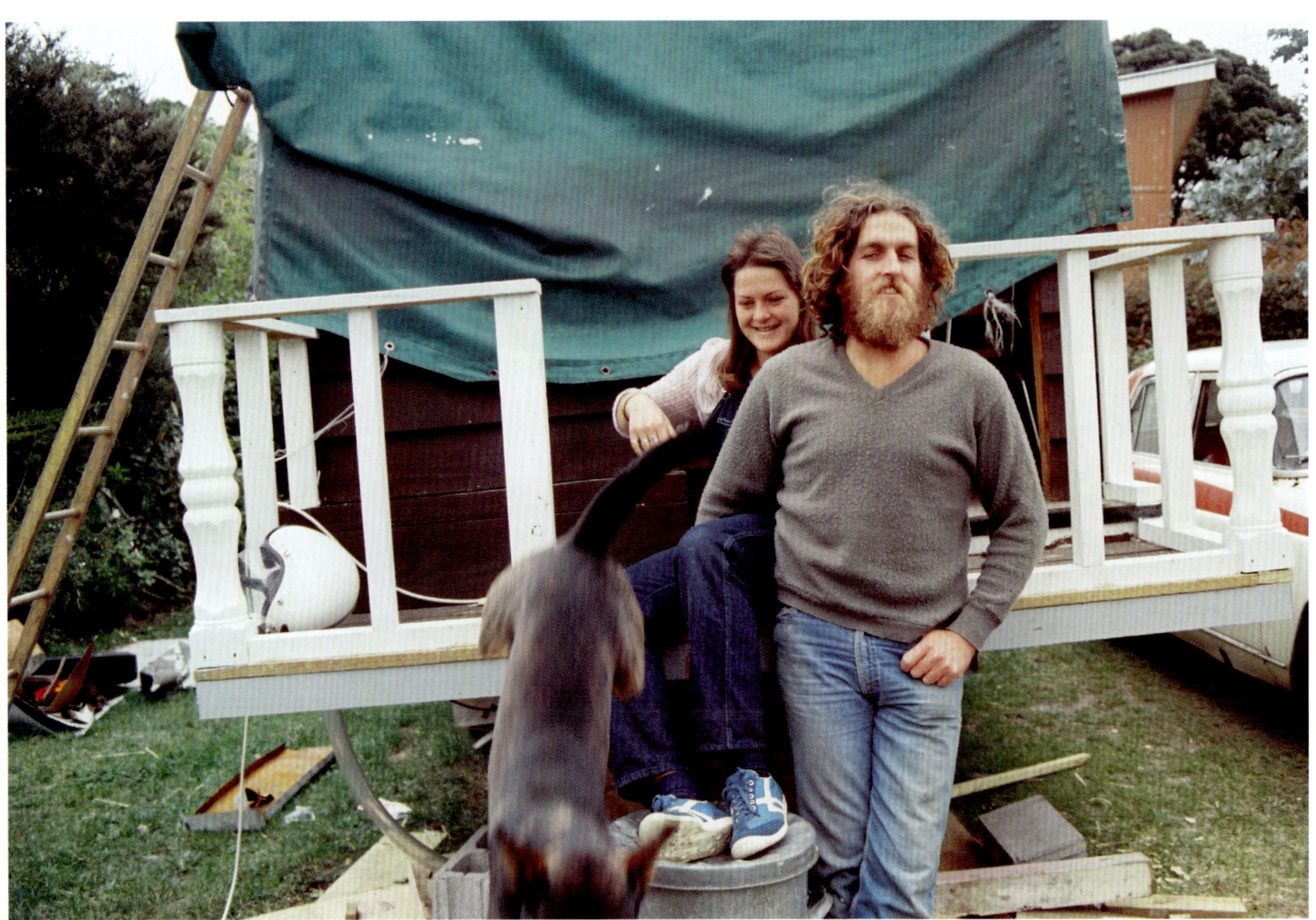

Harry and Rose's truck in the making, Glenfield, Auckland, c.1983

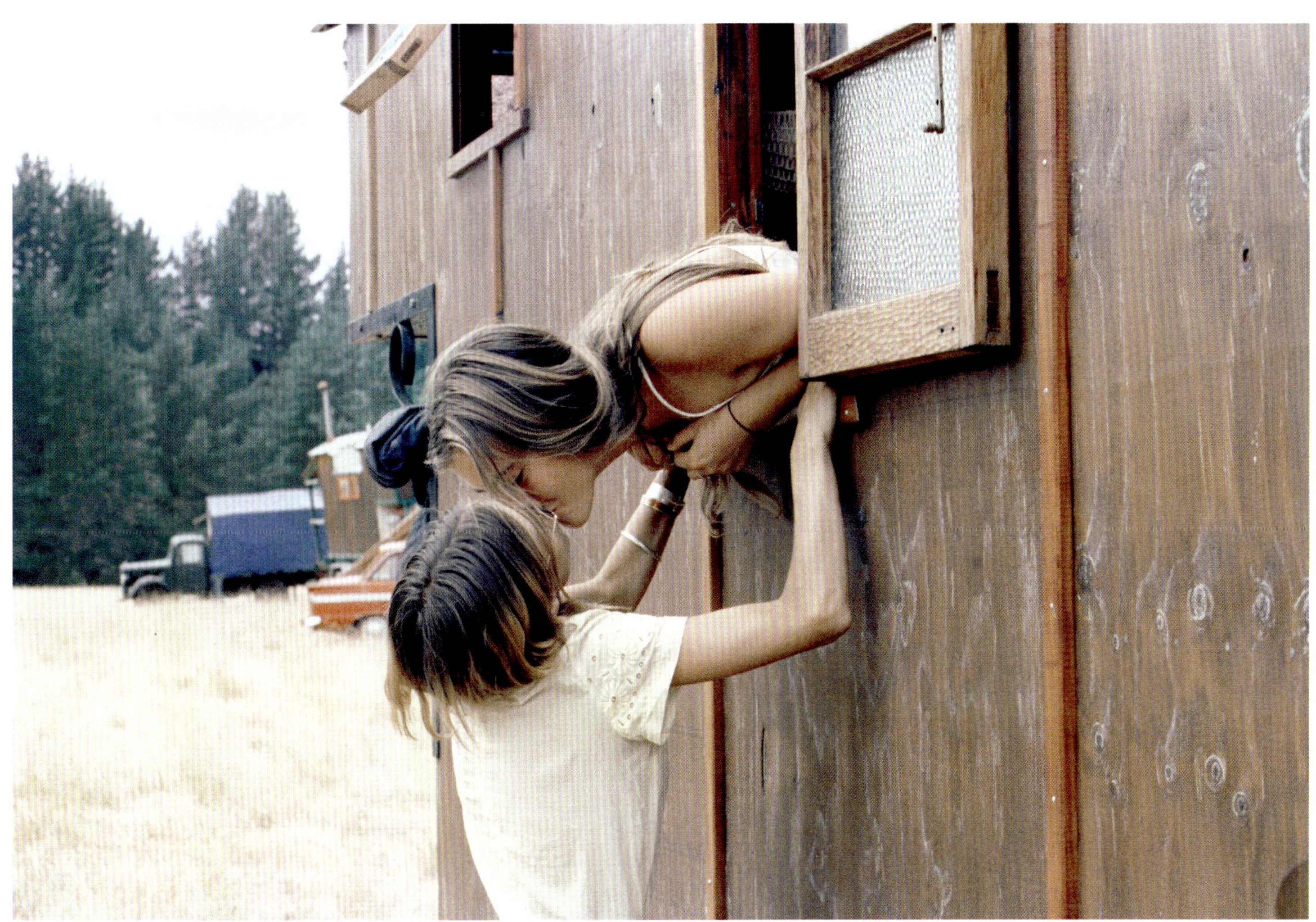

Sue and sister Claire saying hello, Aratiatia, 1984

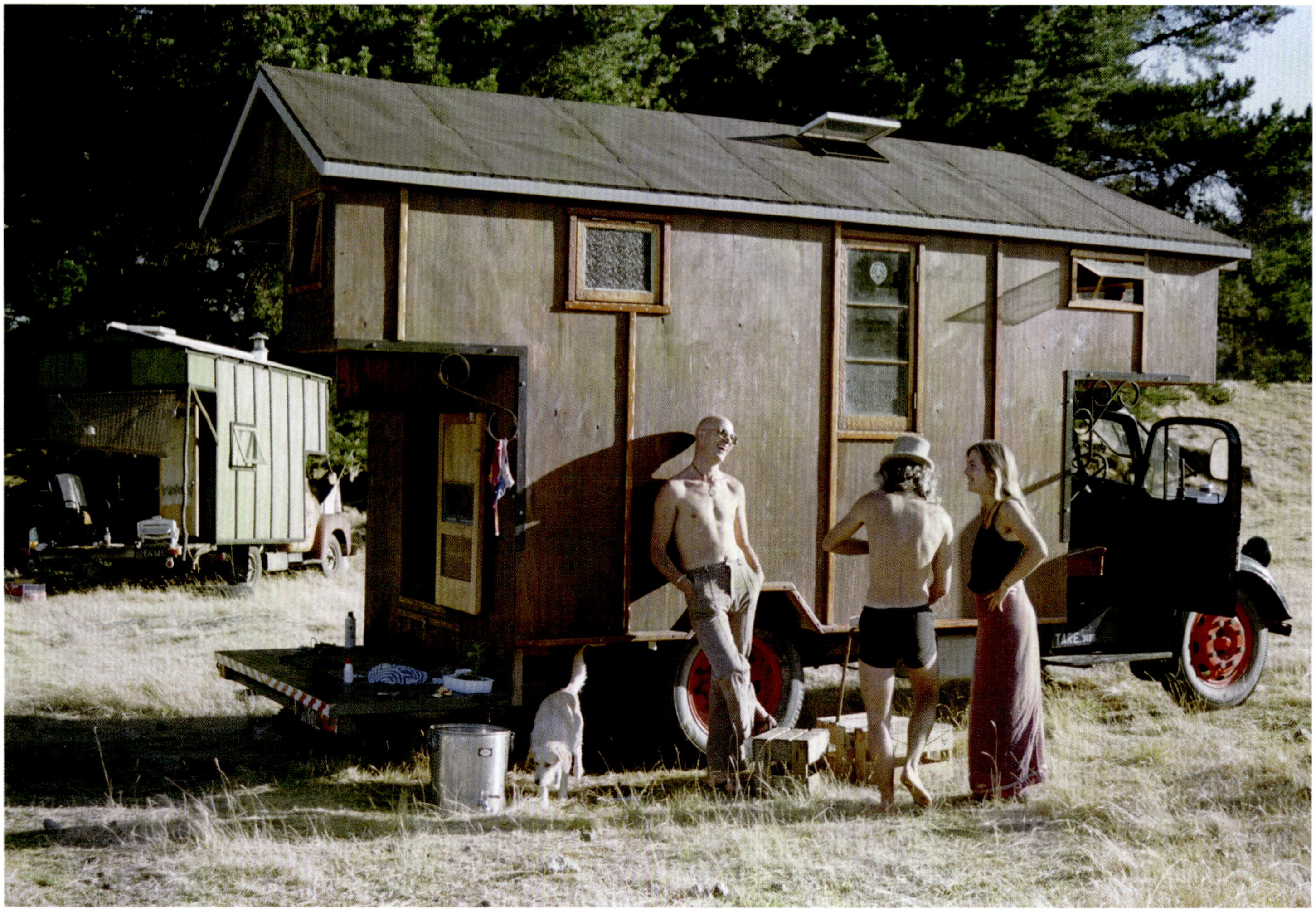

Neelie (the dog), David Sheridan, Paul the 'Grey Wizard' (with hat) and Claire Honeyman in front of the Wizard's truck, Aratiatia, 1984

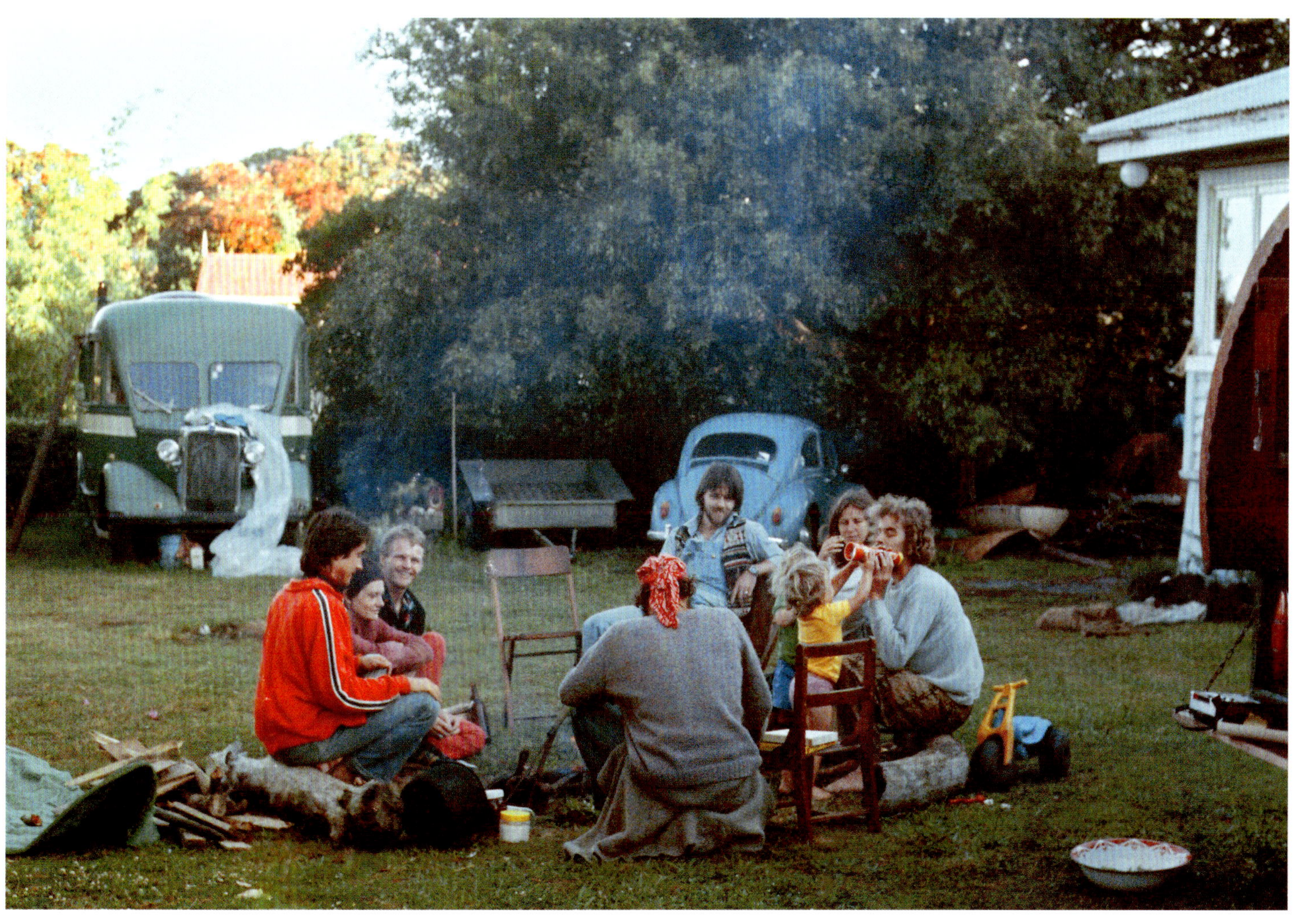

John Tucker, Allison and Wayne, Gypsy Jeff, Bruce Gilbert and unidentified family, Clifton Road, c.1980

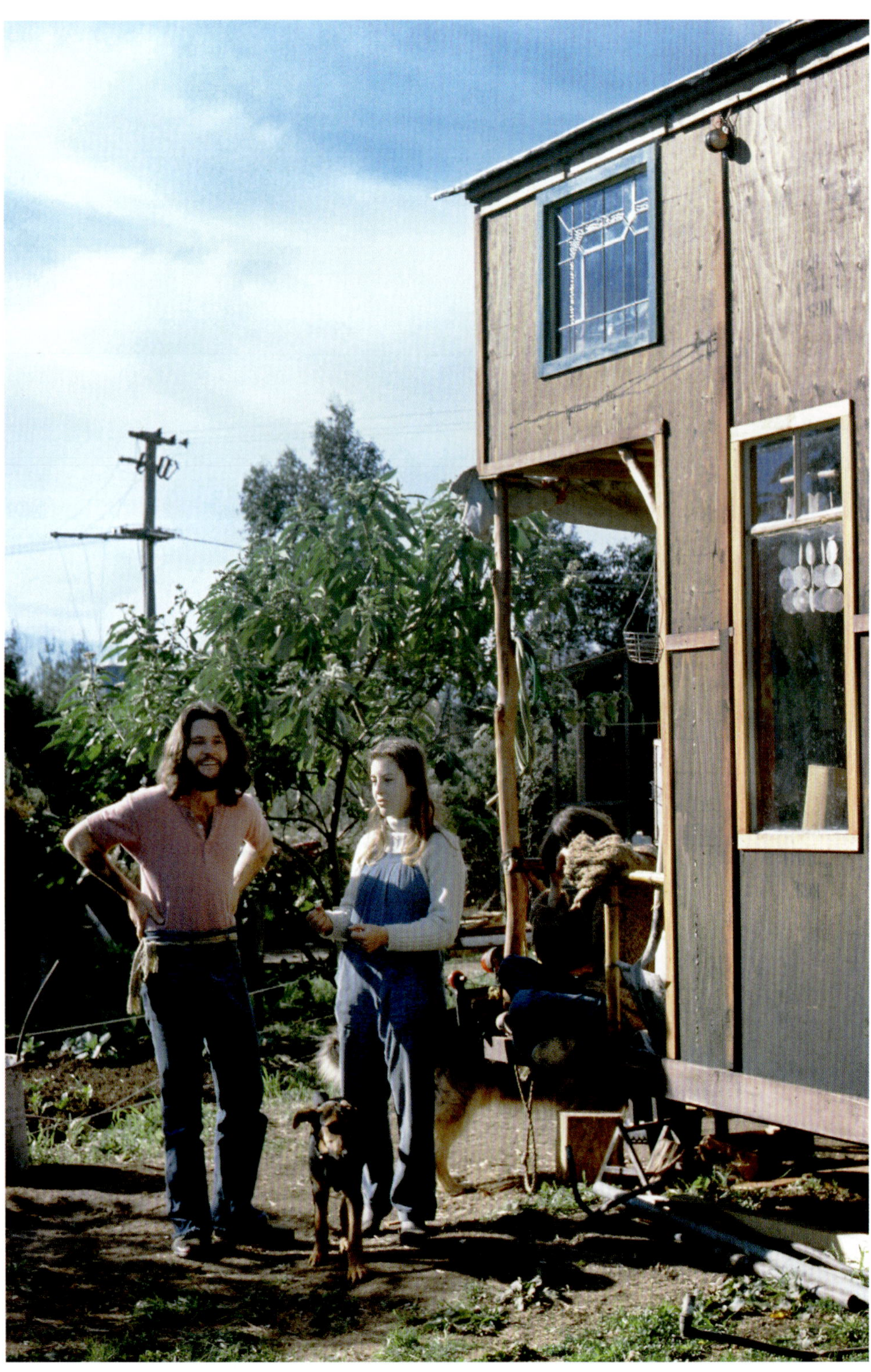

Erin and Claire, Littlewood, Katikati, c.1982

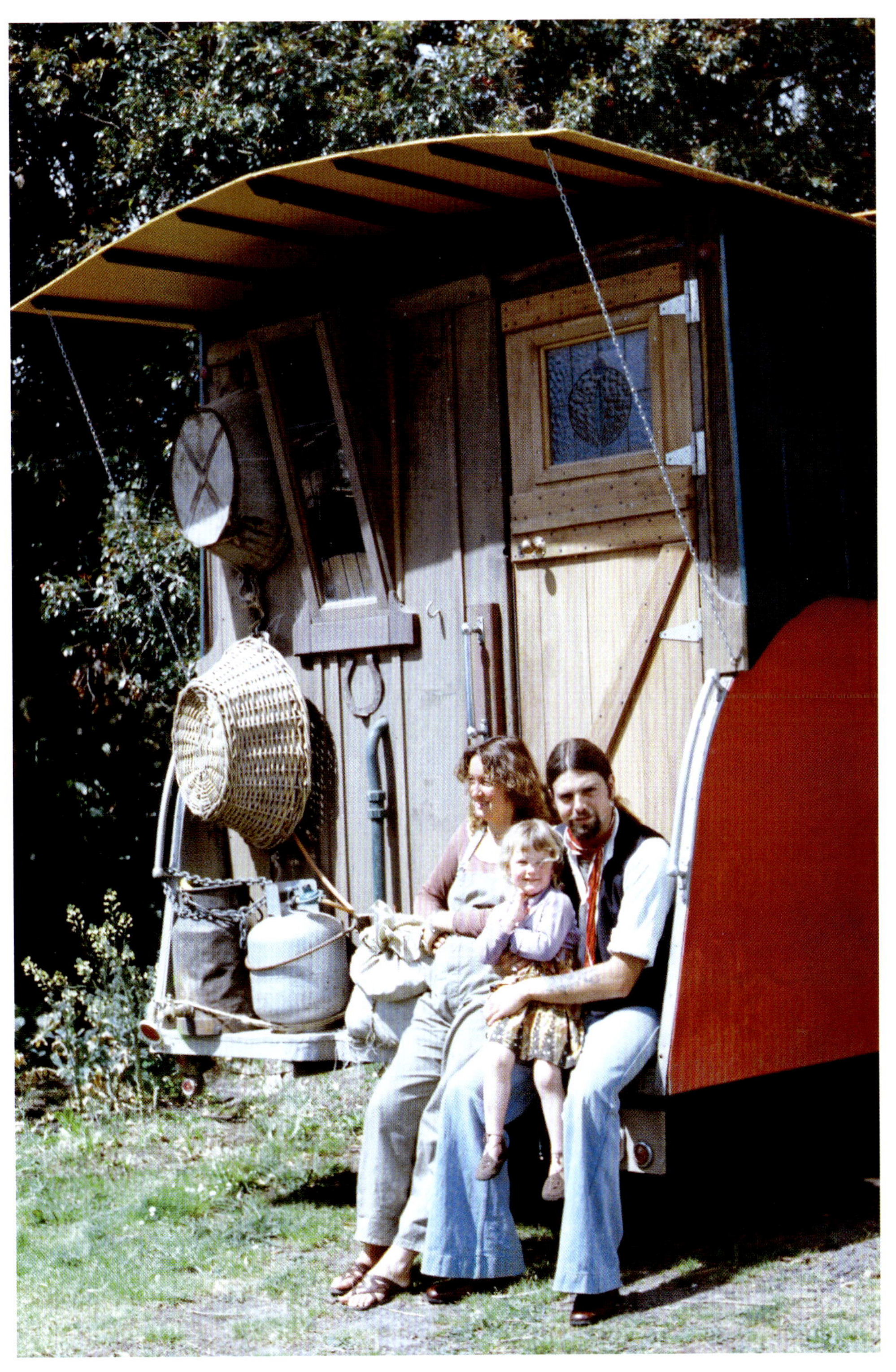

John M Cameron and Linda Barry with Narani Henson, on the 'The Old Fire Engine', Clifton Road, 1979

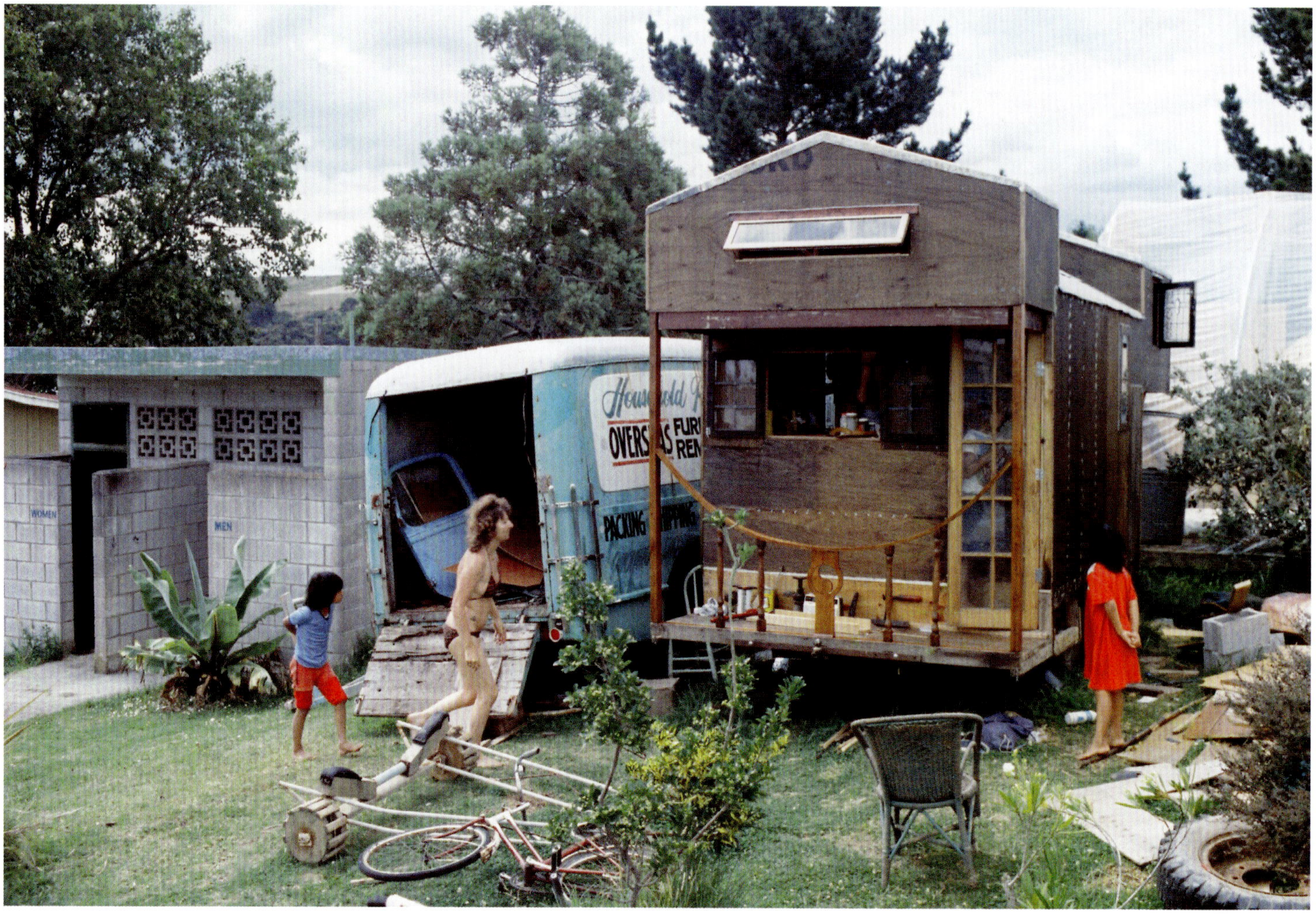

Ewan finishing work on Jade's truck at Te Atatu, Auckland, 1979

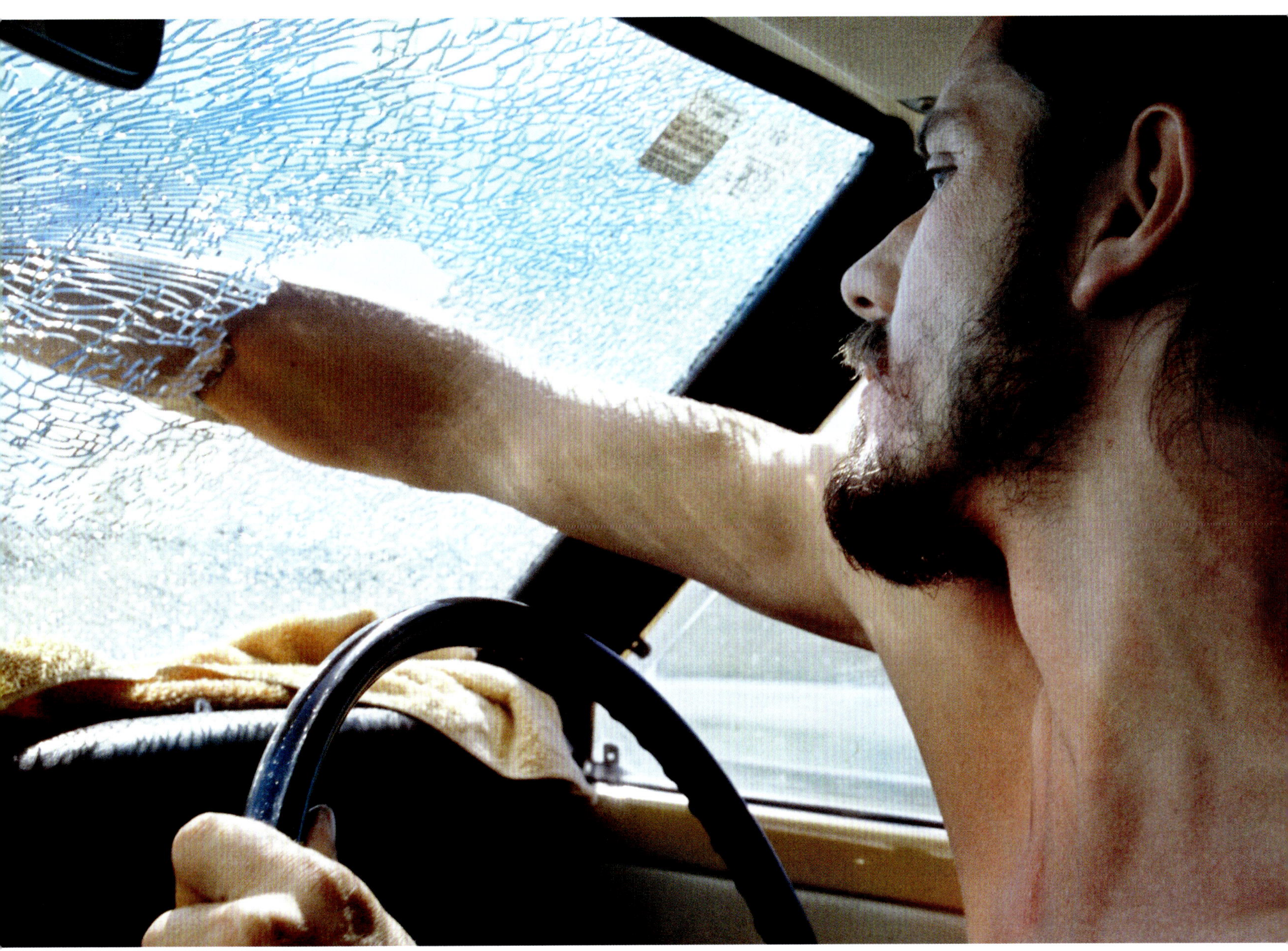

John Tucker driving, n.d.

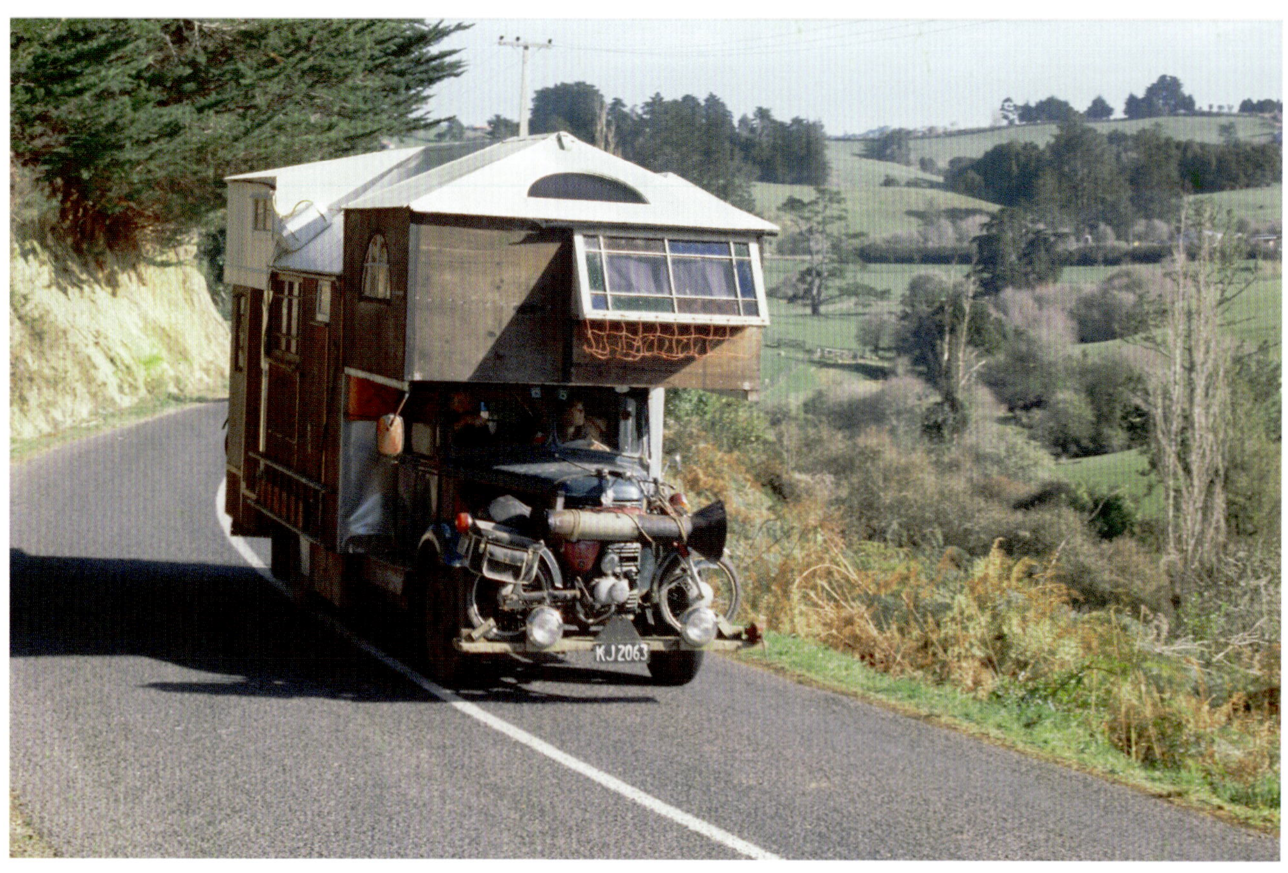

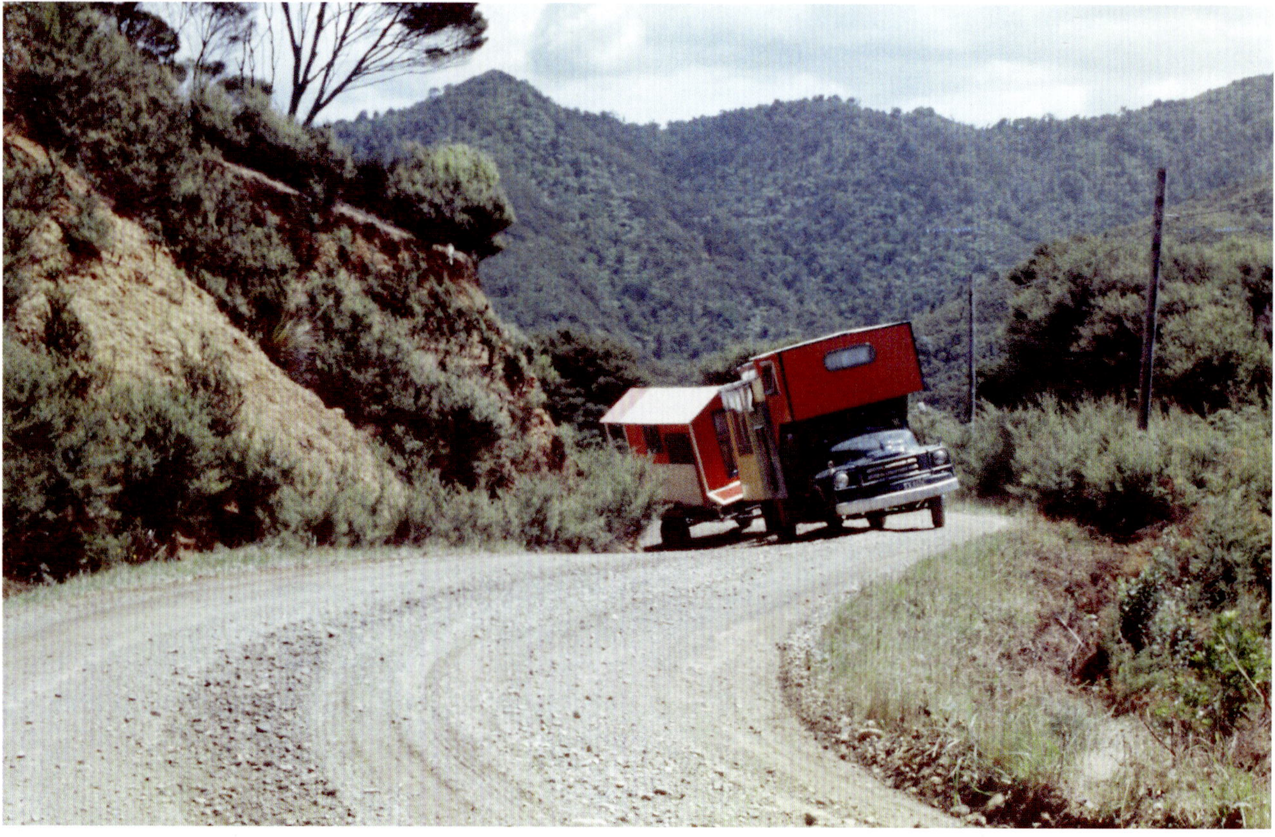

Ewan and Helen's truck leaving Brookby, Twilight Road, 1983

Acorn's Puppet Theatre truck and caravan, Coromandel, 1979

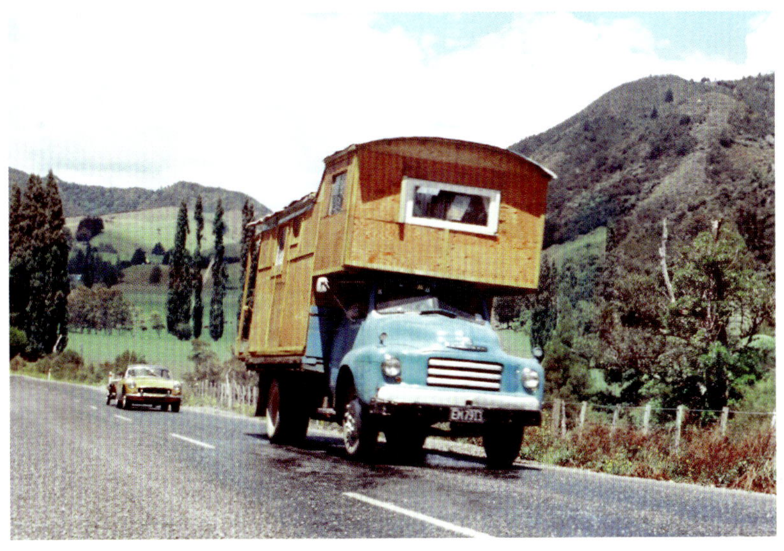

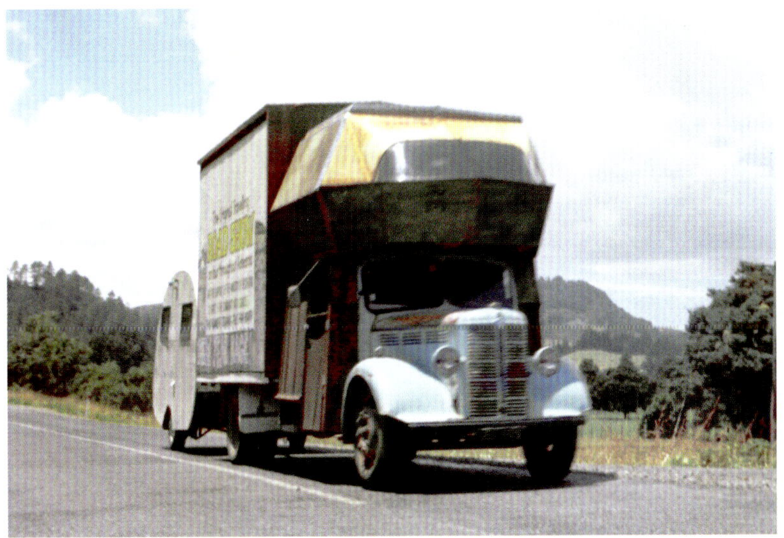

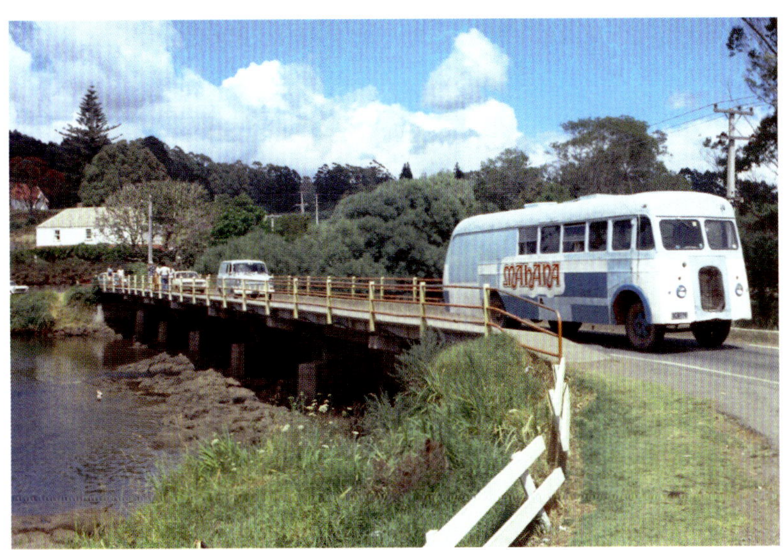

The Original Travelling Road Show on tour, Far North, 1978/79

Stage-truck on tour, Far North, 1978/79

Mahana's bus at Kerikeri, during their Far North tour, 1978/79

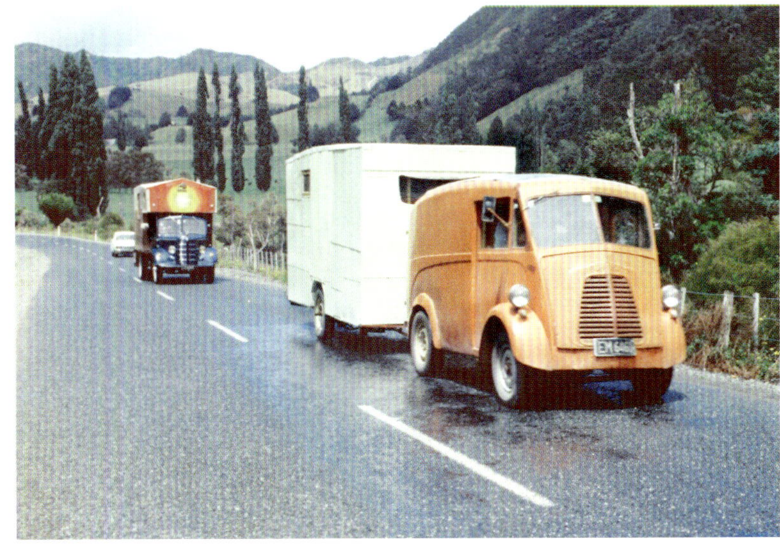

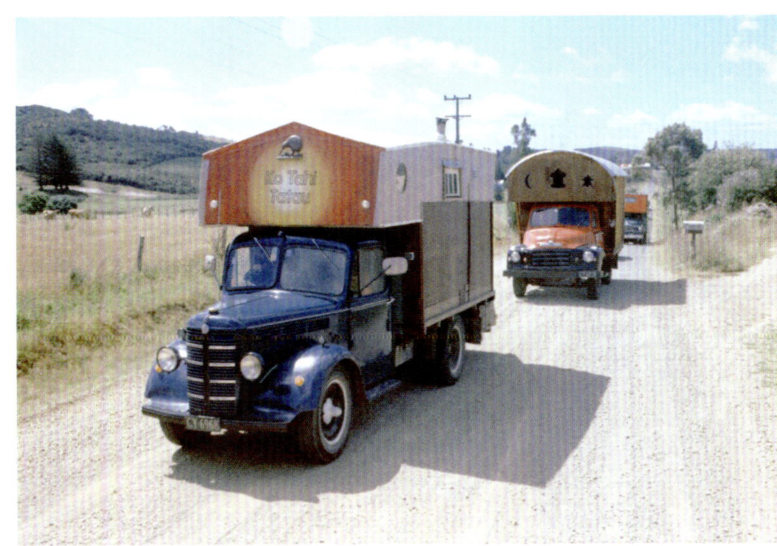

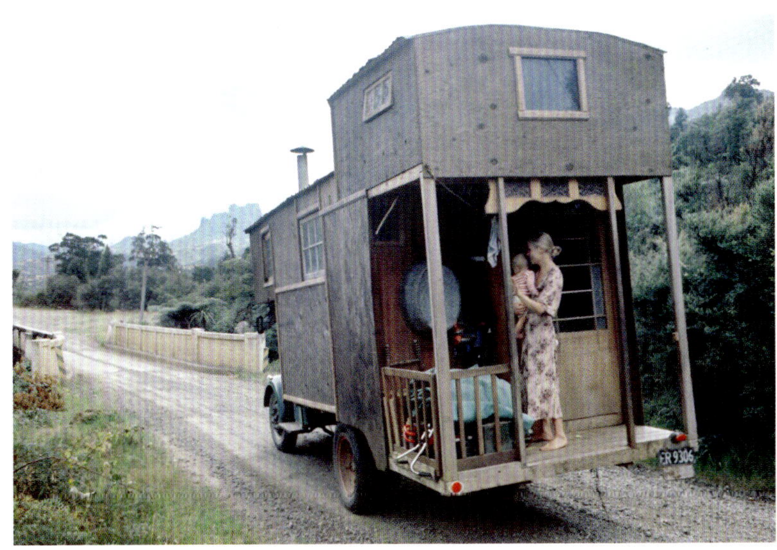

The Original Travelling Road Show on tour, Far North, 1978/79

Goober's first truck leading the tour, Far North, 1978/79

Jimmy and Yvonne's truck, 309 Road, near Waiau, Coromandel, c.1979

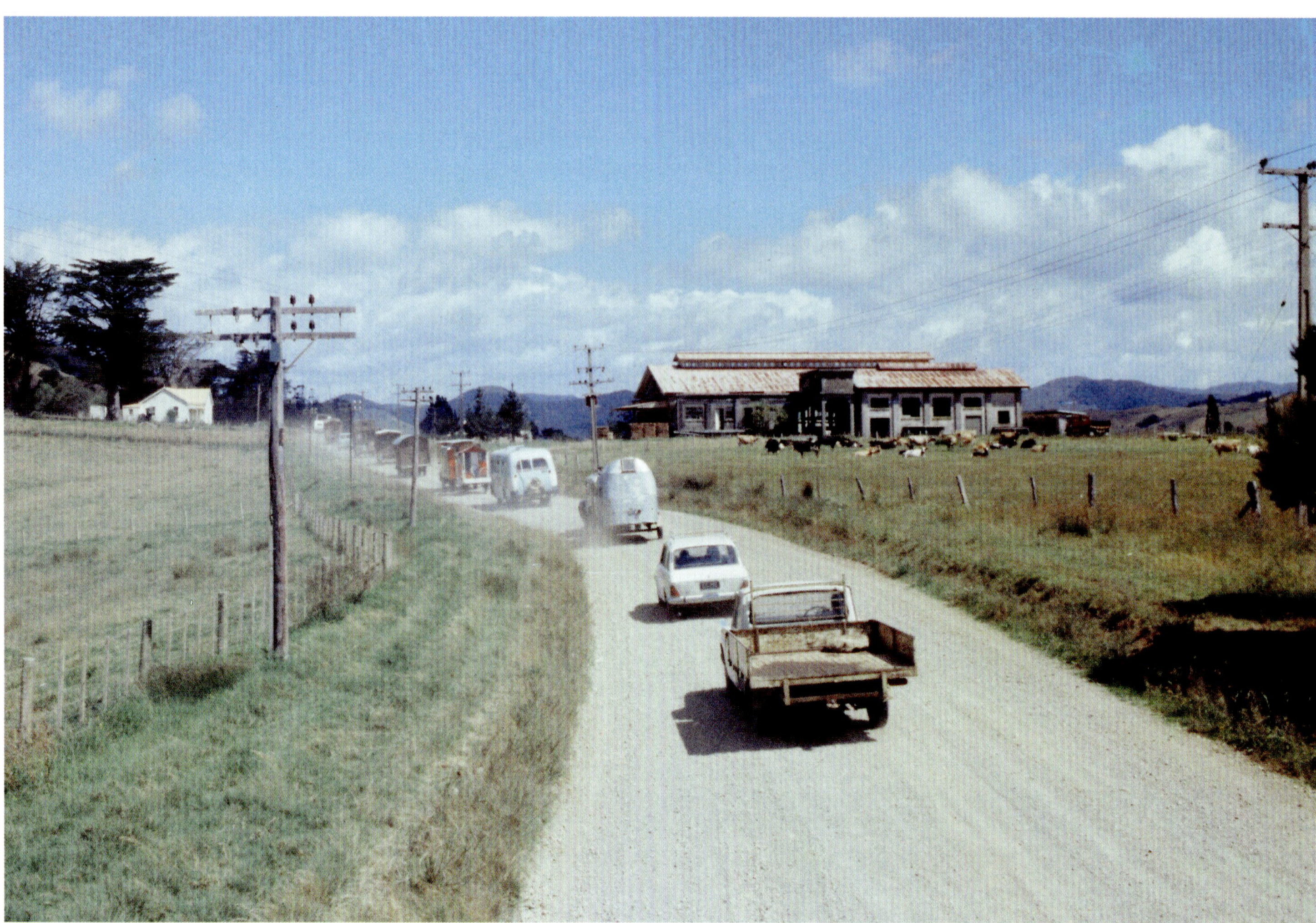

The Original Travelling Road Show on tour, on the road to Fern Flat, Peria, Far North, 1978

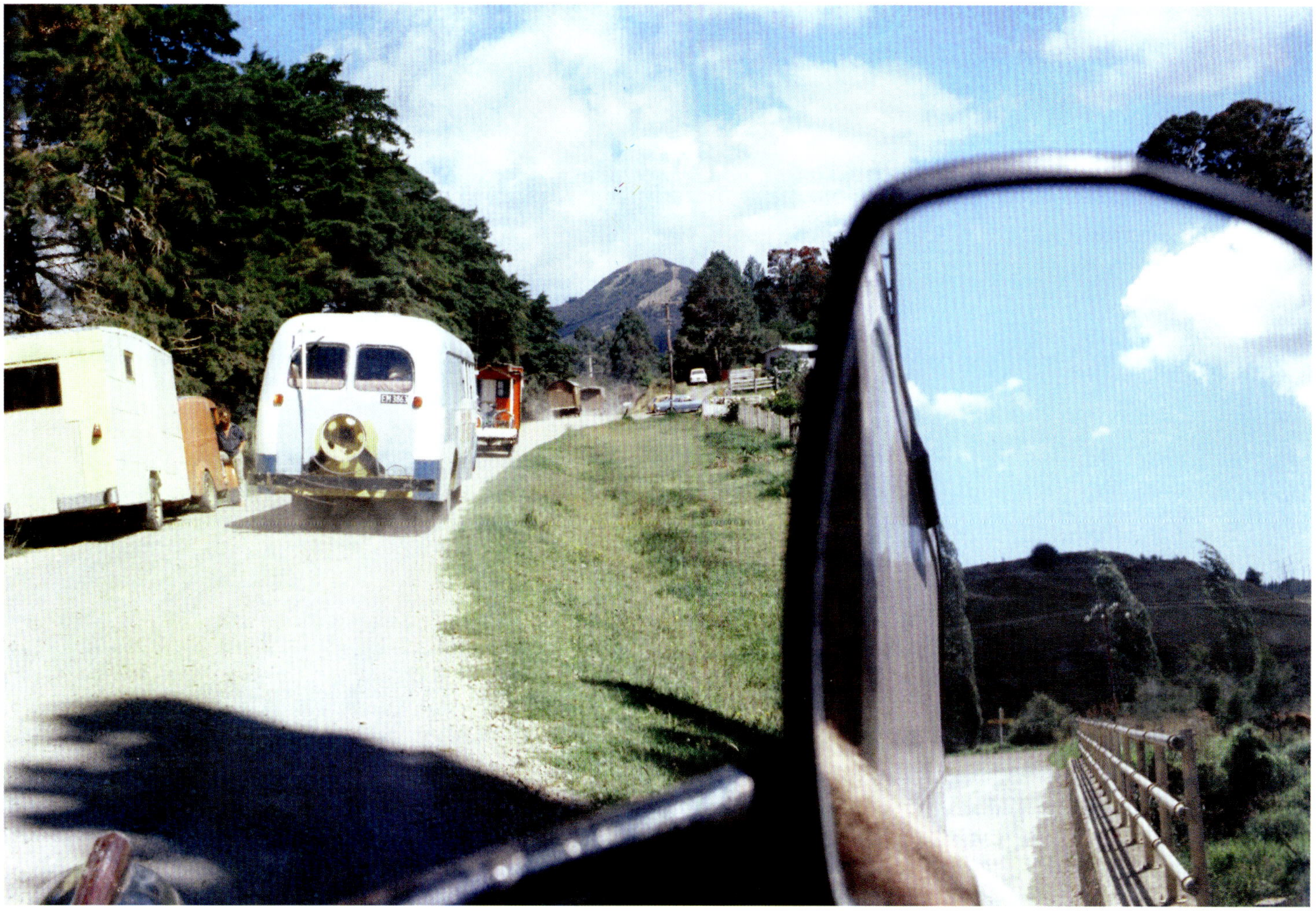

The Original Travelling Road Show on tour, on the road to Fern Flat, Peria, Far North, 1978

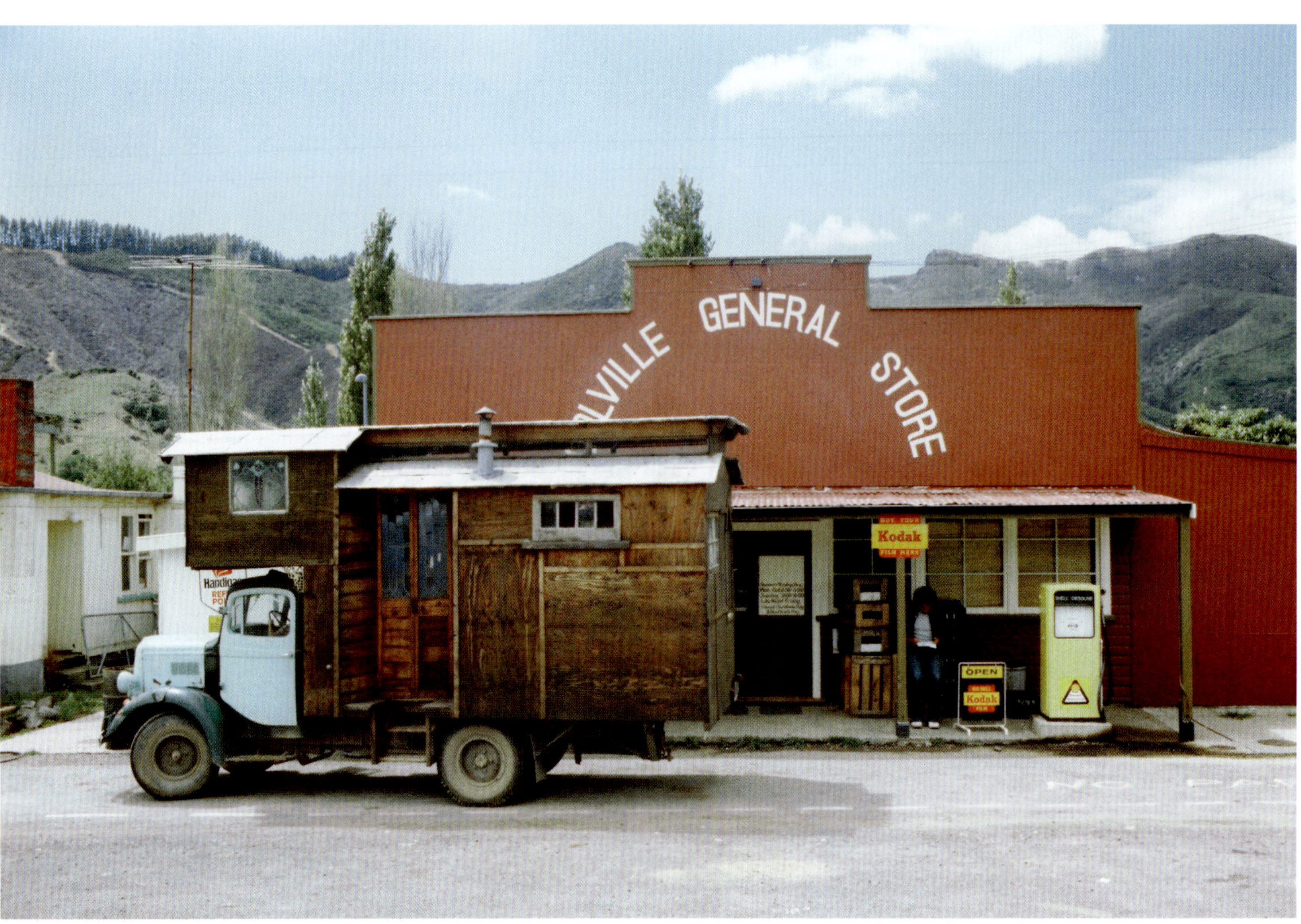

Marie's truck, Colville, Coromandel, 1979

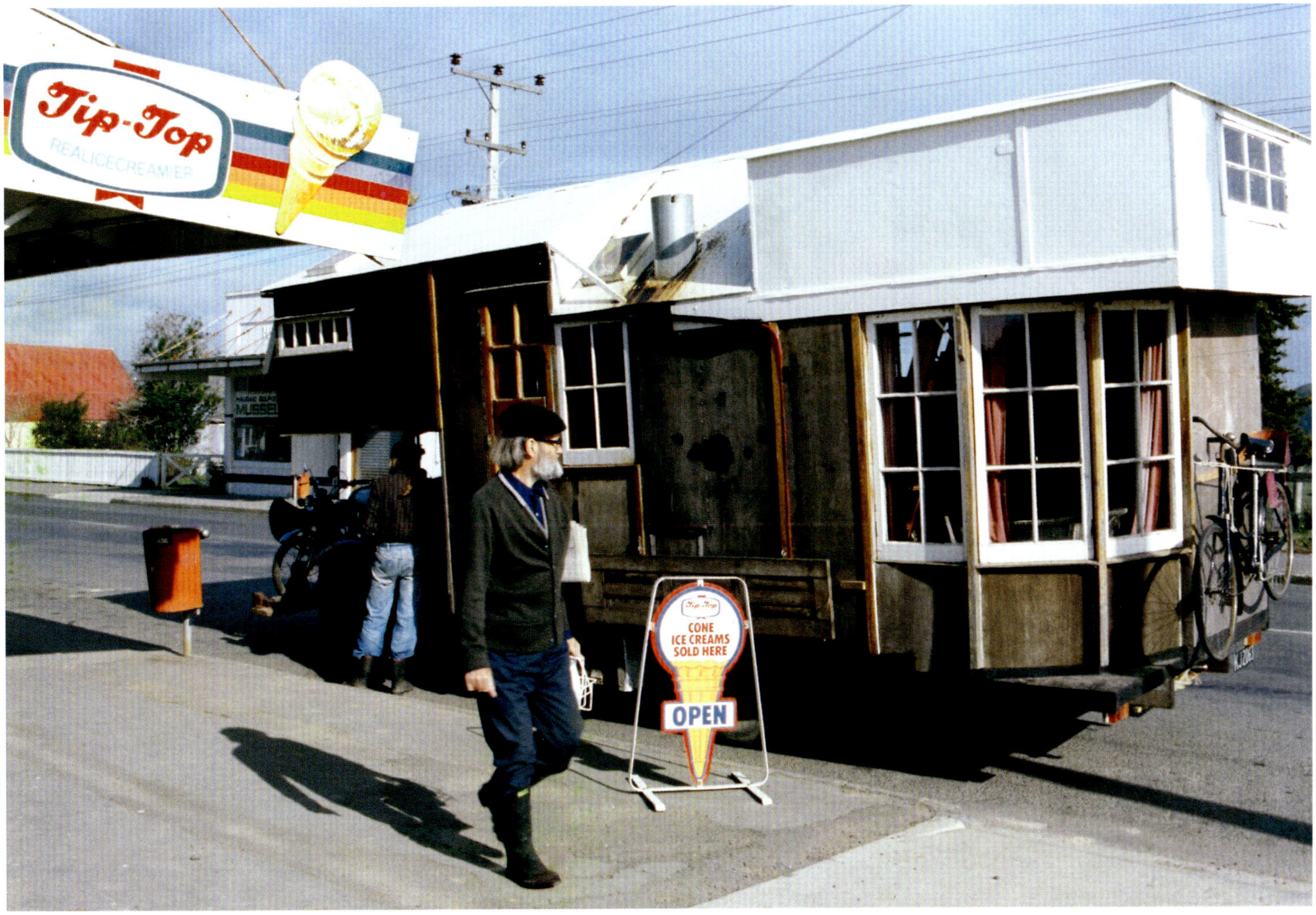

Ewan and Helen's truck, Clevedon, Franklin, Auckland, 1983

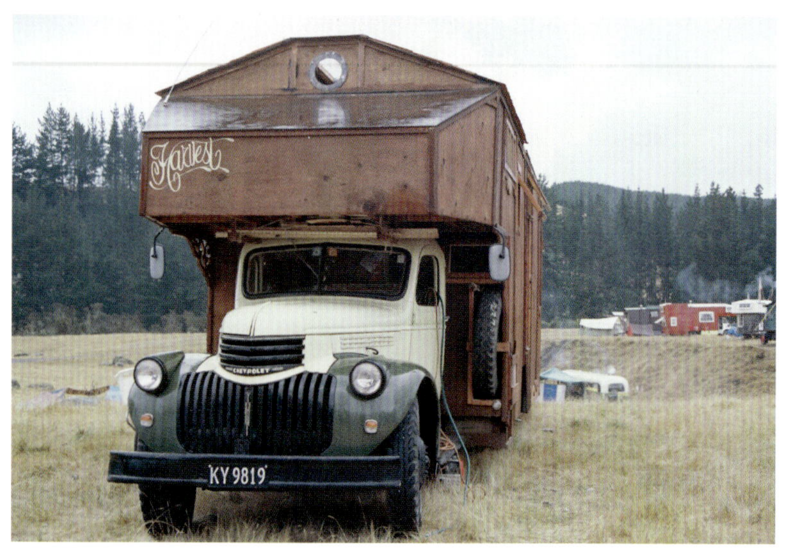

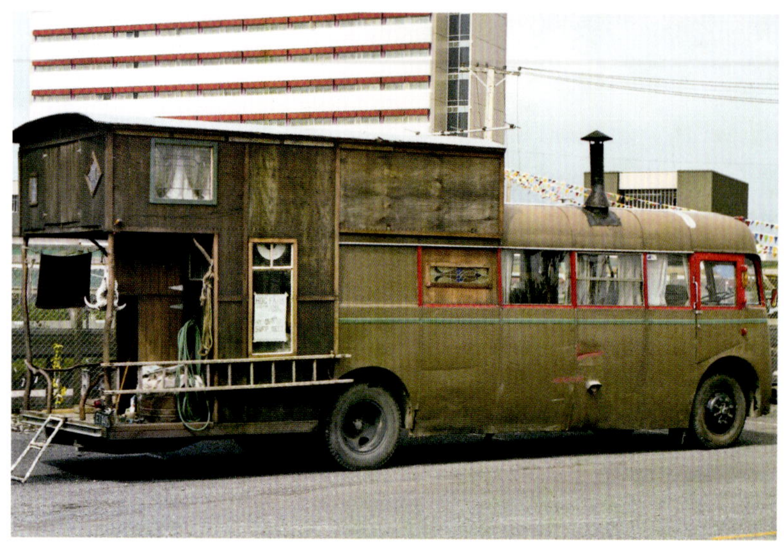

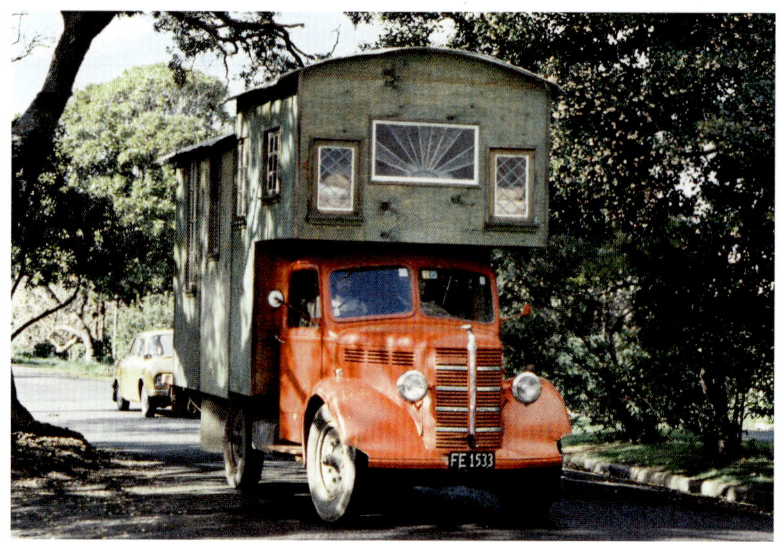

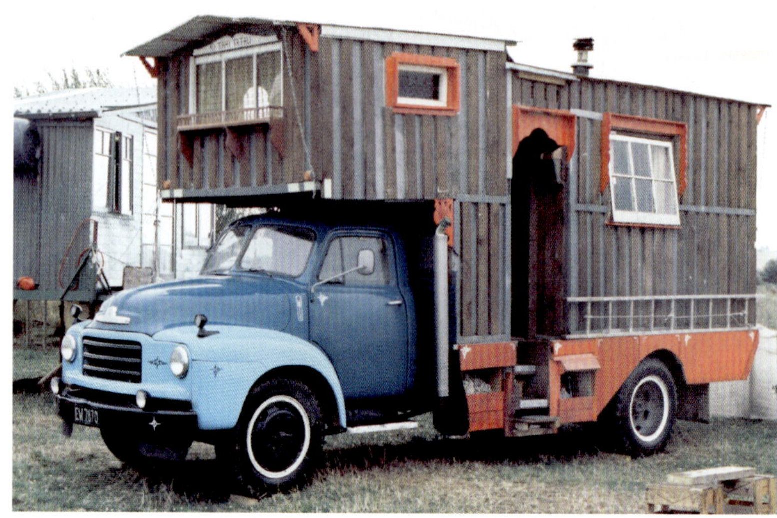

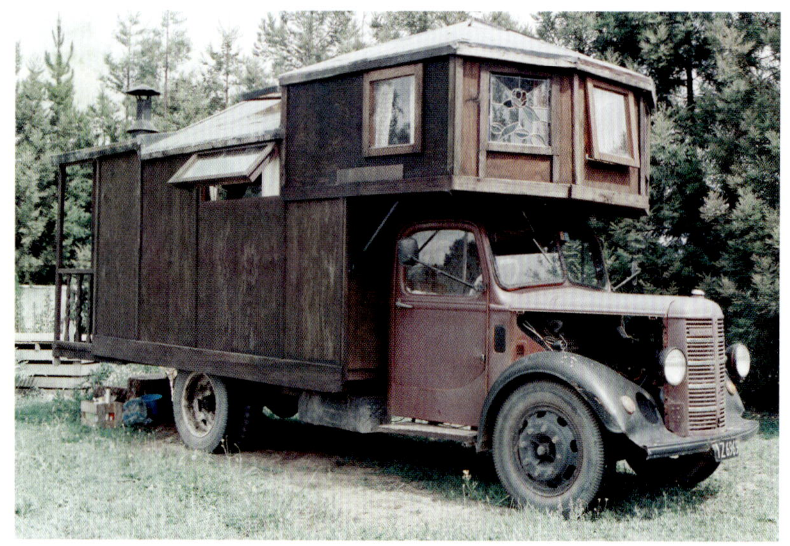

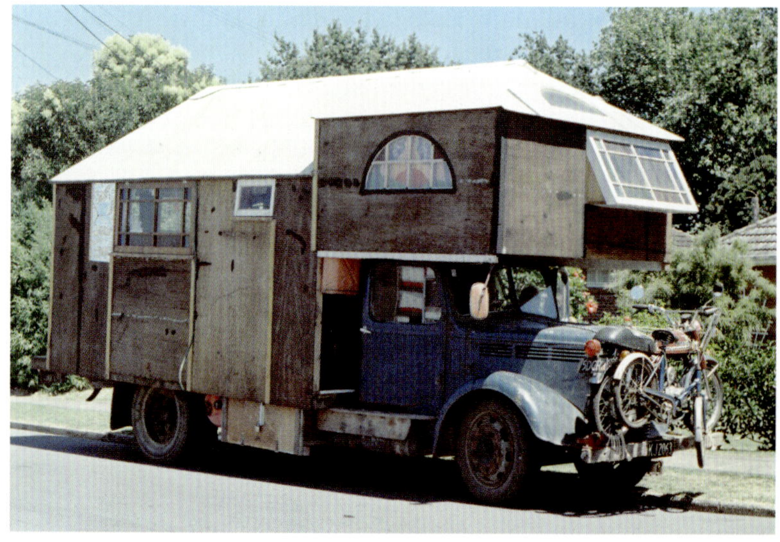

Kathy and Larry's truck, Aratiatia, 1984

Unidentified truck, c.1980

Unidentified truck, c.1984

Eccles' bus, Shakespeare Road, Auckland, n.d.

Goober's #2 truck, Sweetwaters site pre-festival, 1980

Ewan and Helen's truck, c.1983

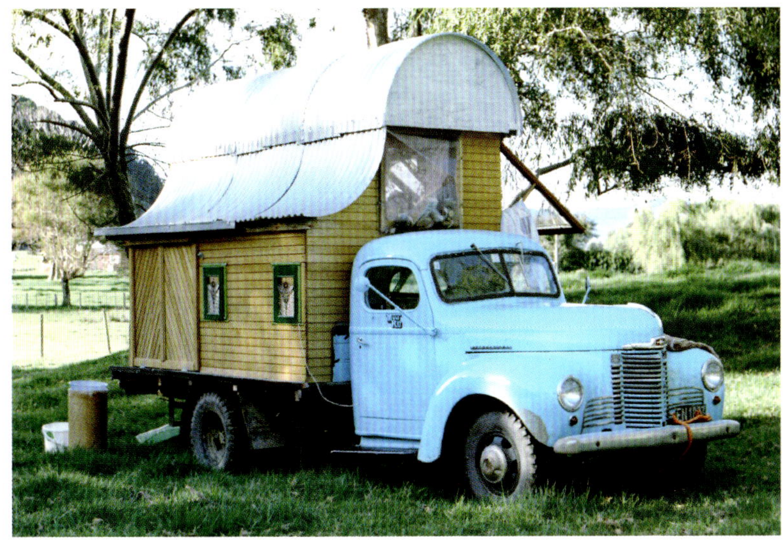

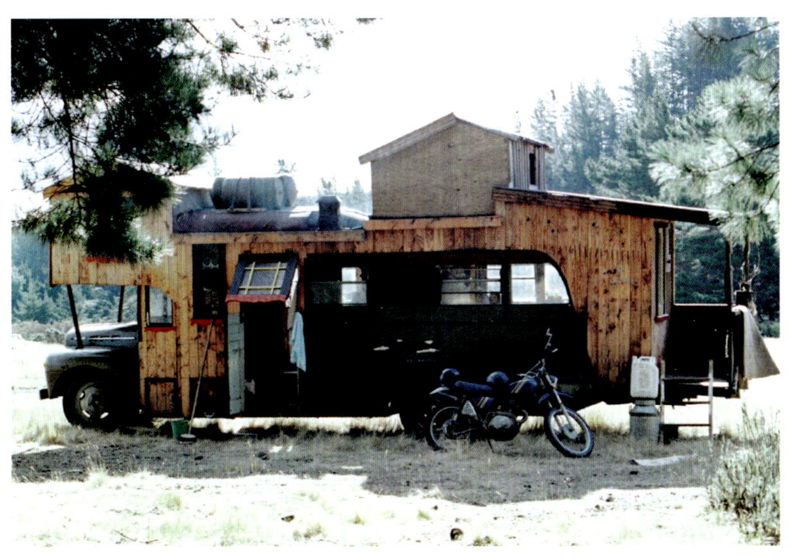

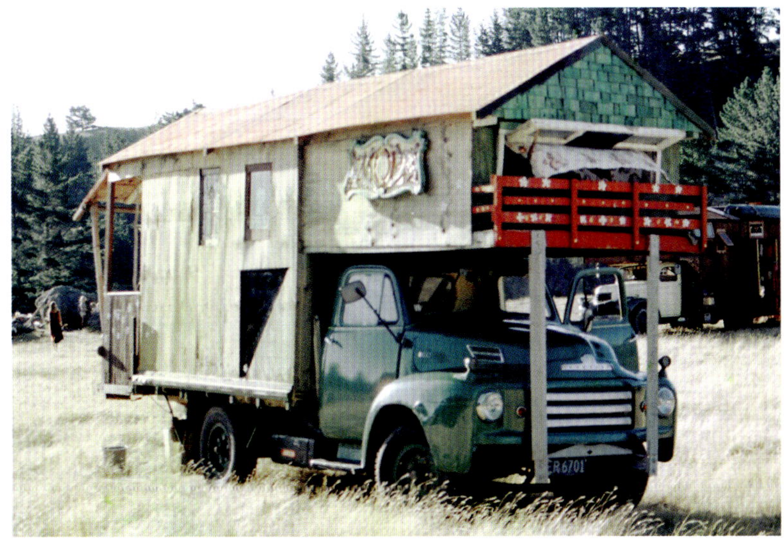

Terry St George's truck 'Mole Rat' at Moehau Community, Sandy Bay, 1978

Unidentified truck, Aratiatia, 1984

Unidentified truck, Aratiatia, 1984

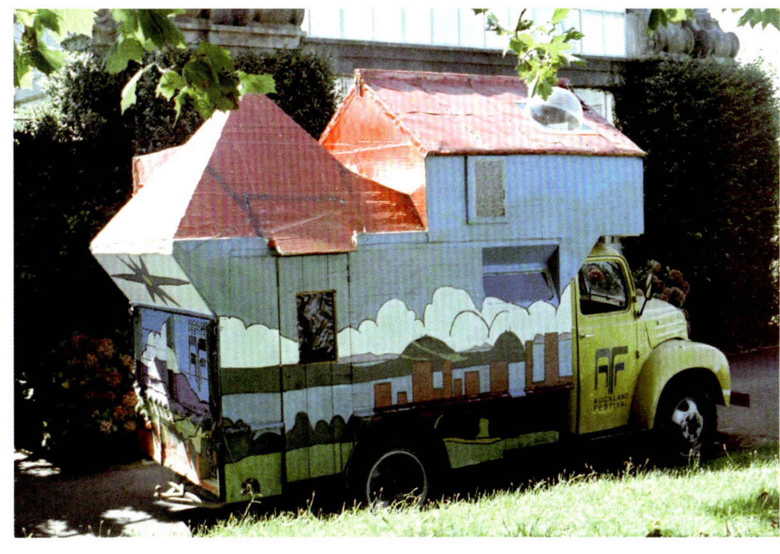

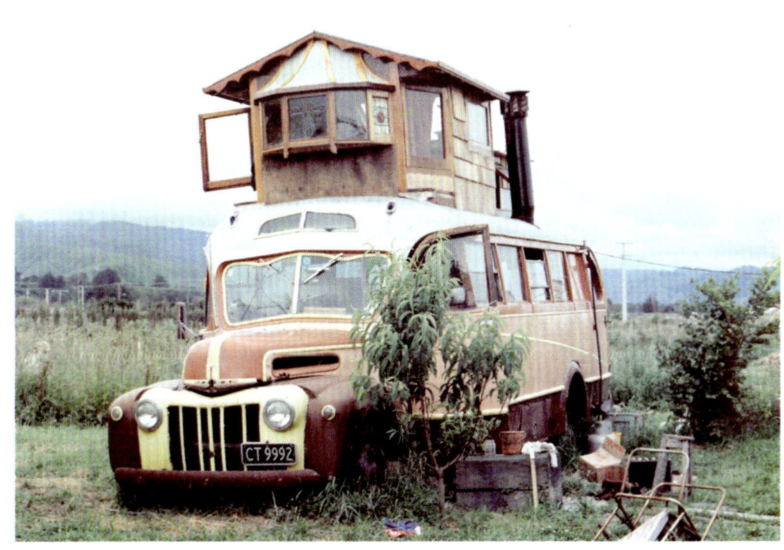

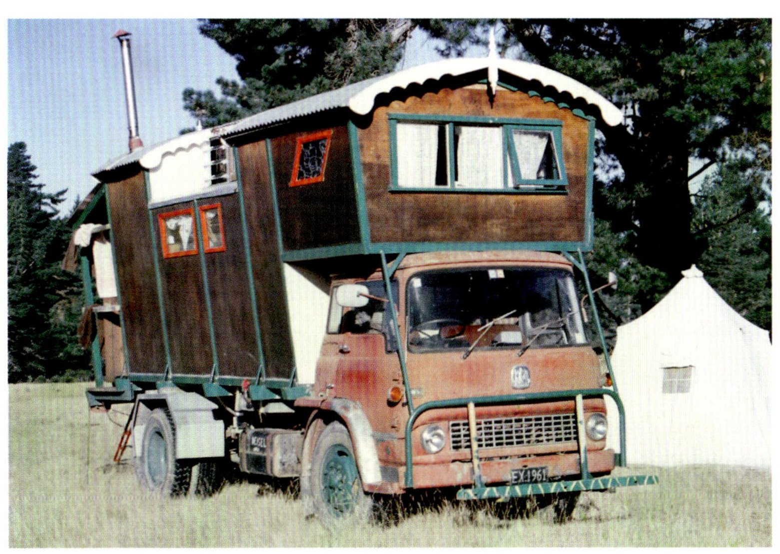

'Betsy', Auckland Festival truck, later adopted by Trunk Fools, 1978

Unidentified bus, c.1979

Unidentified truck, Aratiatia, 1984

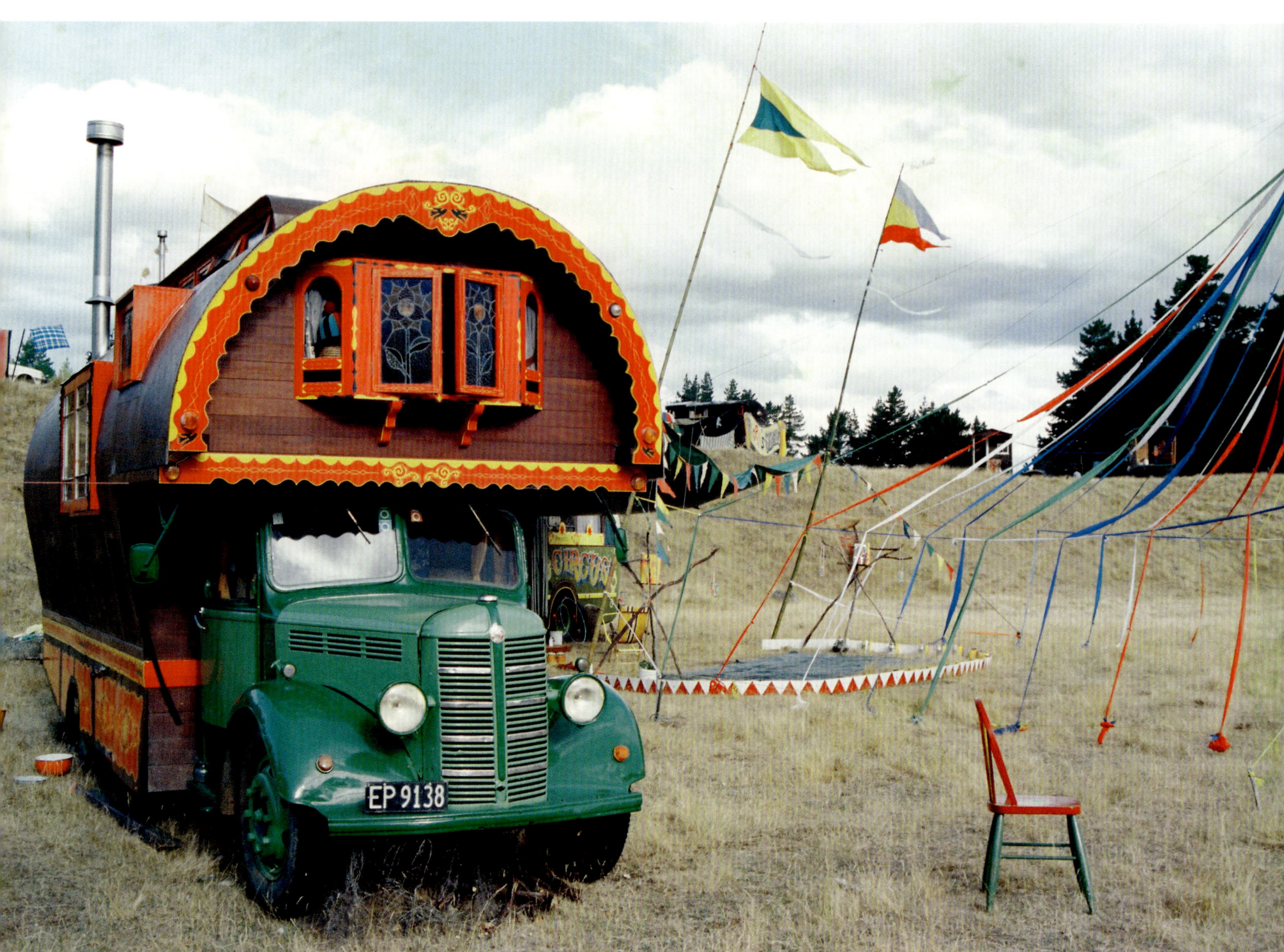

Dom and Jane's truck, People's Circus, House Truck & Mobile Shelter Convention, Aratiatia, January 1984

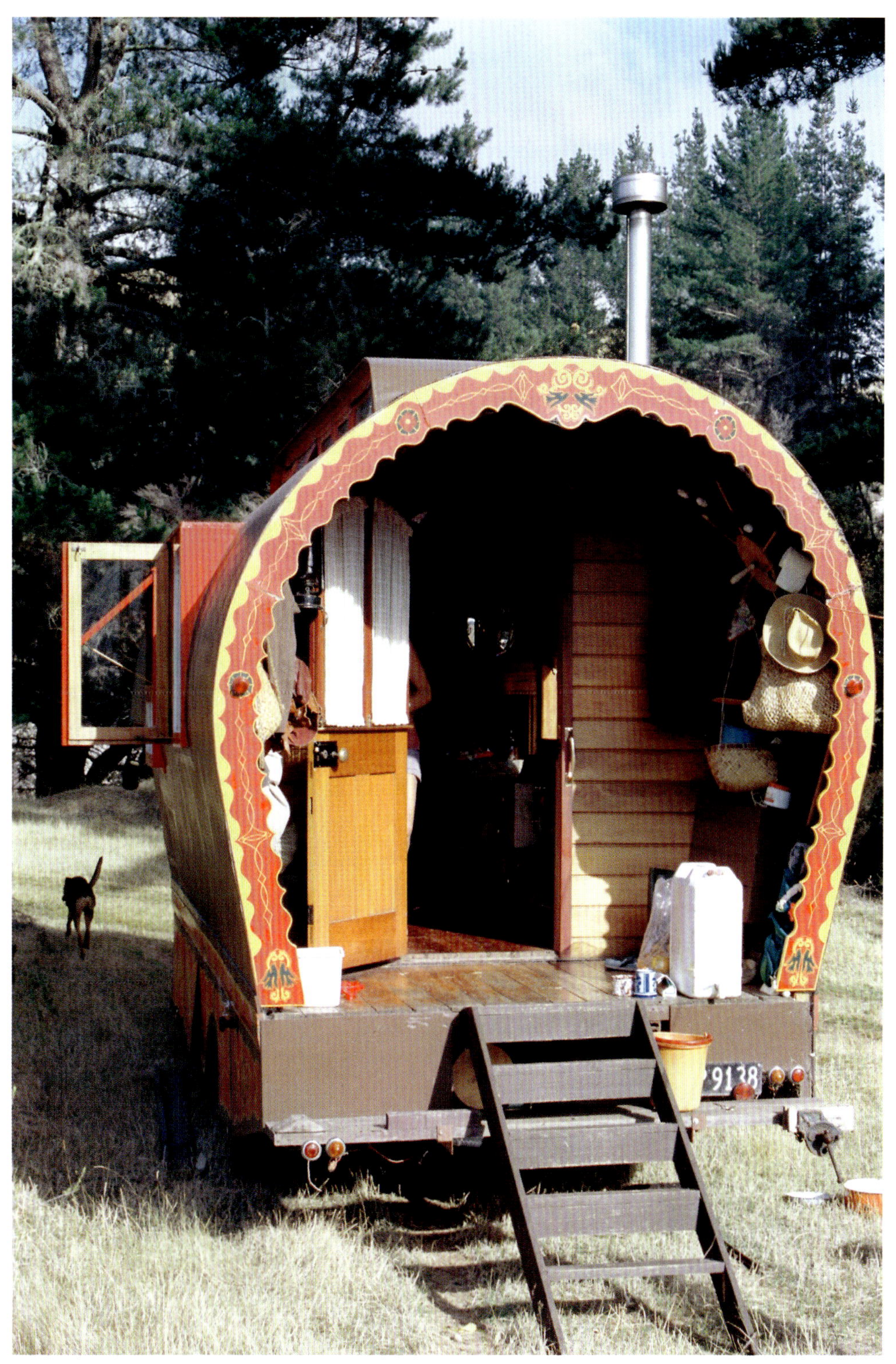

‘Front’ door of Dom and Jane’s truck, Aratiatia, 1984

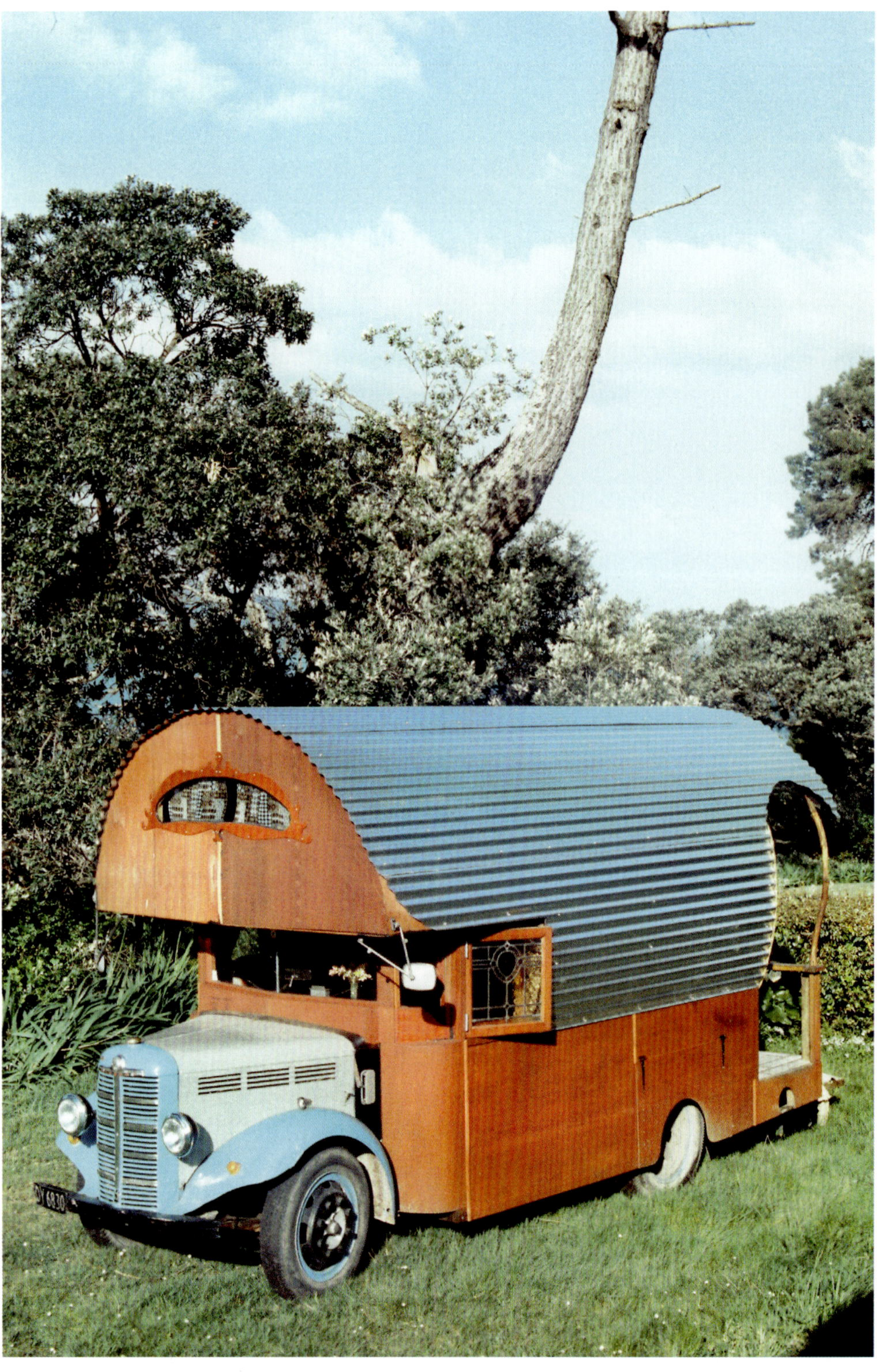

Podge and Diane's truck, Clifton Road, c.1979

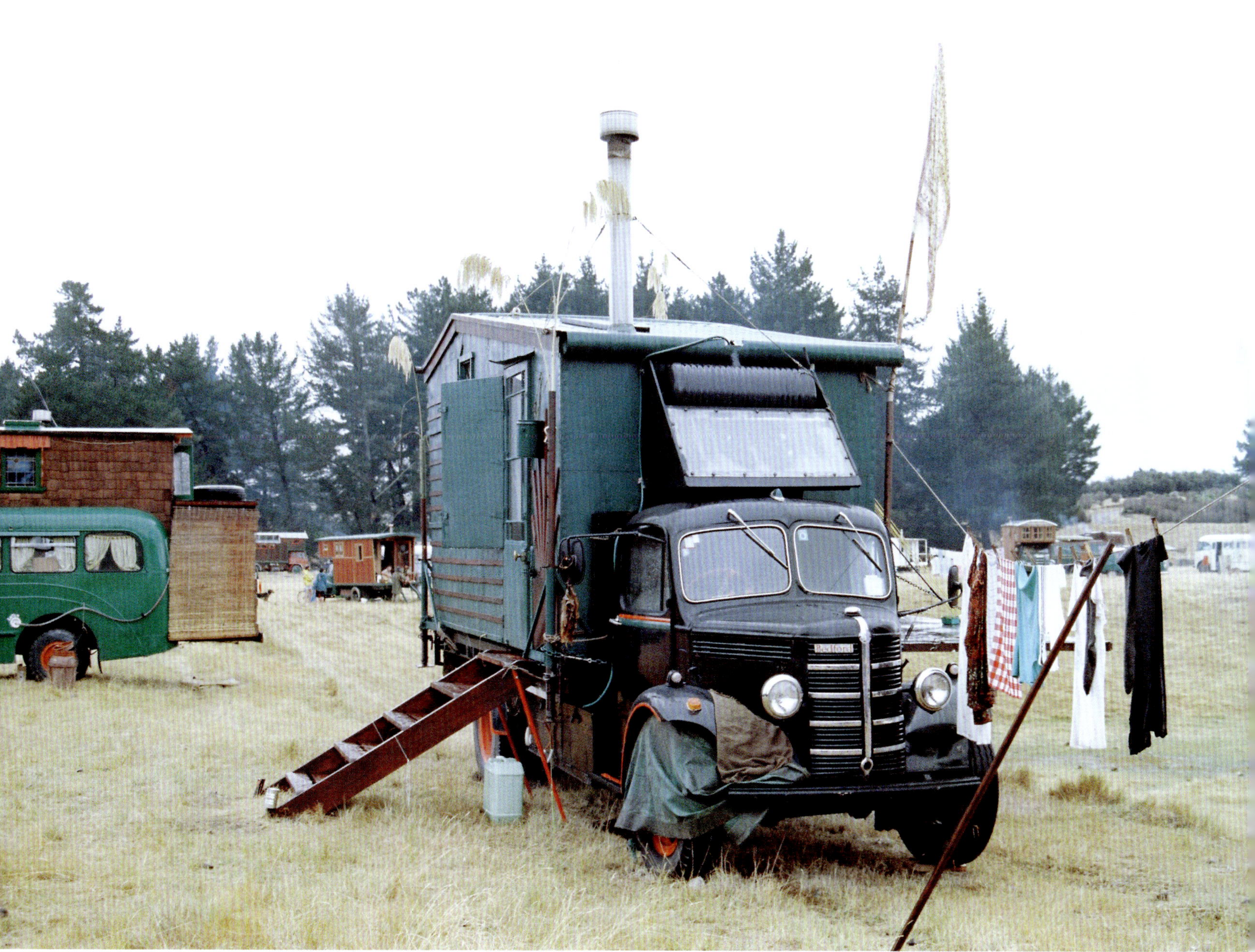

Gypsy Jeff's 'Stealth Truck', Aratiatia, 1984

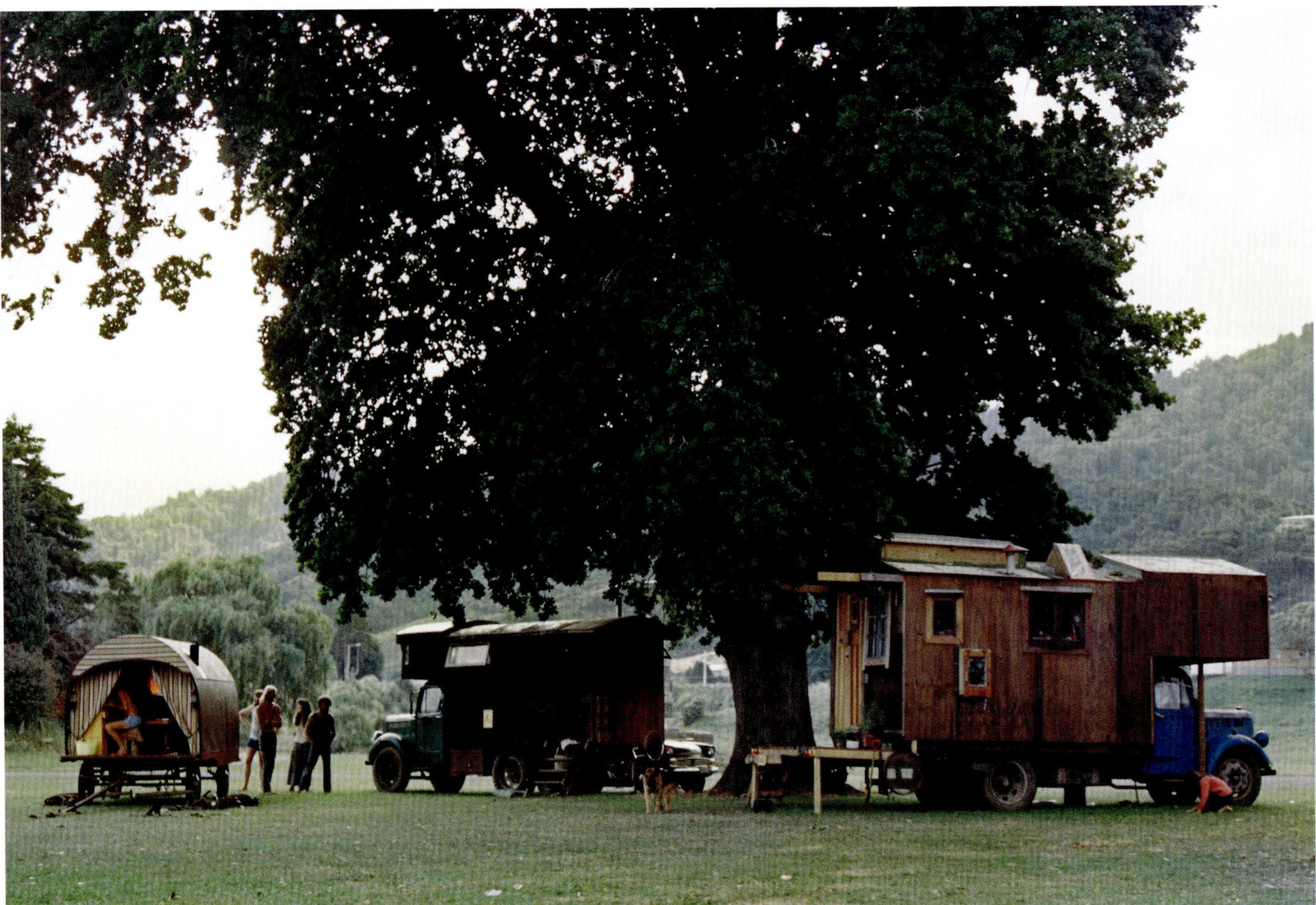

Unidentified trucks at unknown location, n.d

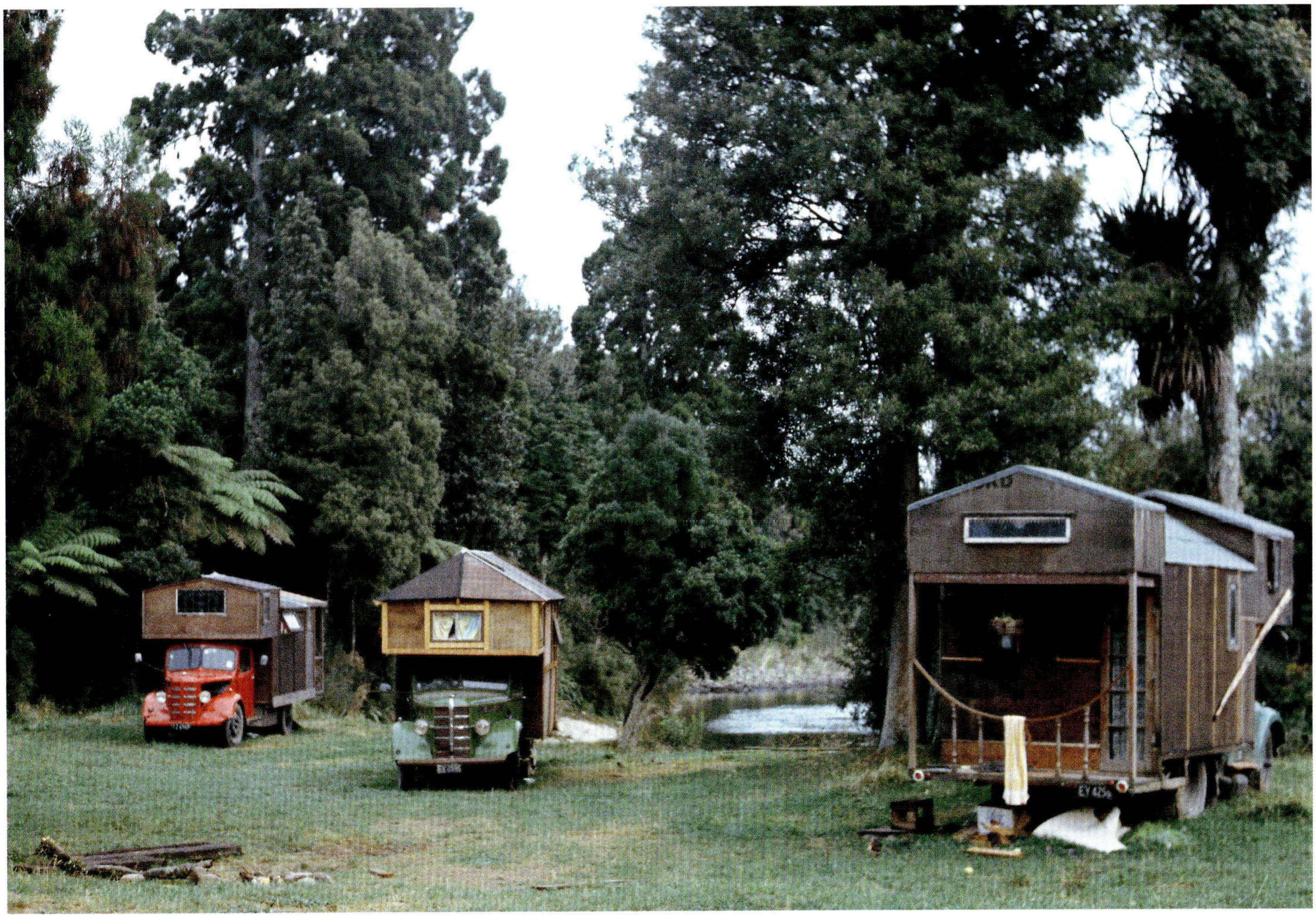

First arrivals at Nambassa Five-day Celebration of Music, Crafts and Alternative Lifestyles Culture, Waitawheta Valley near Waihi, 1981

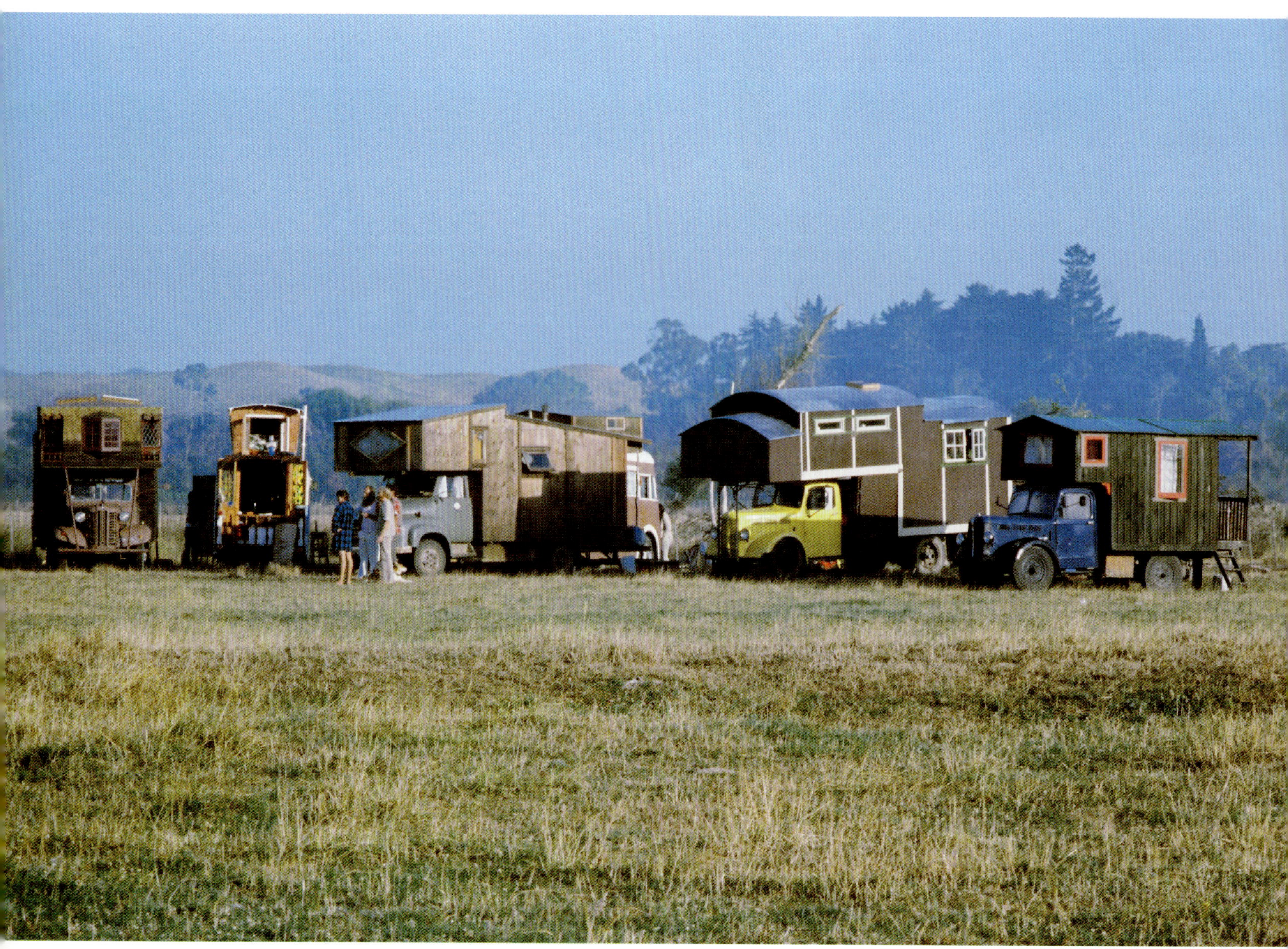

Sweetwaters Music Festival, Pukekawa, Lower Waikato River, 1982

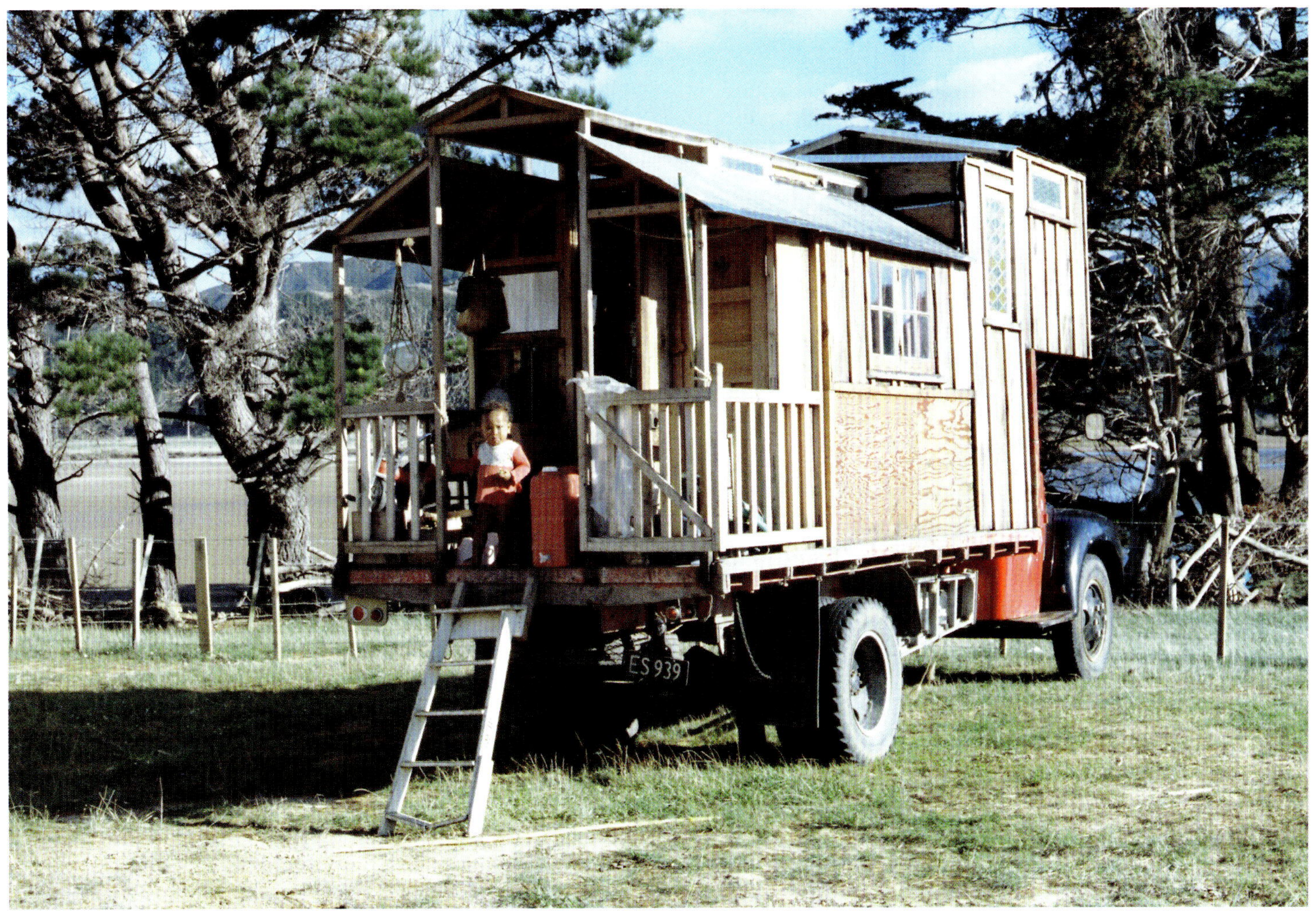

Trevor and Rangi McGlinchey's truck with Mauriri on veranda, Waikawau Bay, Coromandel, 1979

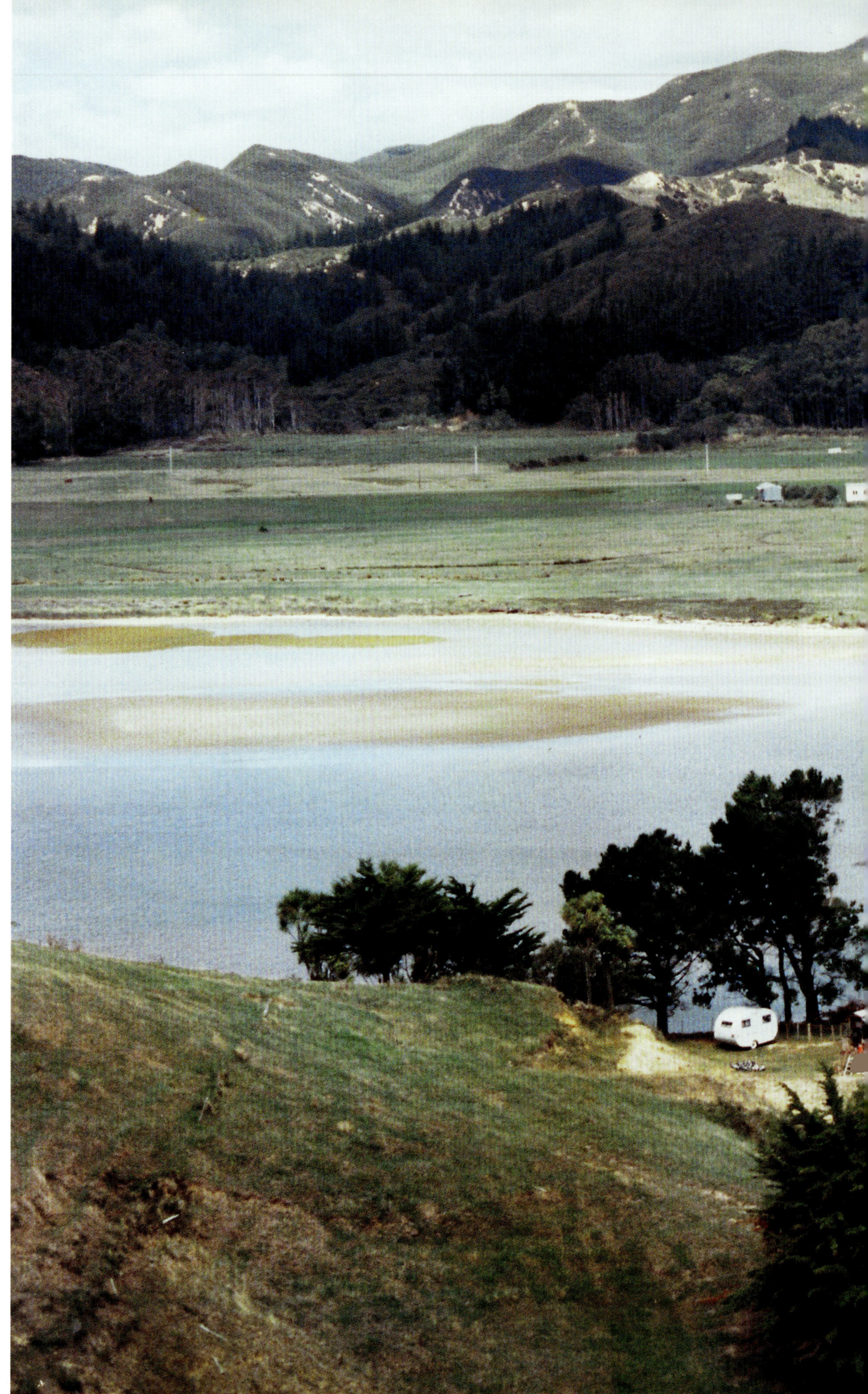

Waikawau Bay,
Coromandel, 1979

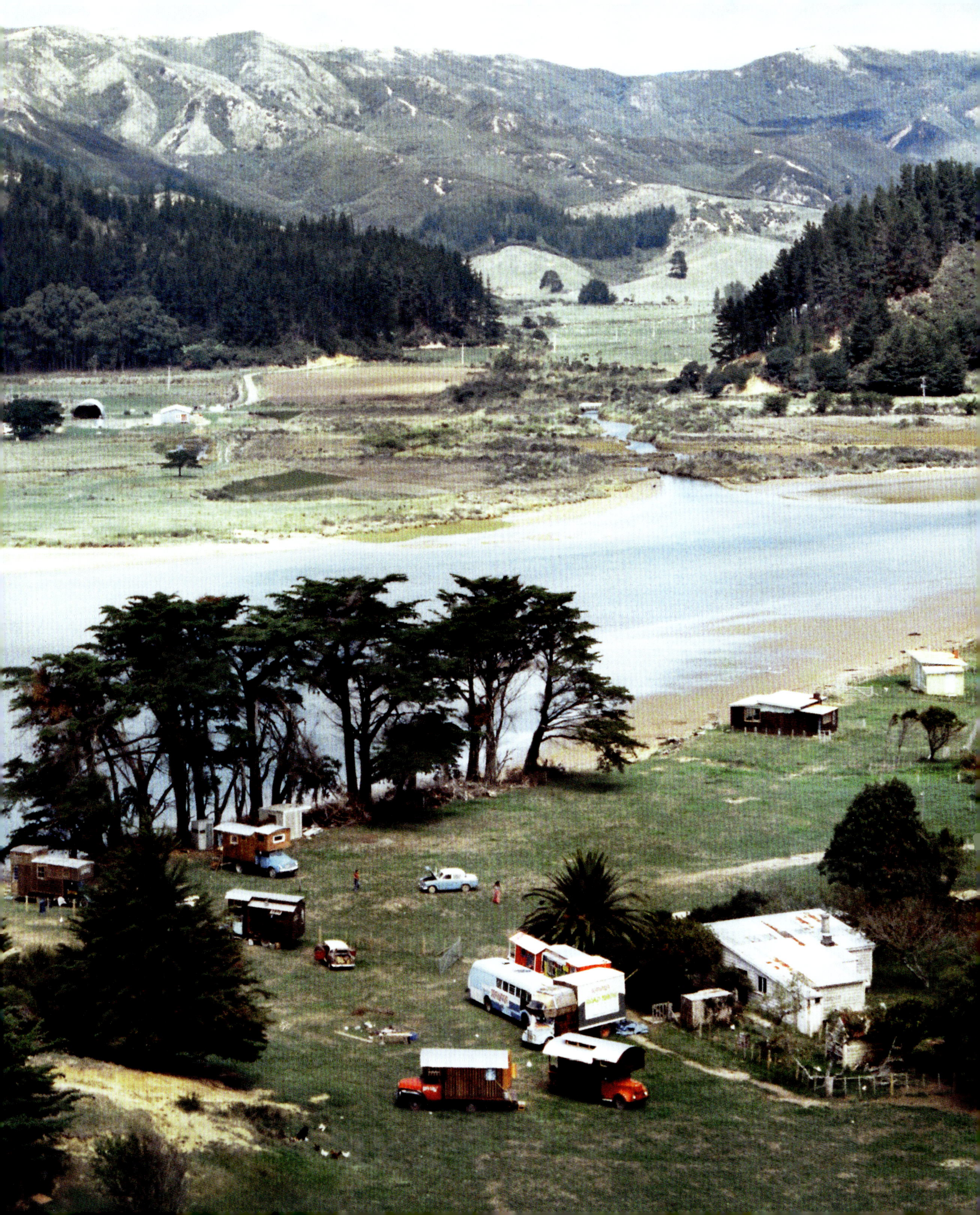

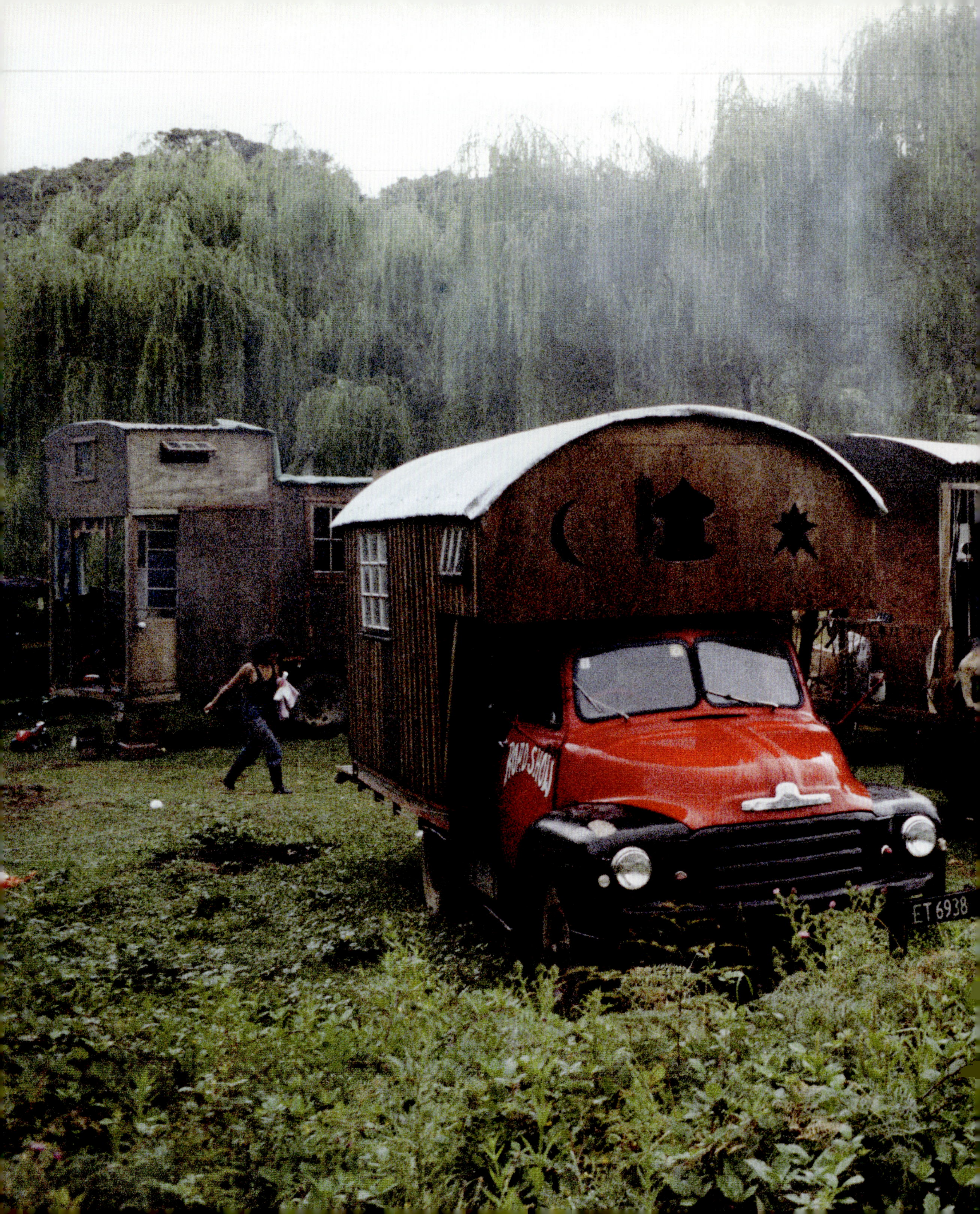
ROAD SHOW
ET 6938

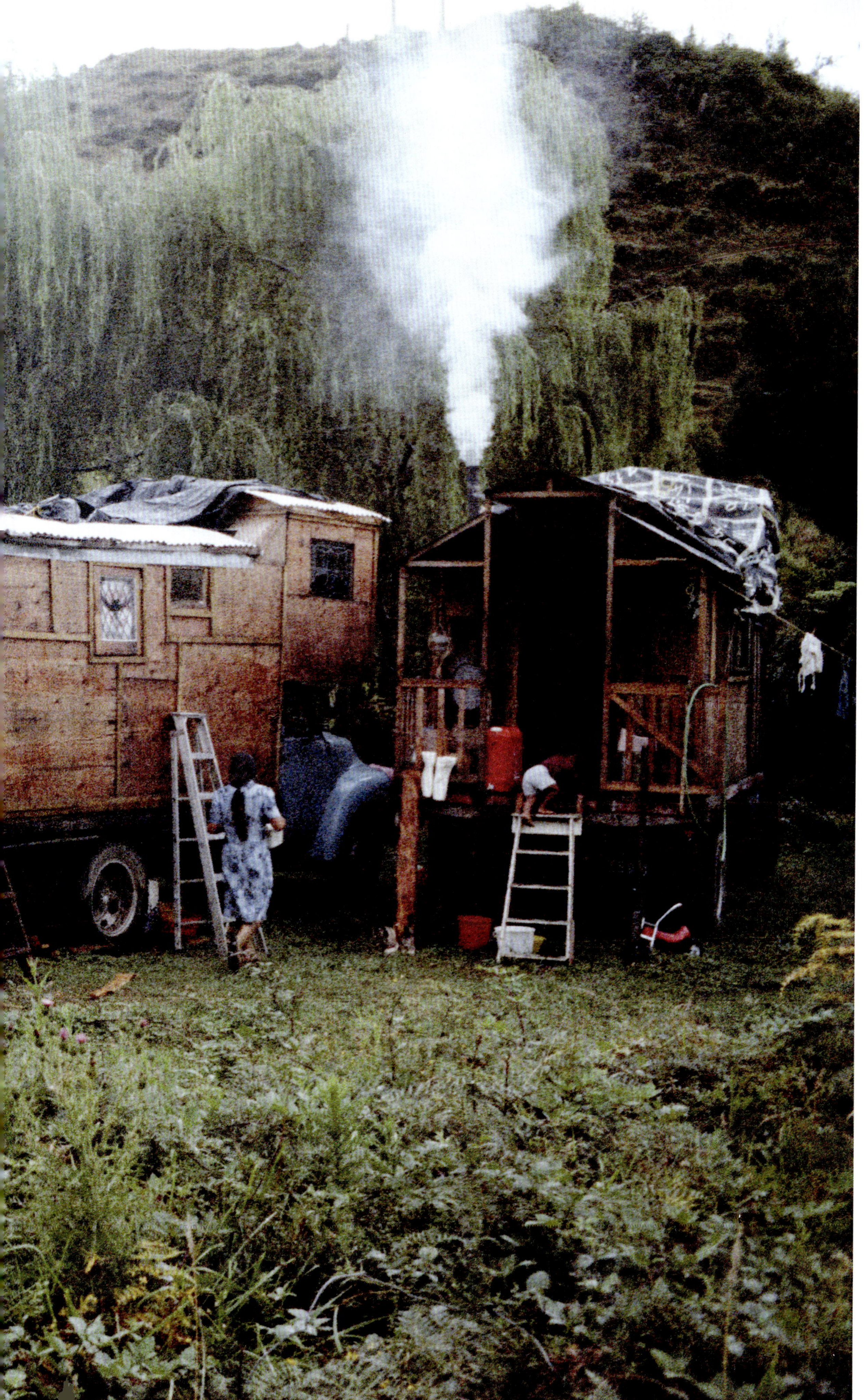

Road Show camp, 309 Road, Coromandel, after Nambassa, 1979

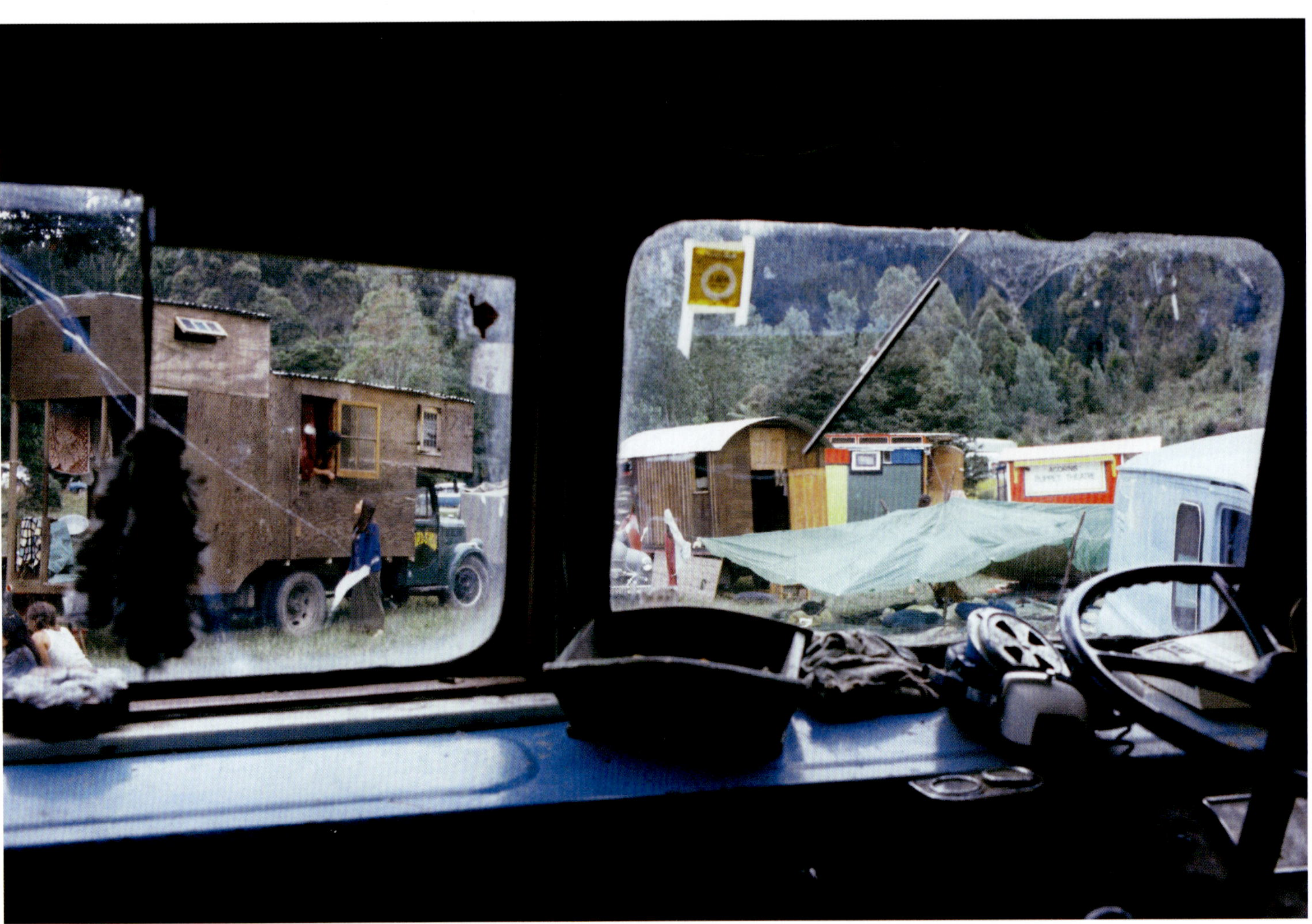

Fern Flat camp, Northland, 1978/79

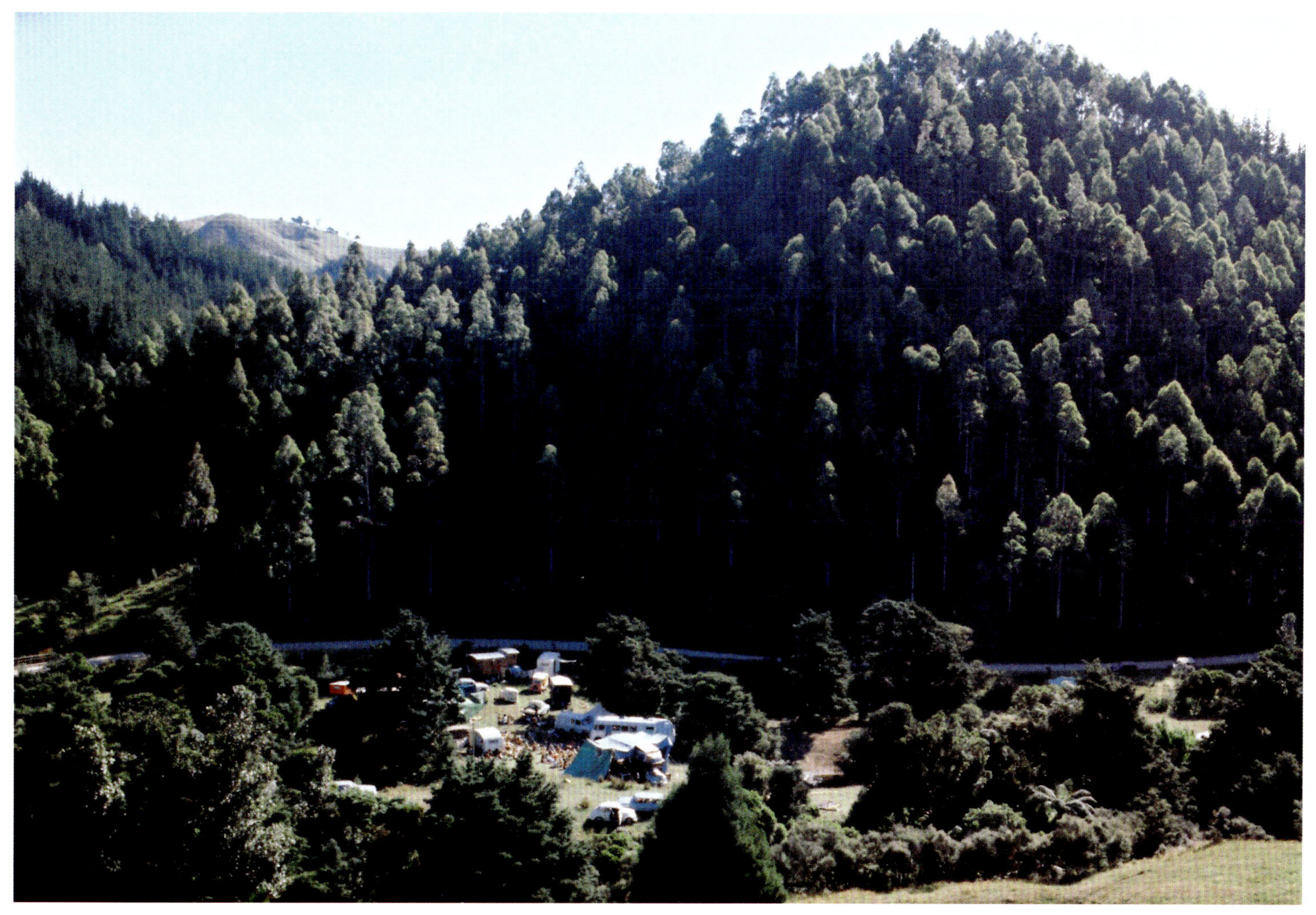

The Road Show on tour, Fern Flat camp, Northland, 1978/79

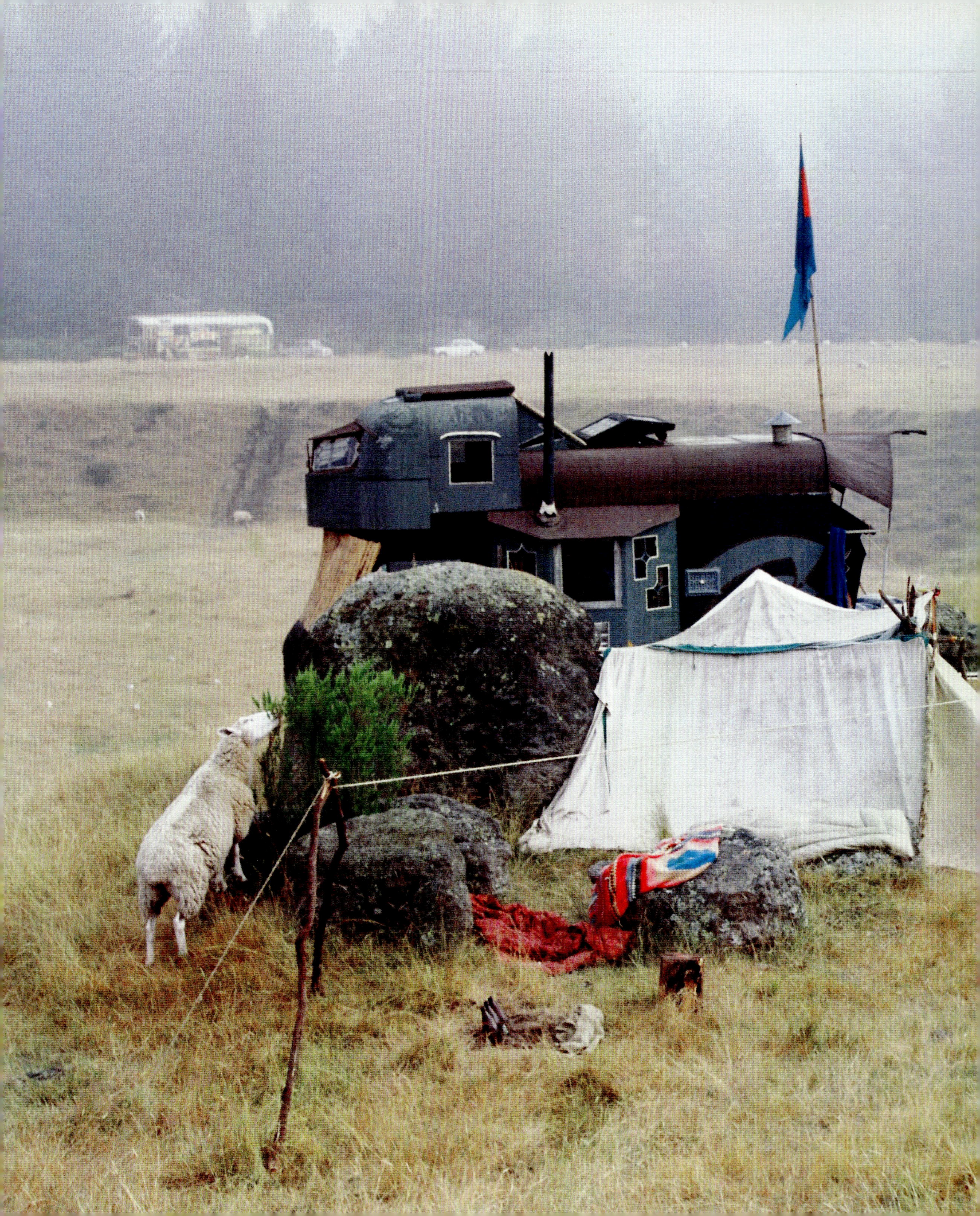

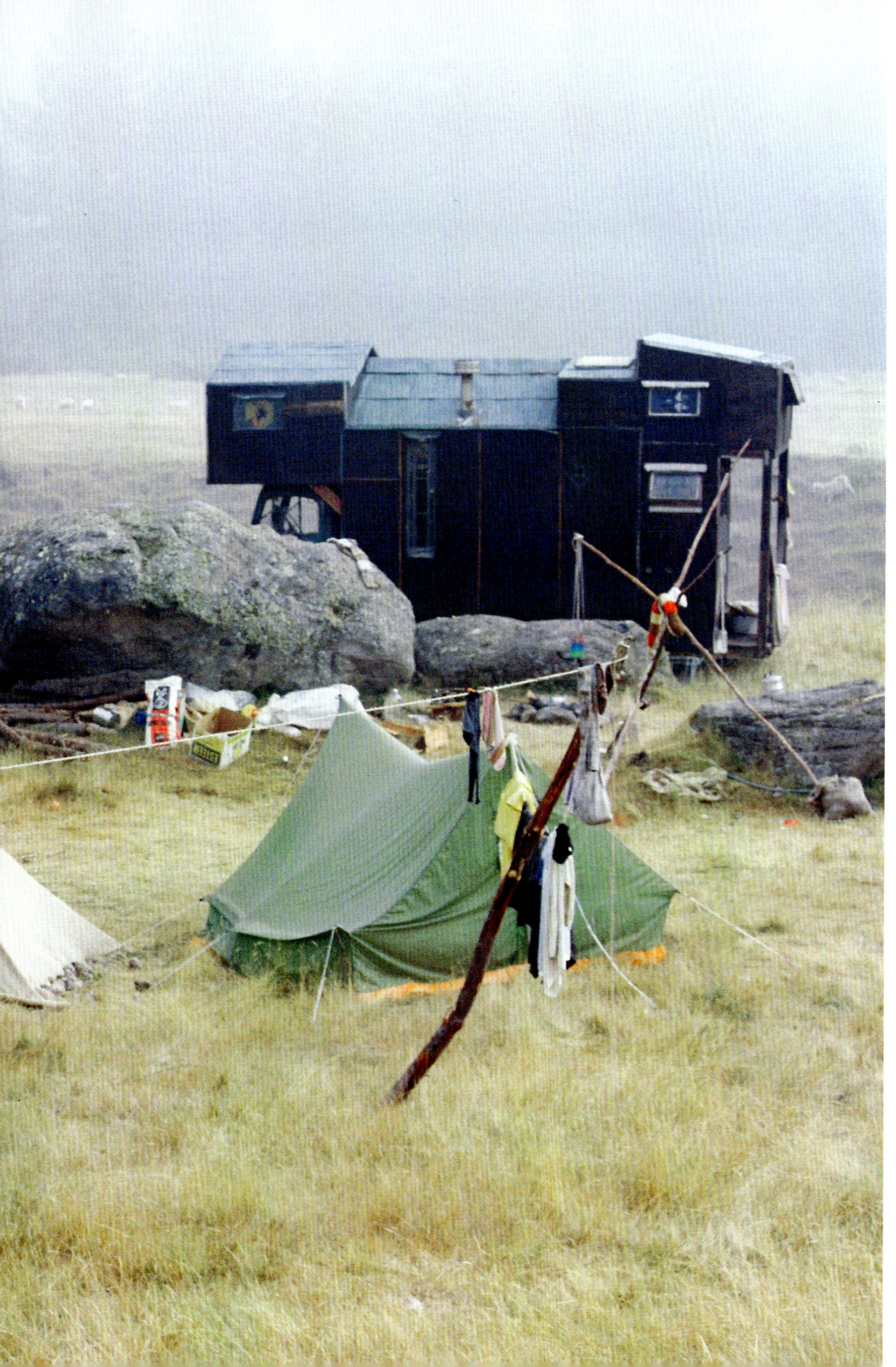

Rob Tee's van-home and Jade's truck, Aratiatia, 1984

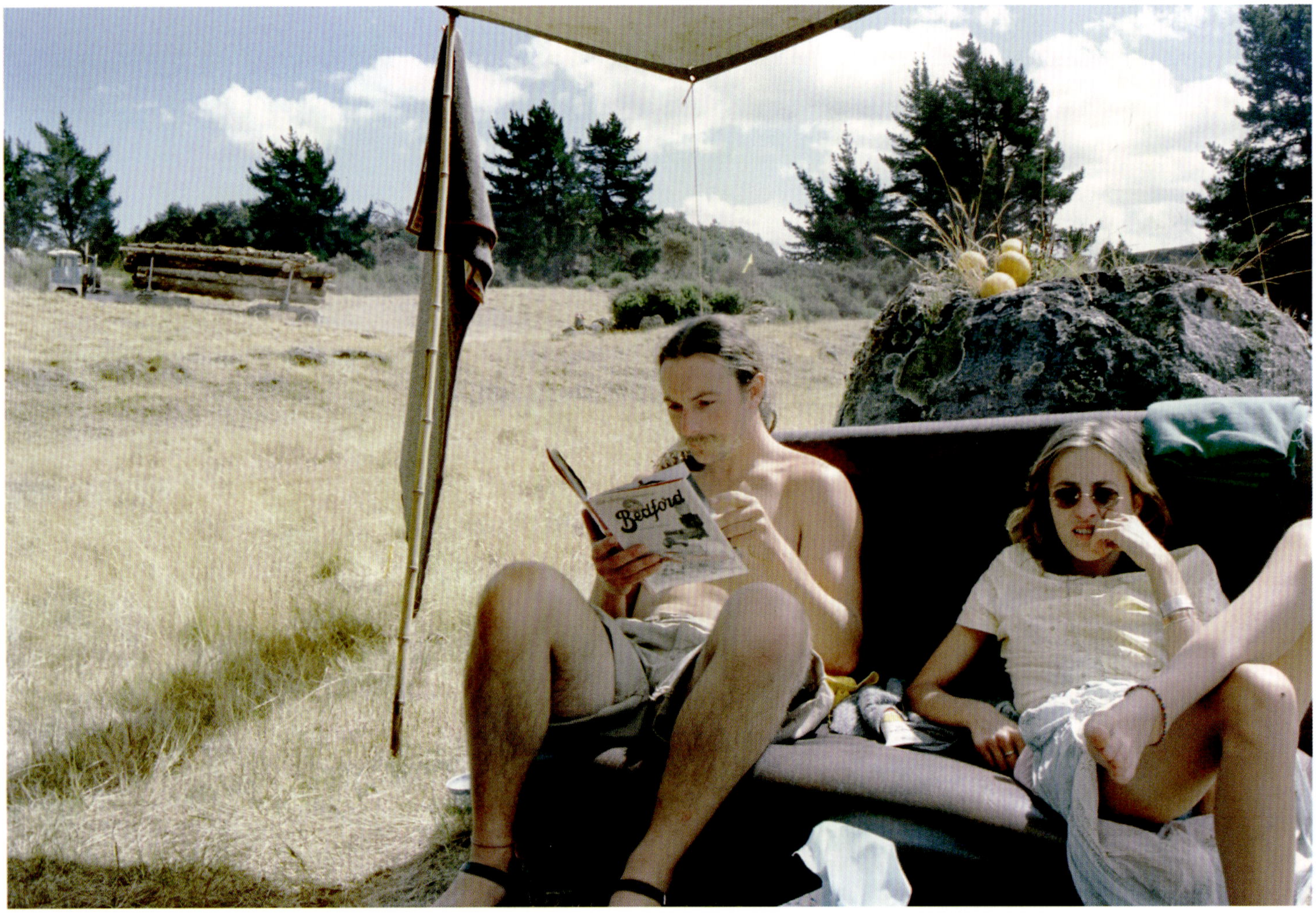

Graham Cairns and Sue Honeyman, Aratiatia, 1984

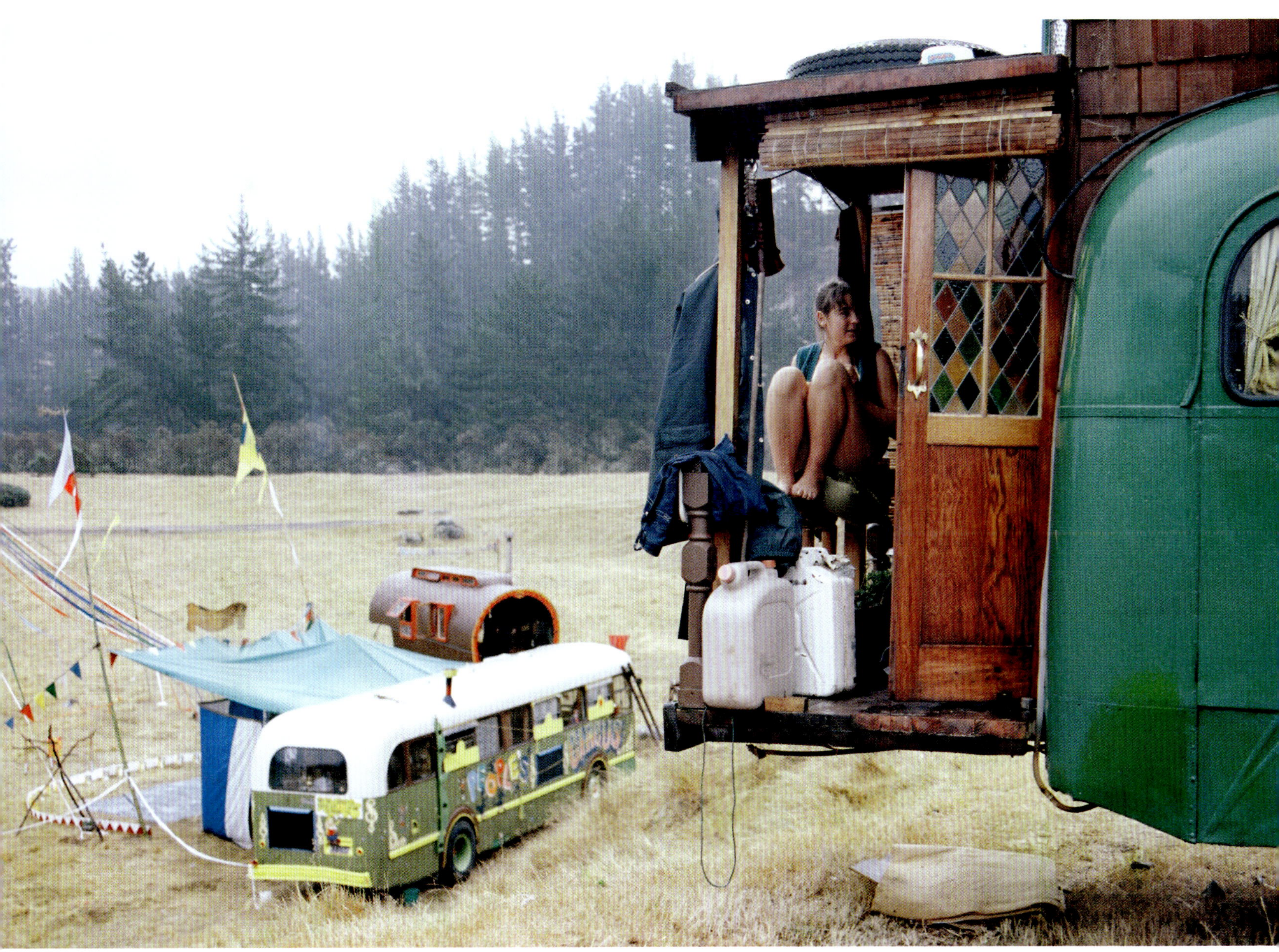

The People's Circus stage, Aratiatia, 1984

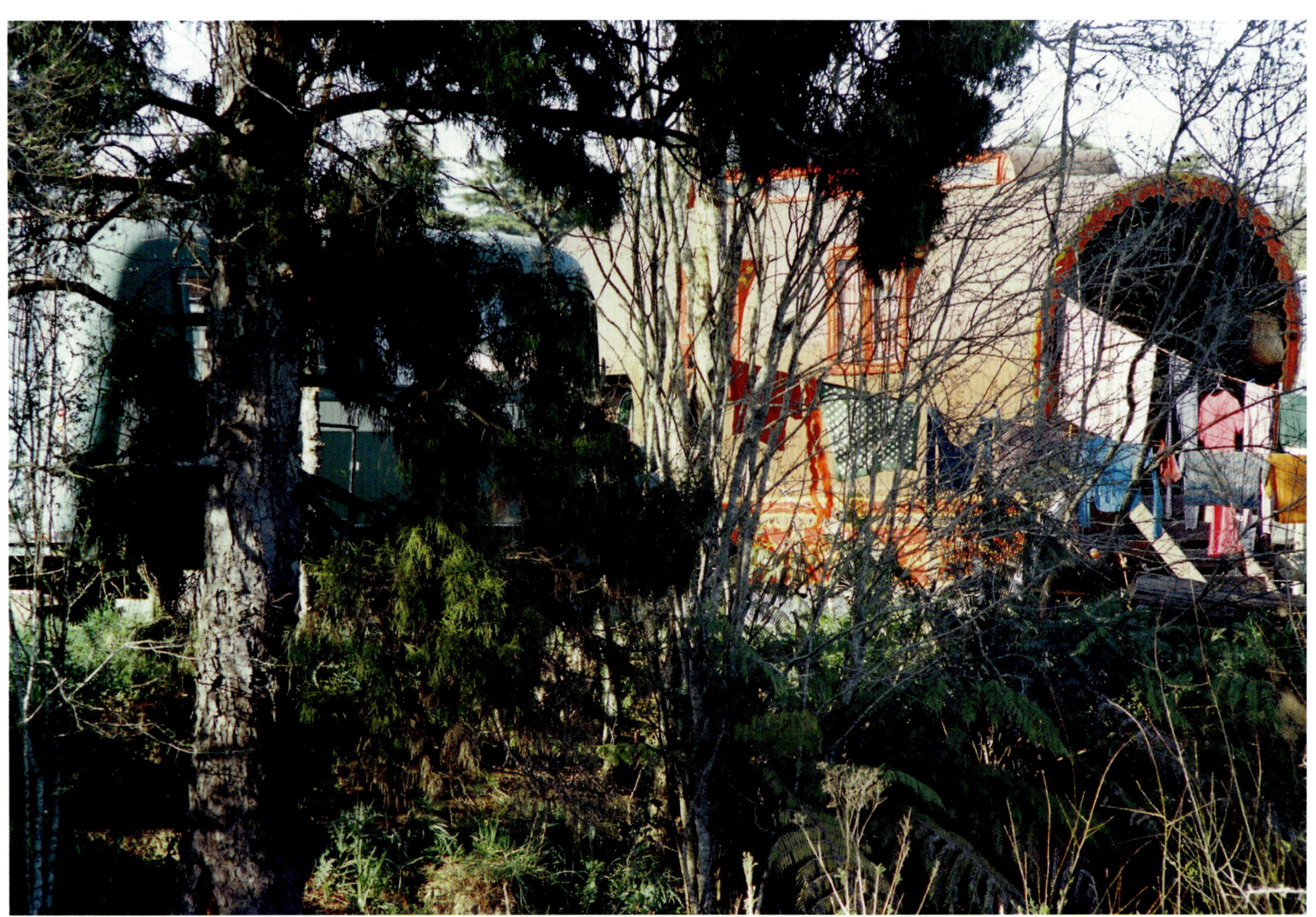

Paul's bus 'Iris' and Dom and Jane's #2 truck, at Brookby, 1984

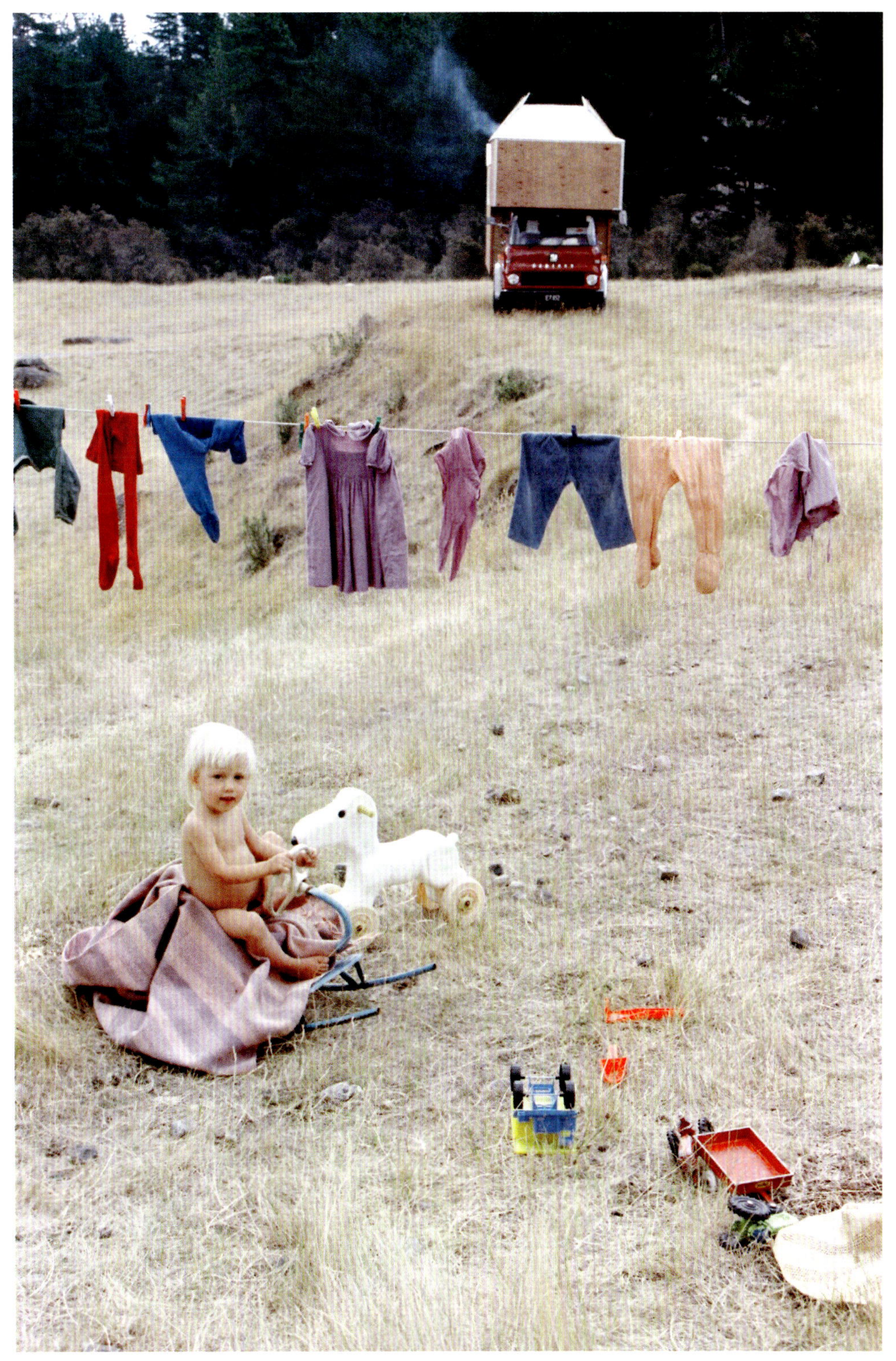

Aya, wash day, Aratiatia, 1984

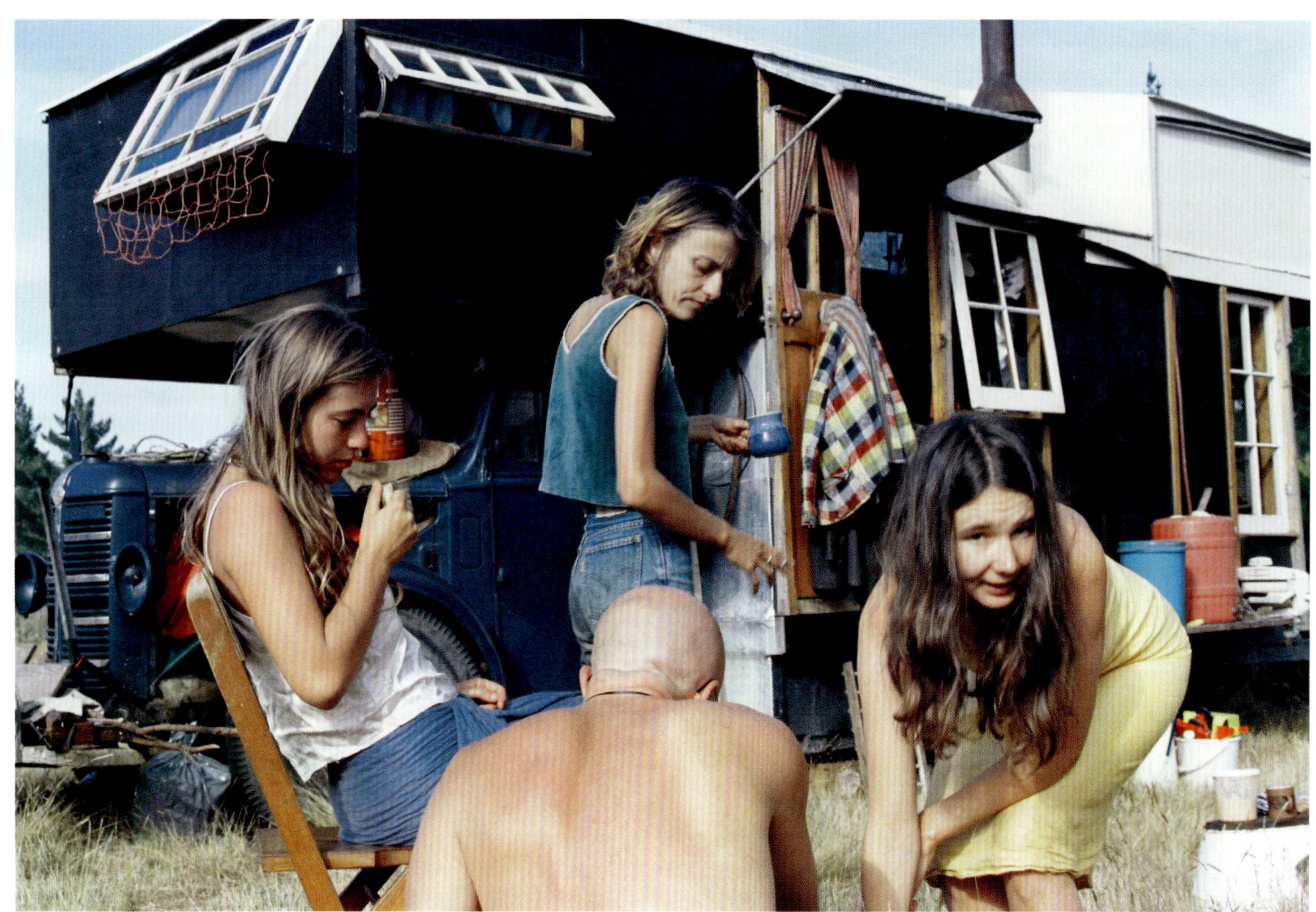

Claire, David Sheridan, Sue and Helen Driver with Ewan and Helen's truck, Aratiatia, 1984

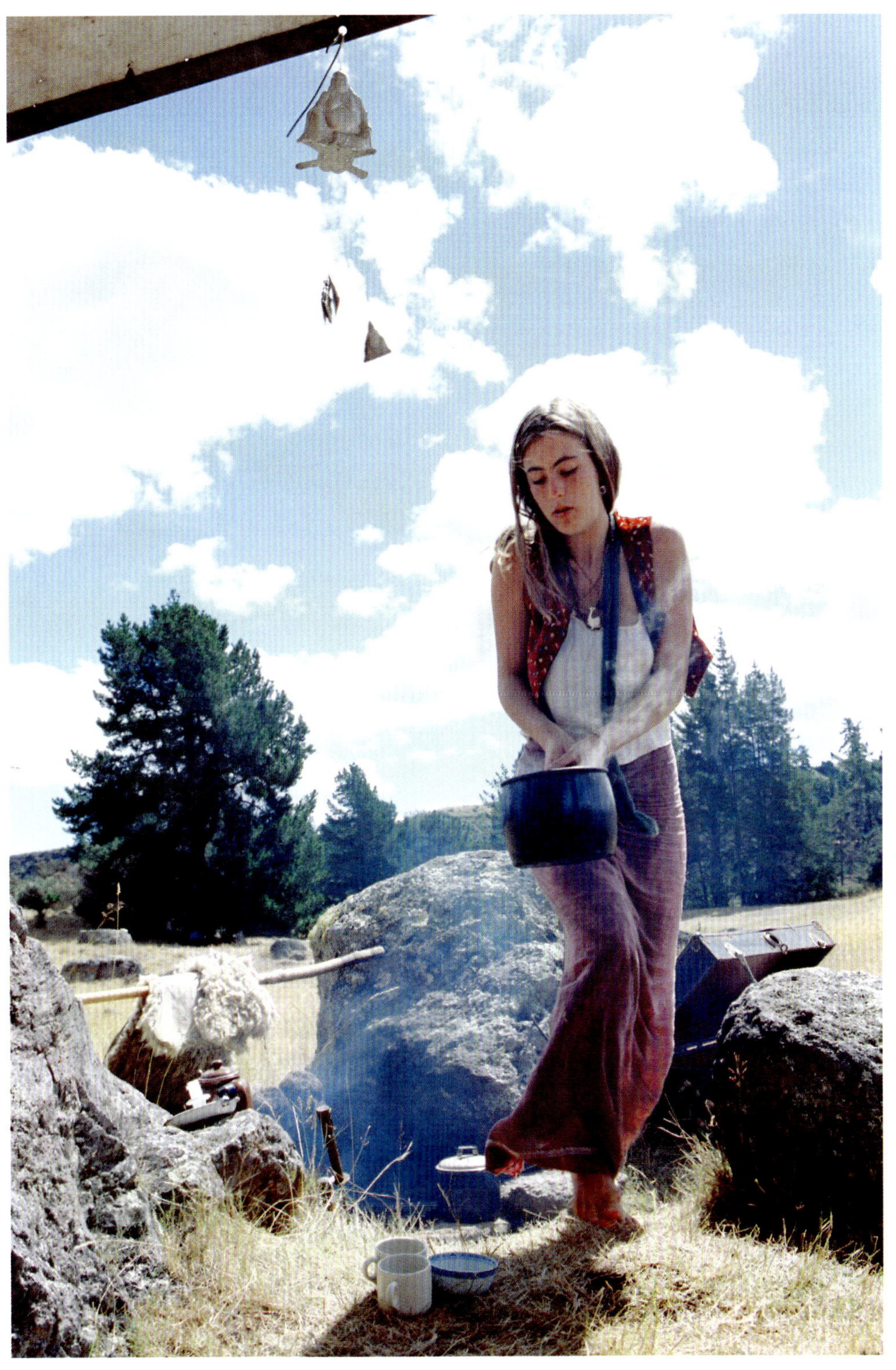

Claire making tea, Aratiatia, 1984

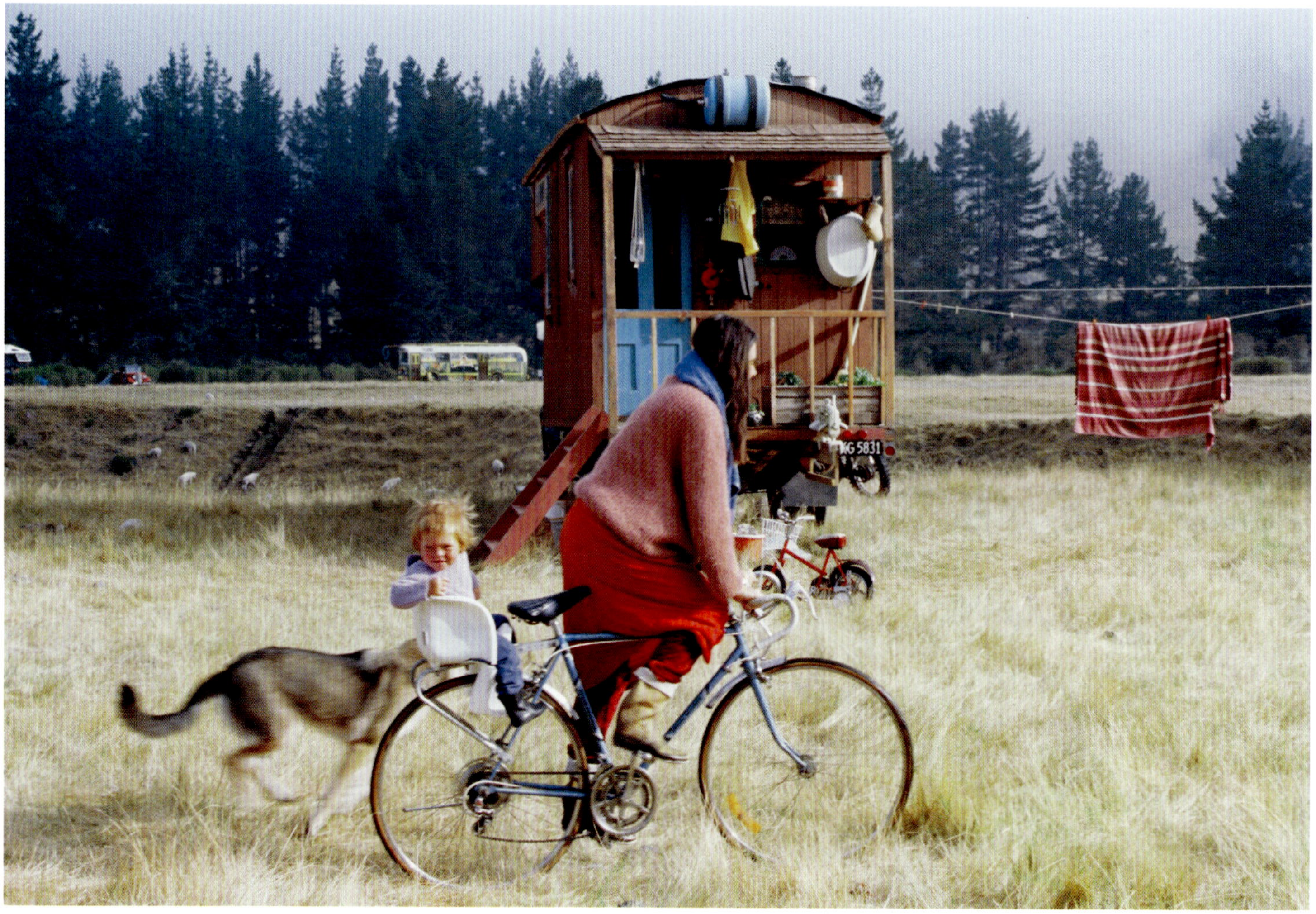

Steve and Shona Smith's truck, post gathering, Aratiatia 1984

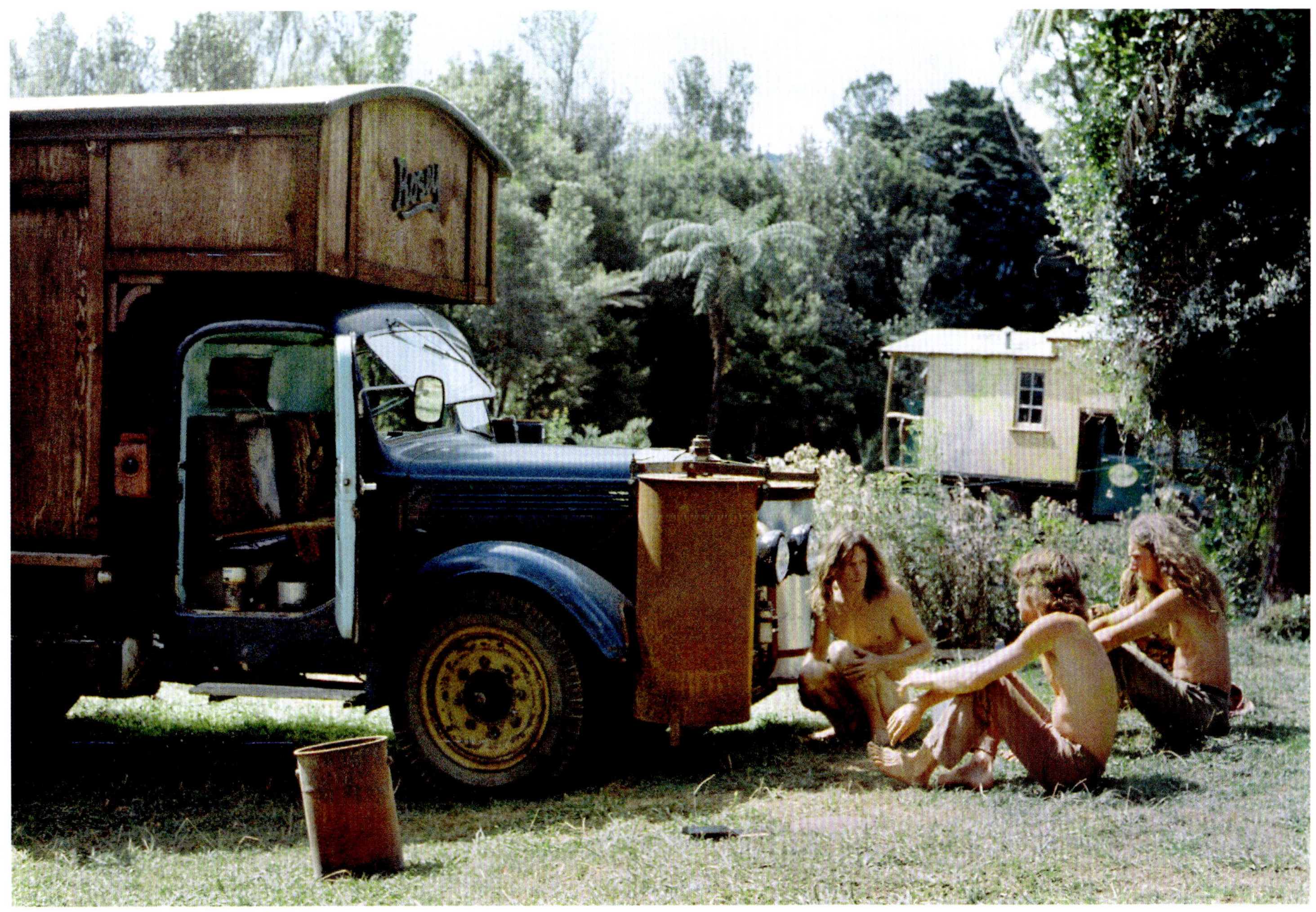

Rosey, Pete's gasifier truck. 'One sack of coal to get to Hamilton from Auckland', Nambassa, 1981

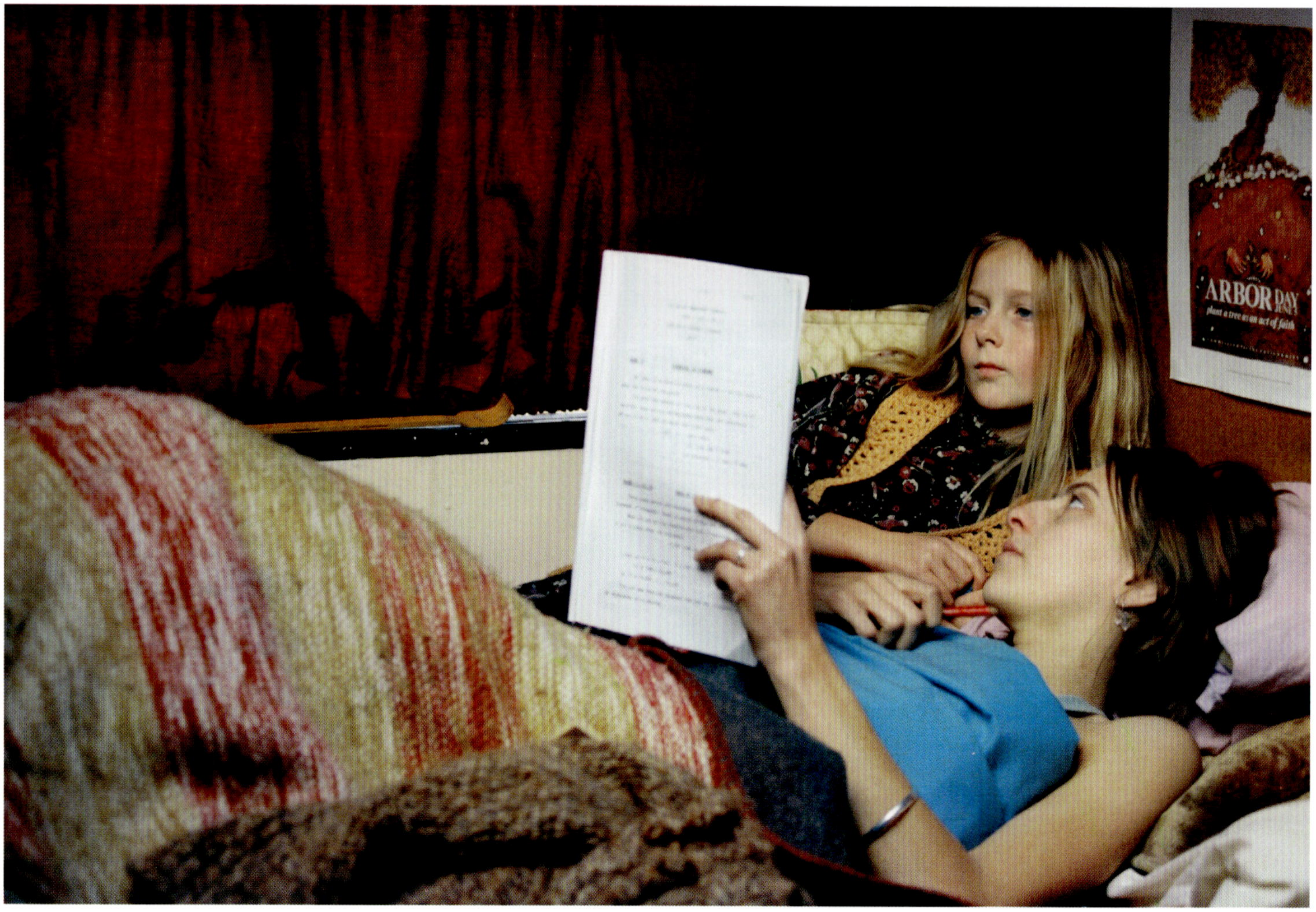

Sue helping Petra with homework in 'Iris', Clifton Road, 1981

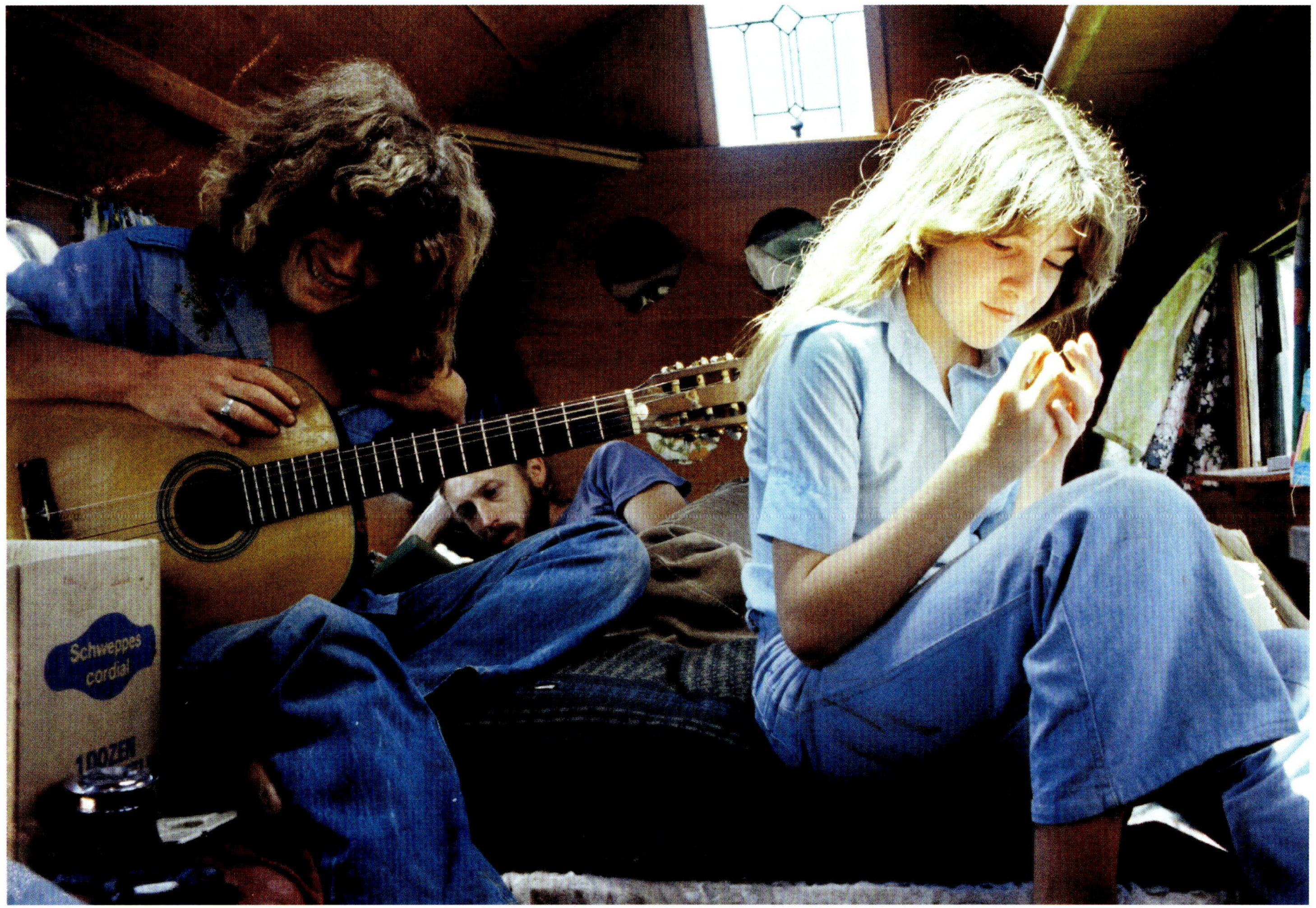

Steve Terry on guitar, with Ewan and Tanya in the aeroplane caravan, 1979

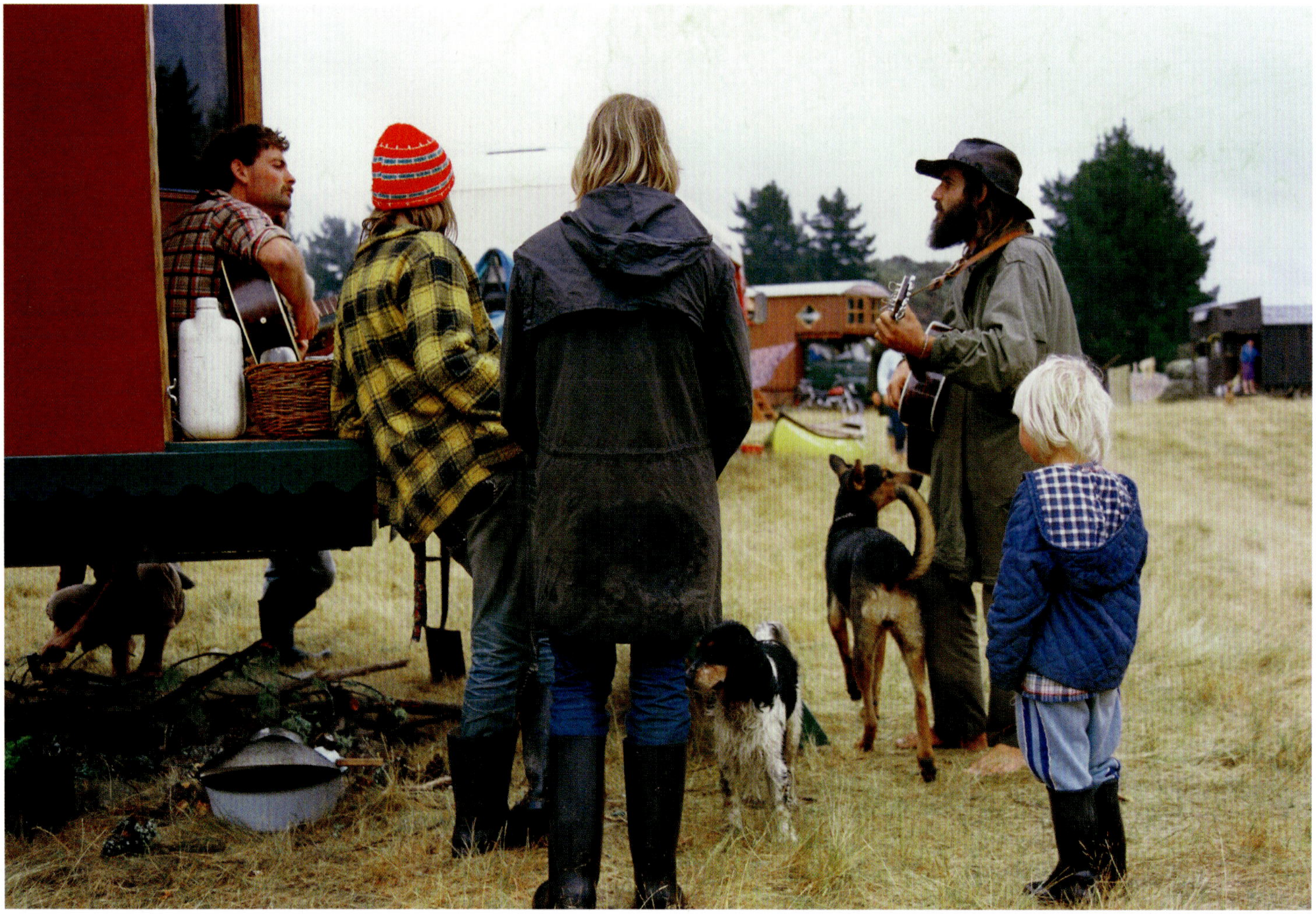

Jamming session with Richard, cane weaver 'Panashé Basket' (wearing big brimmed hat), Aratiatia, 1984

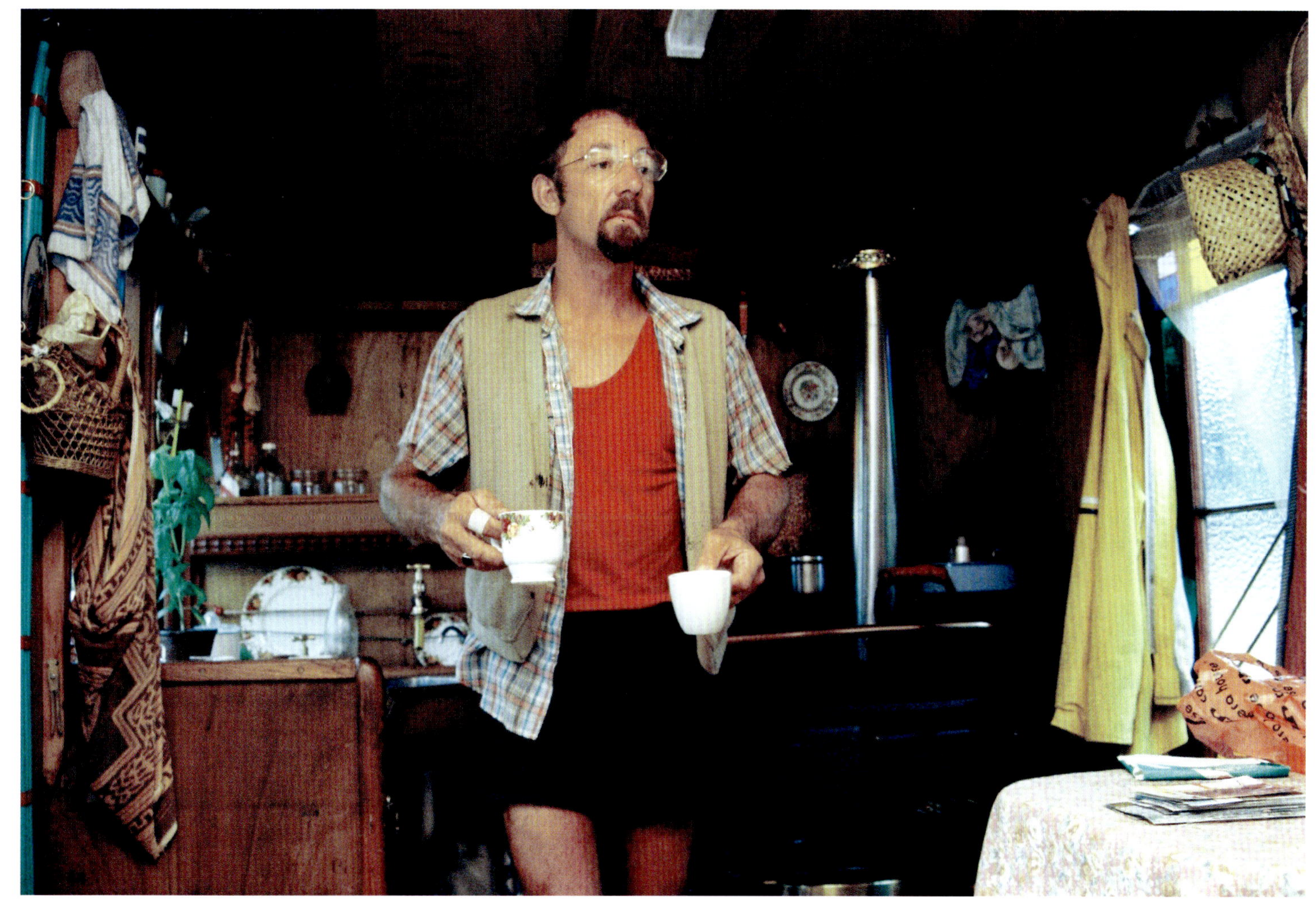

Gypsy Jeff serving tea inside his ‘Stealth Truck’, Aratiatia, 1984

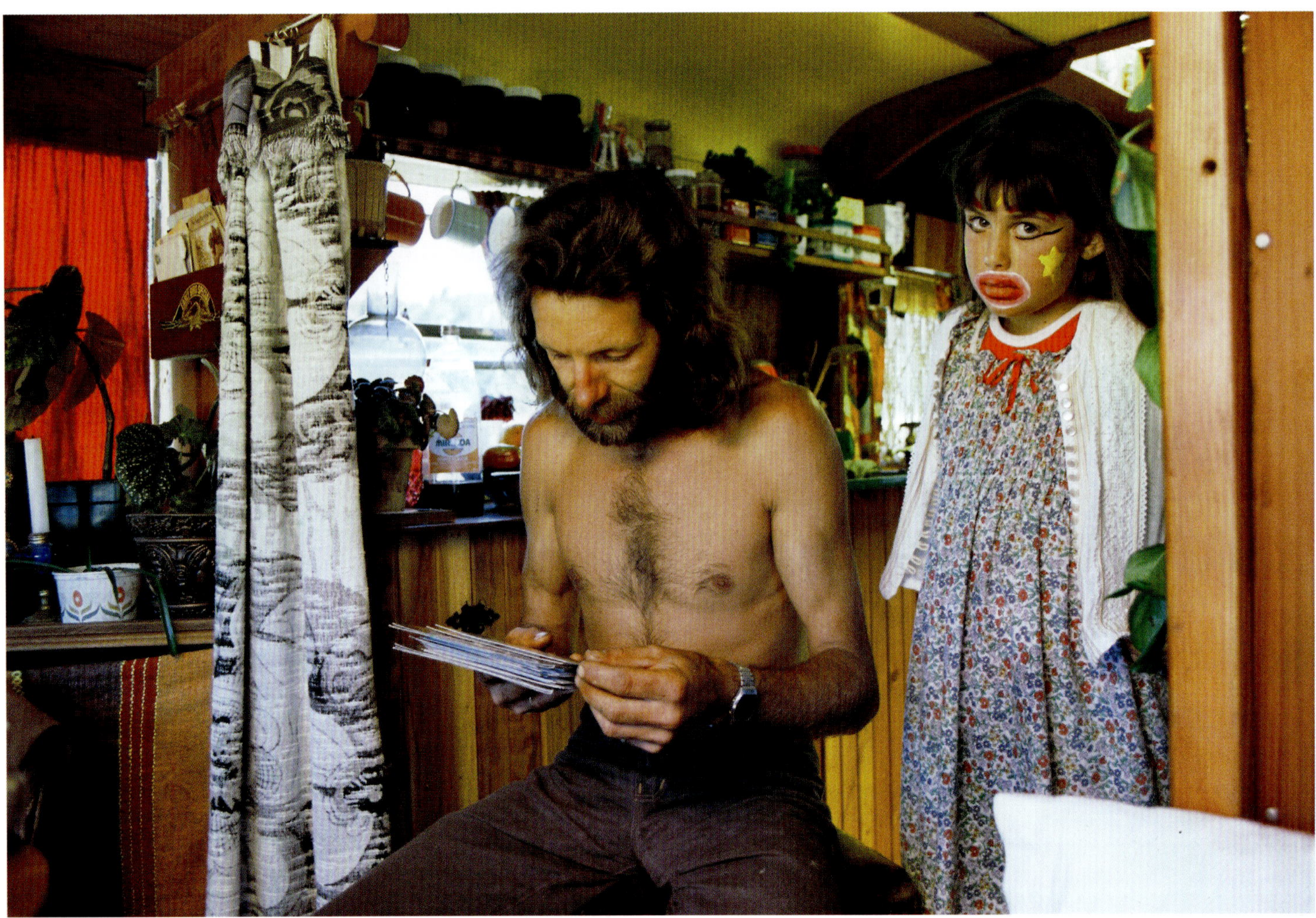

Steve with his daughter Chantal, Aratiatia, 1984

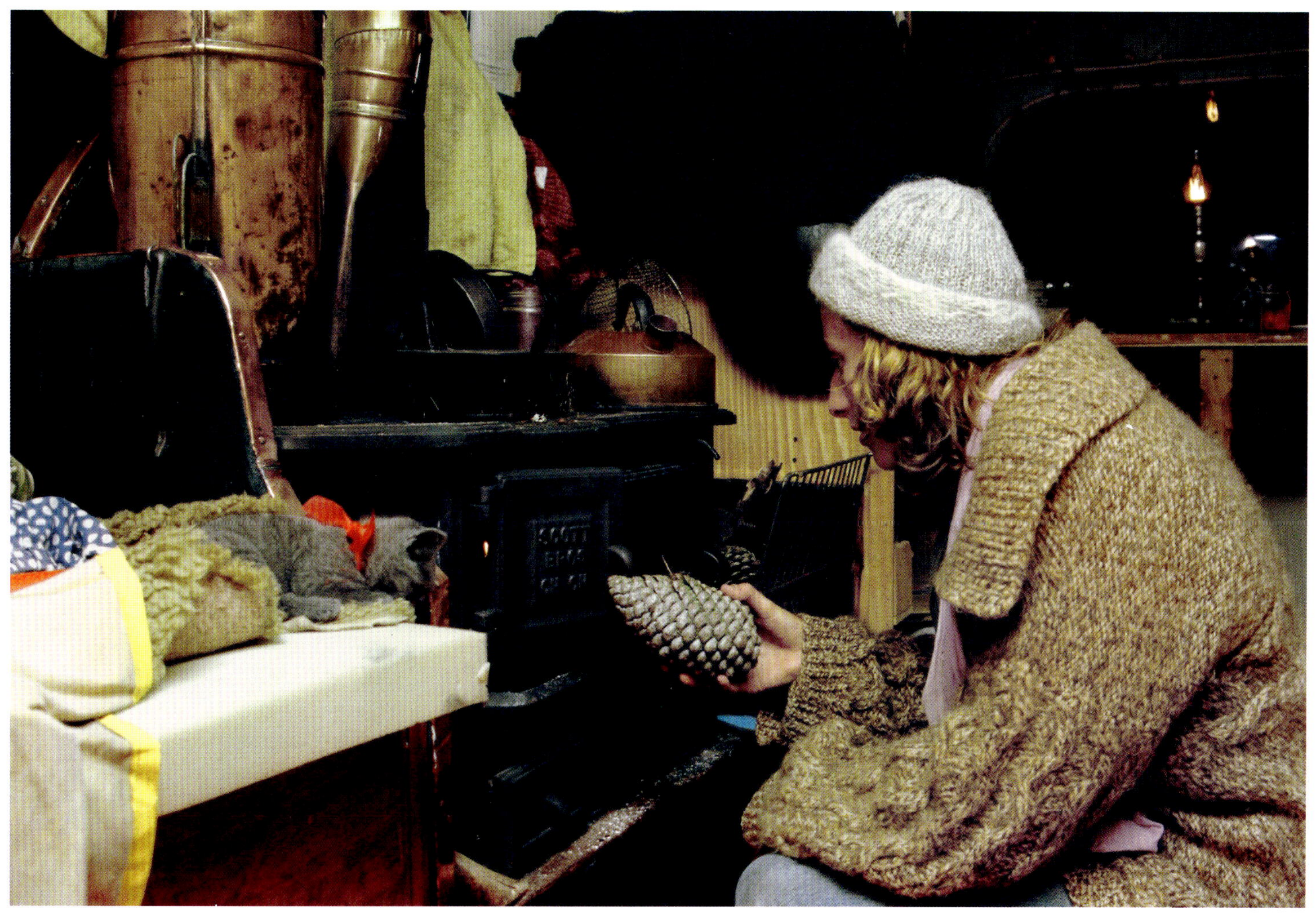

Sue Honeyman, Aratiatia, 1984

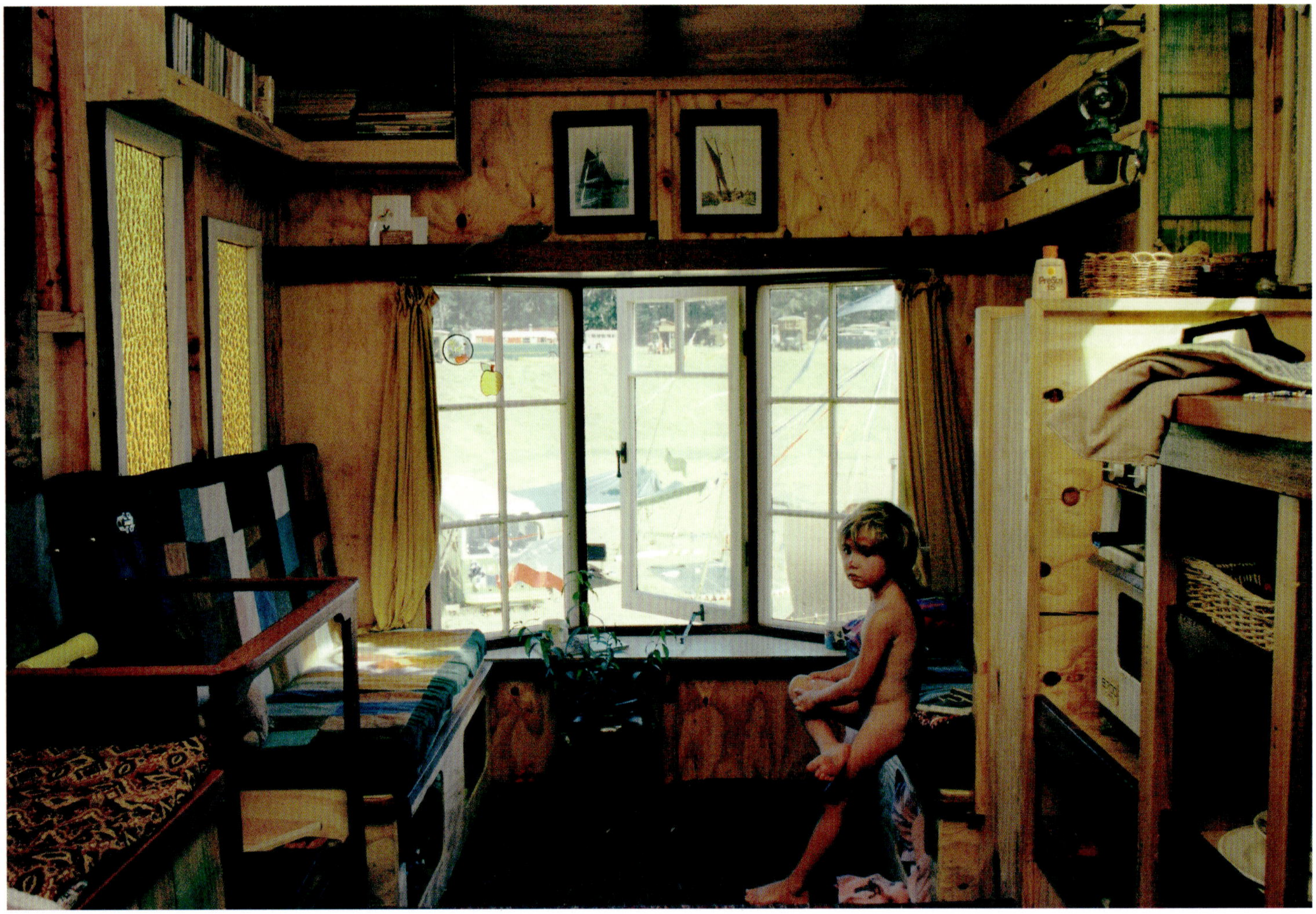

Leigh in Ann and Bill's house truck, Aratiatia, 1984

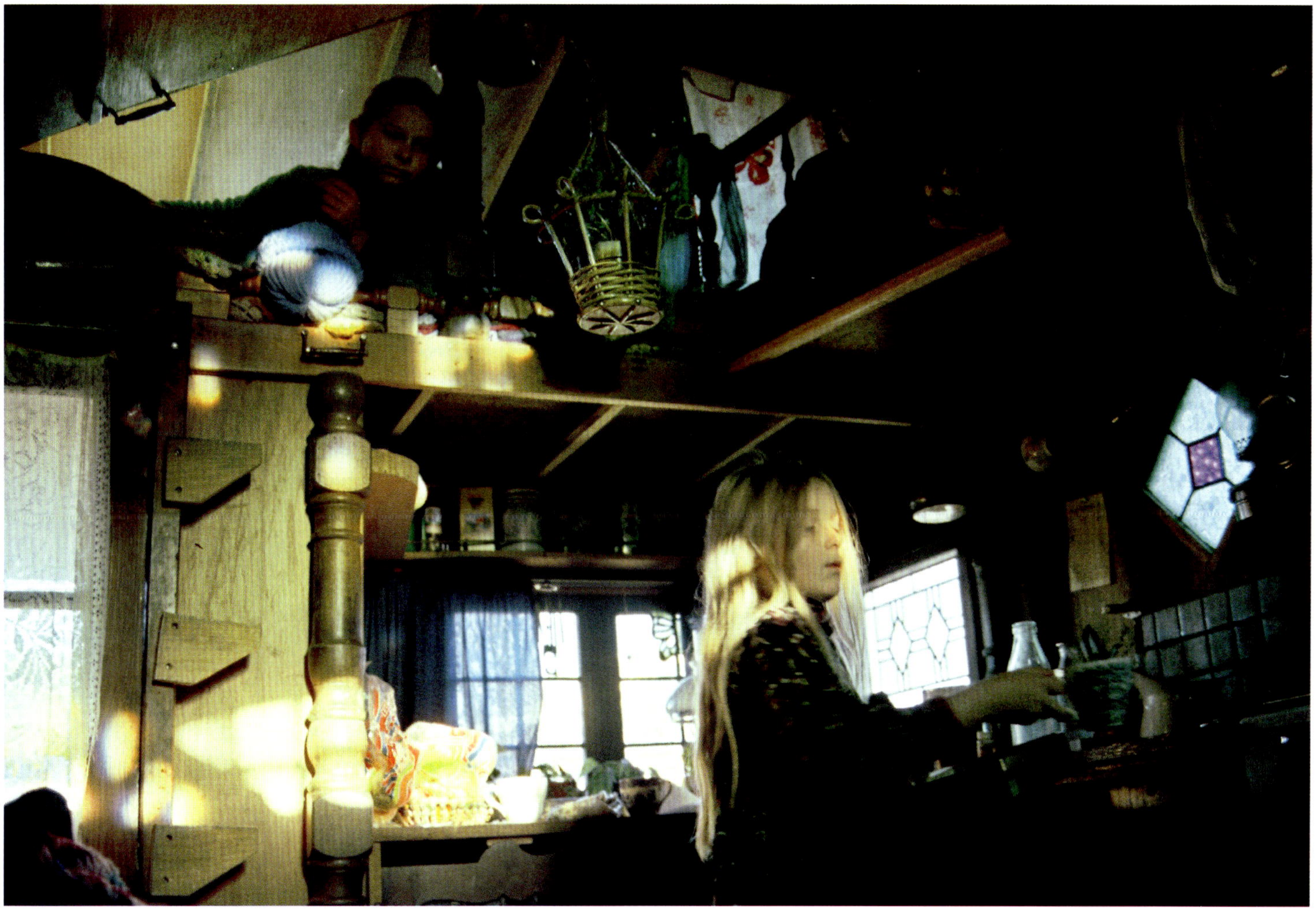

Petra making tea in Jade's house truck, with sister Lisa looking on, Aratiatia, 1984

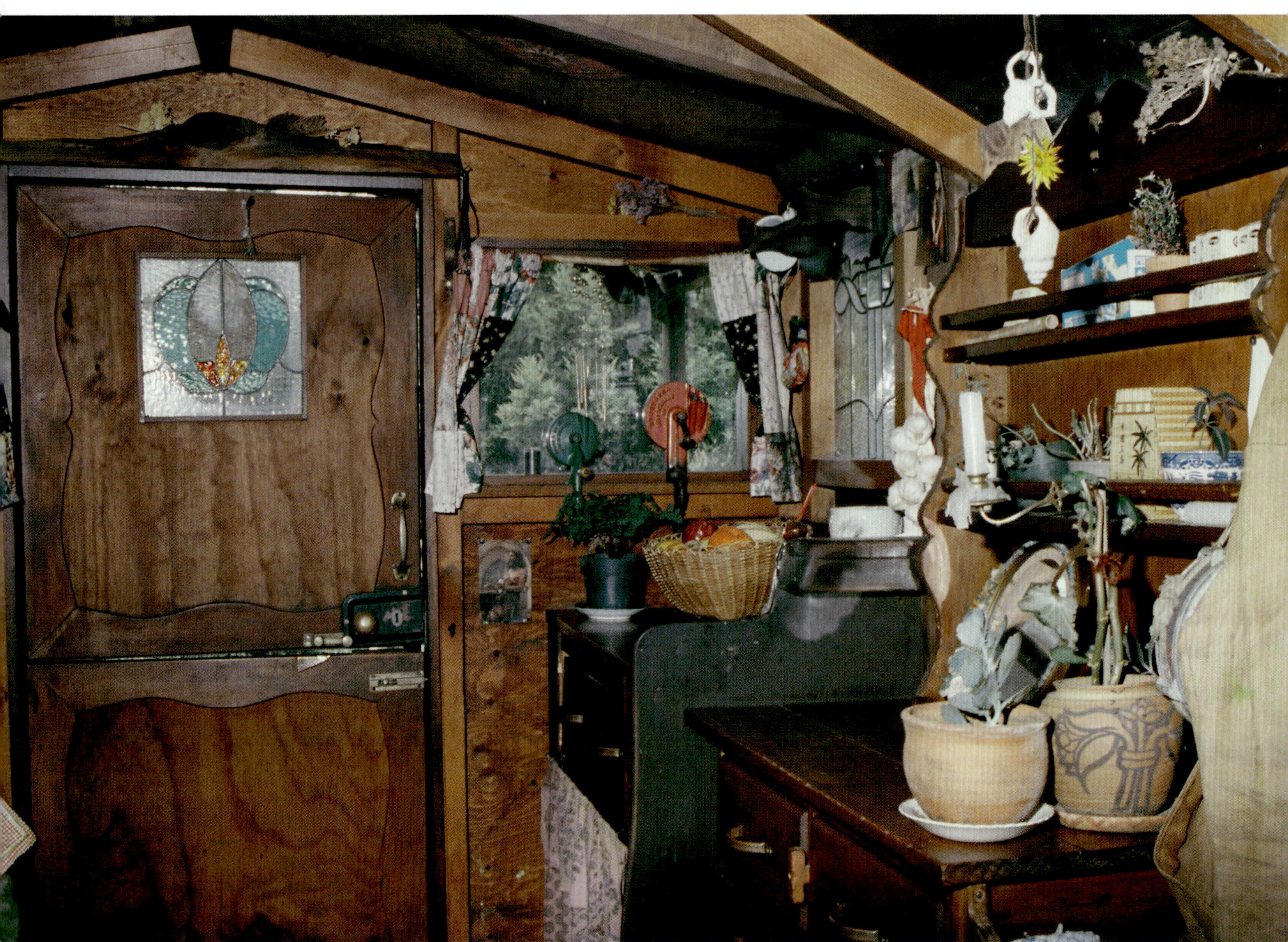

Interior, unidentified truck, Aratiatia, 1984

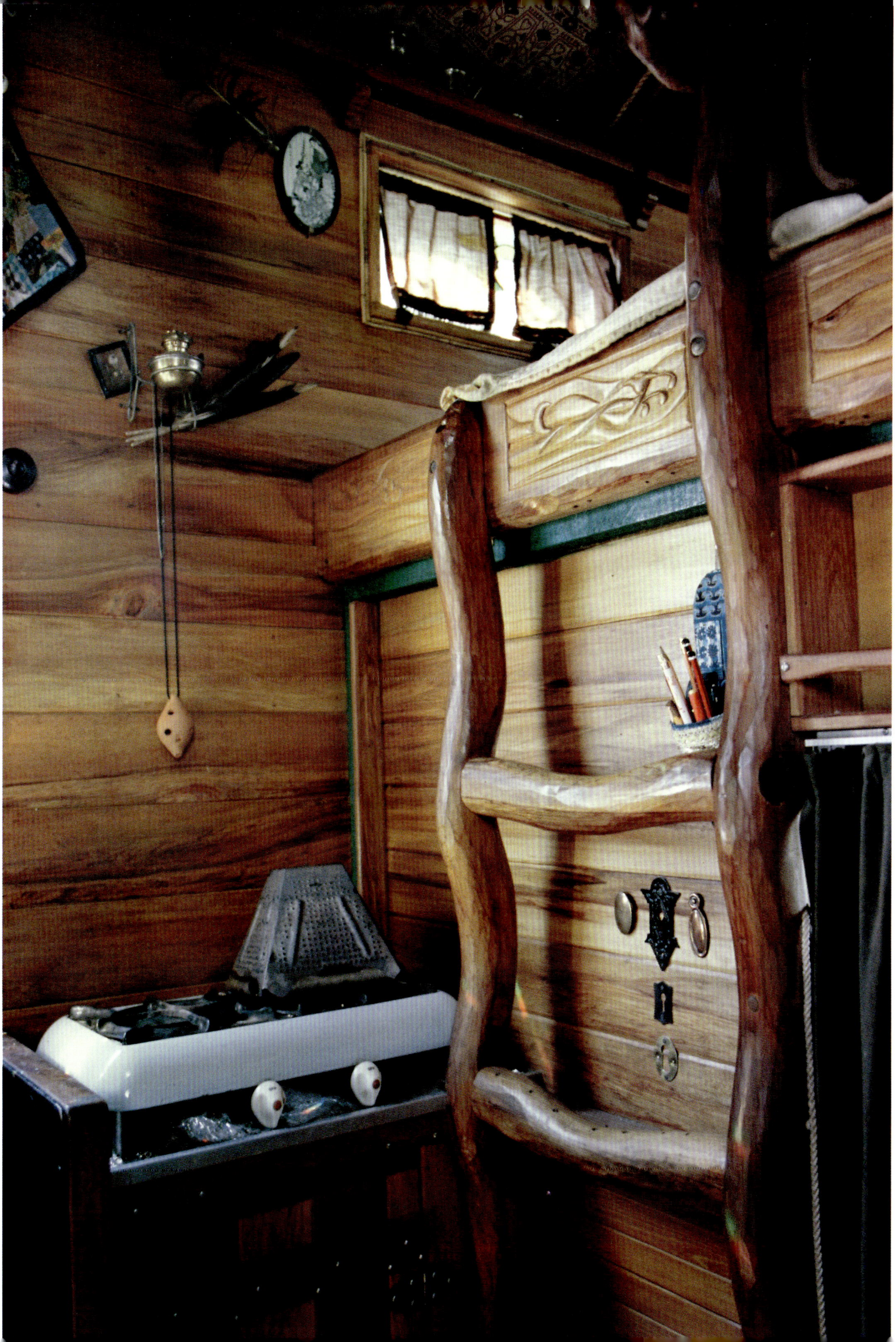

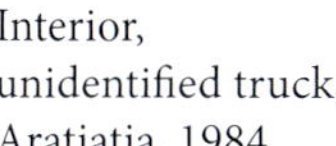

Interior,
unidentified truck,
Aratiatia, 1984

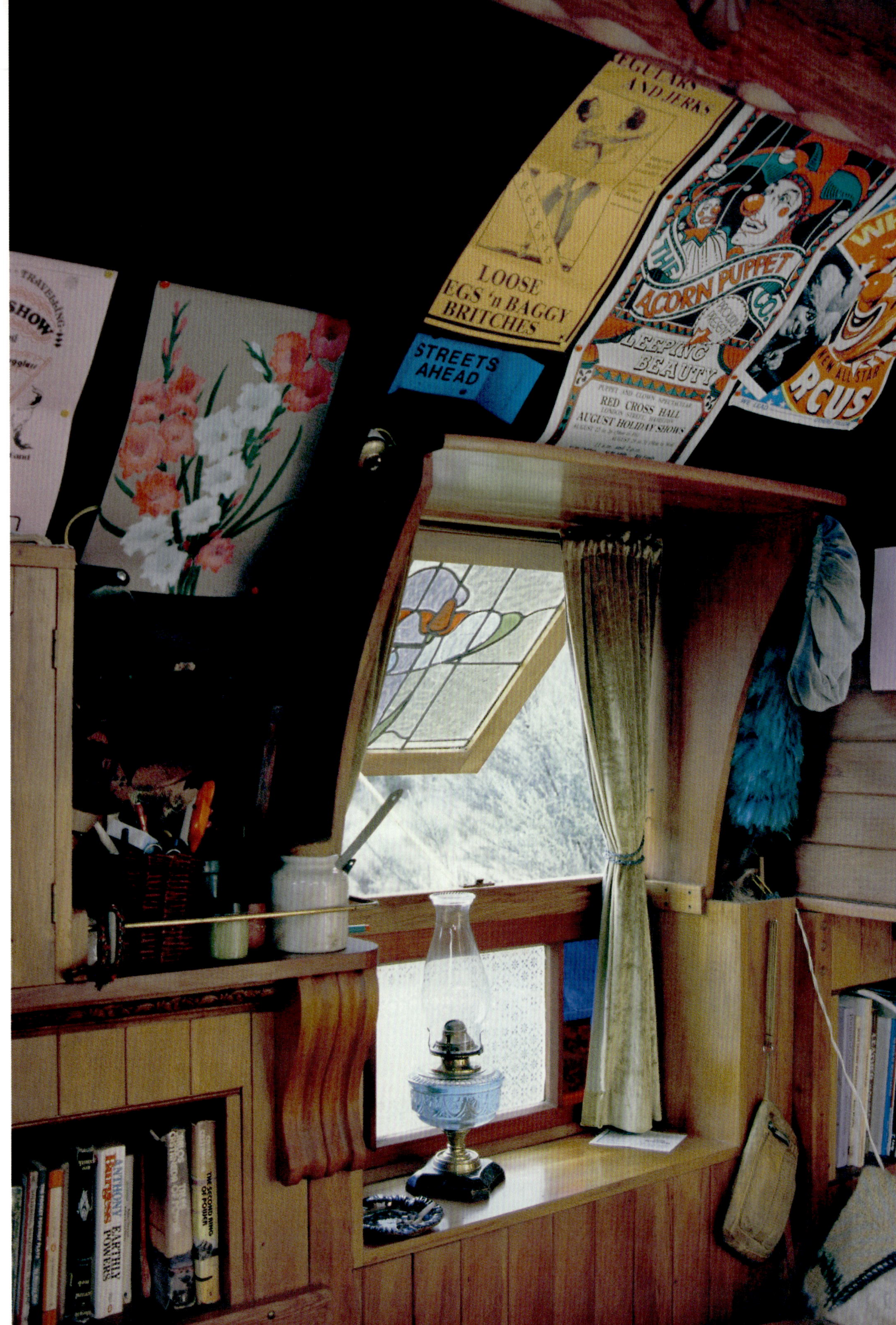

Dom and Jane's house truck, Aratiatia, 1984

Dom and Jane's house truck, Aratiatia, 1984

Dom and Jane's house truck, Aratiatia, 1984

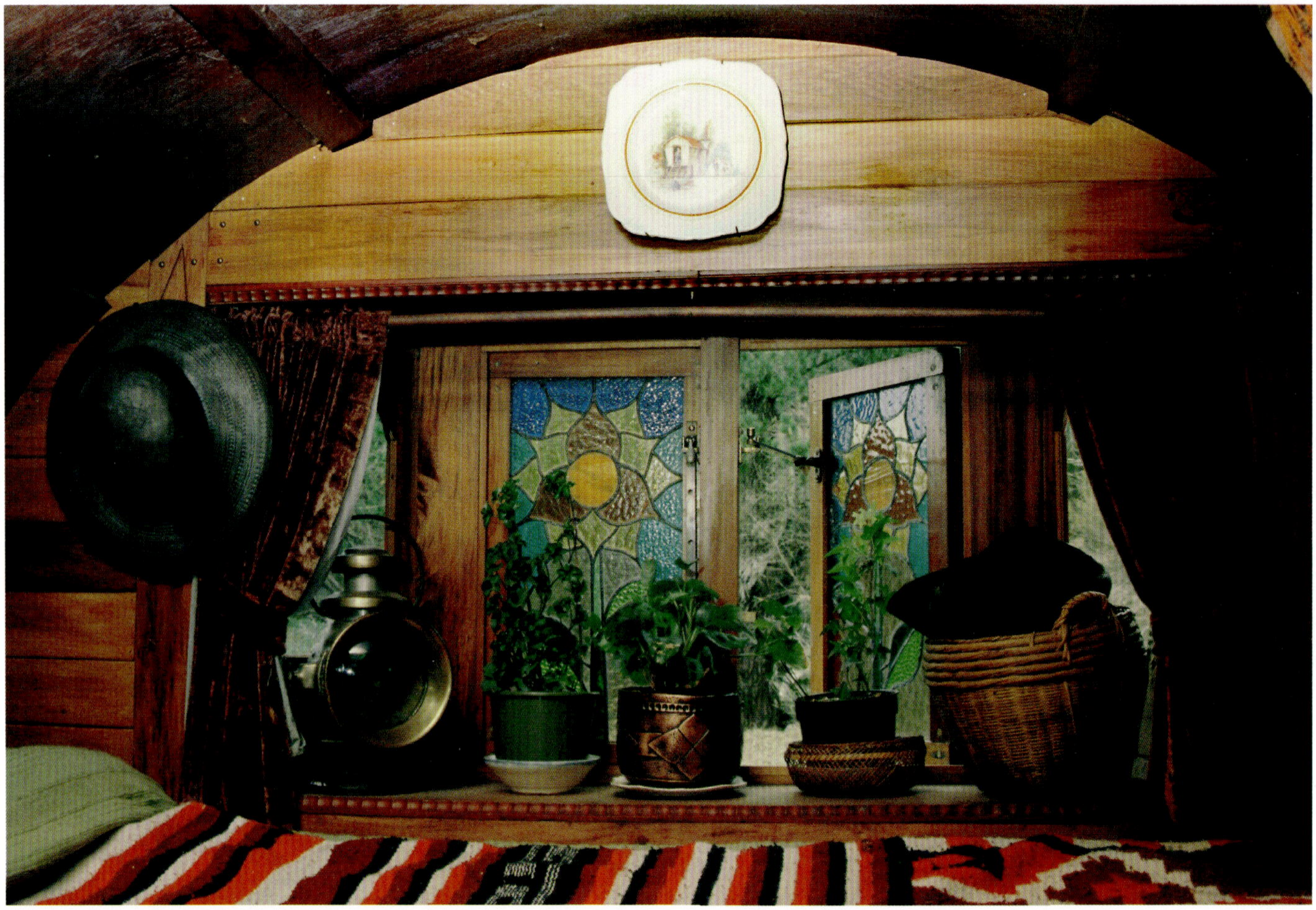

Dom and Jane's house truck, (mirror from Auckland Regional Authority bus), Aratiatia, 1984

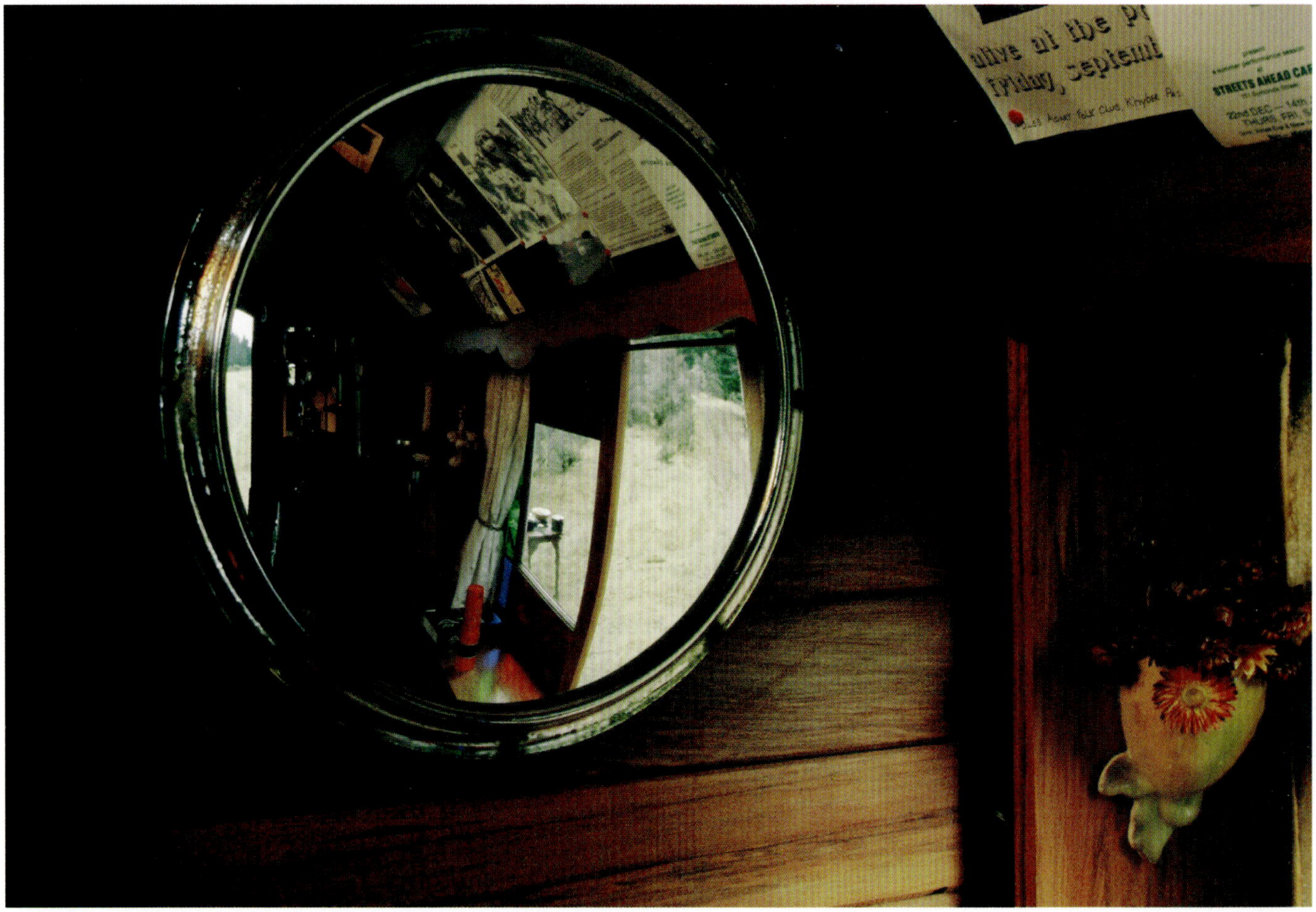

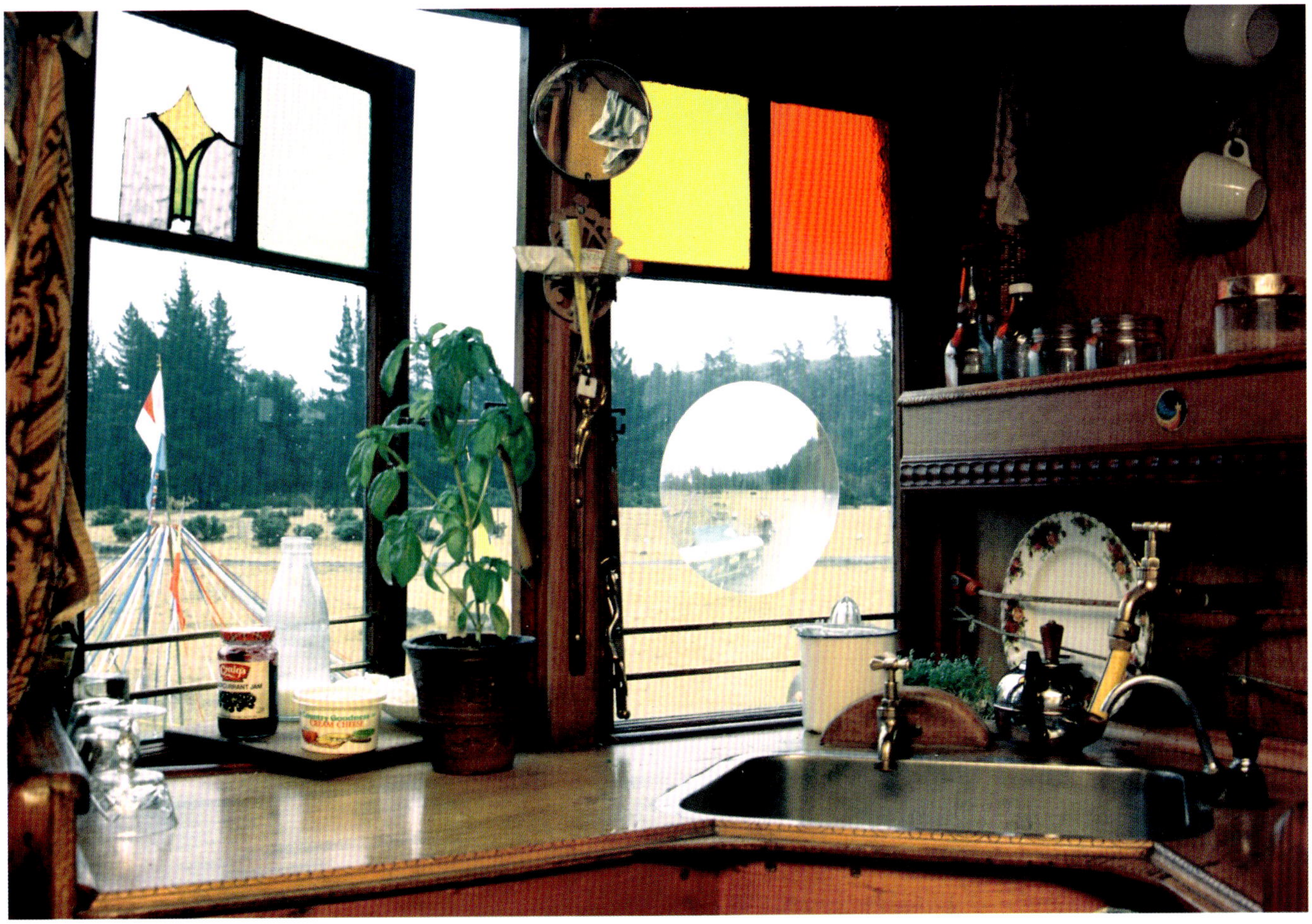

Gypsy Jeff's
'Stealth Truck',
Aratiatia, 1984

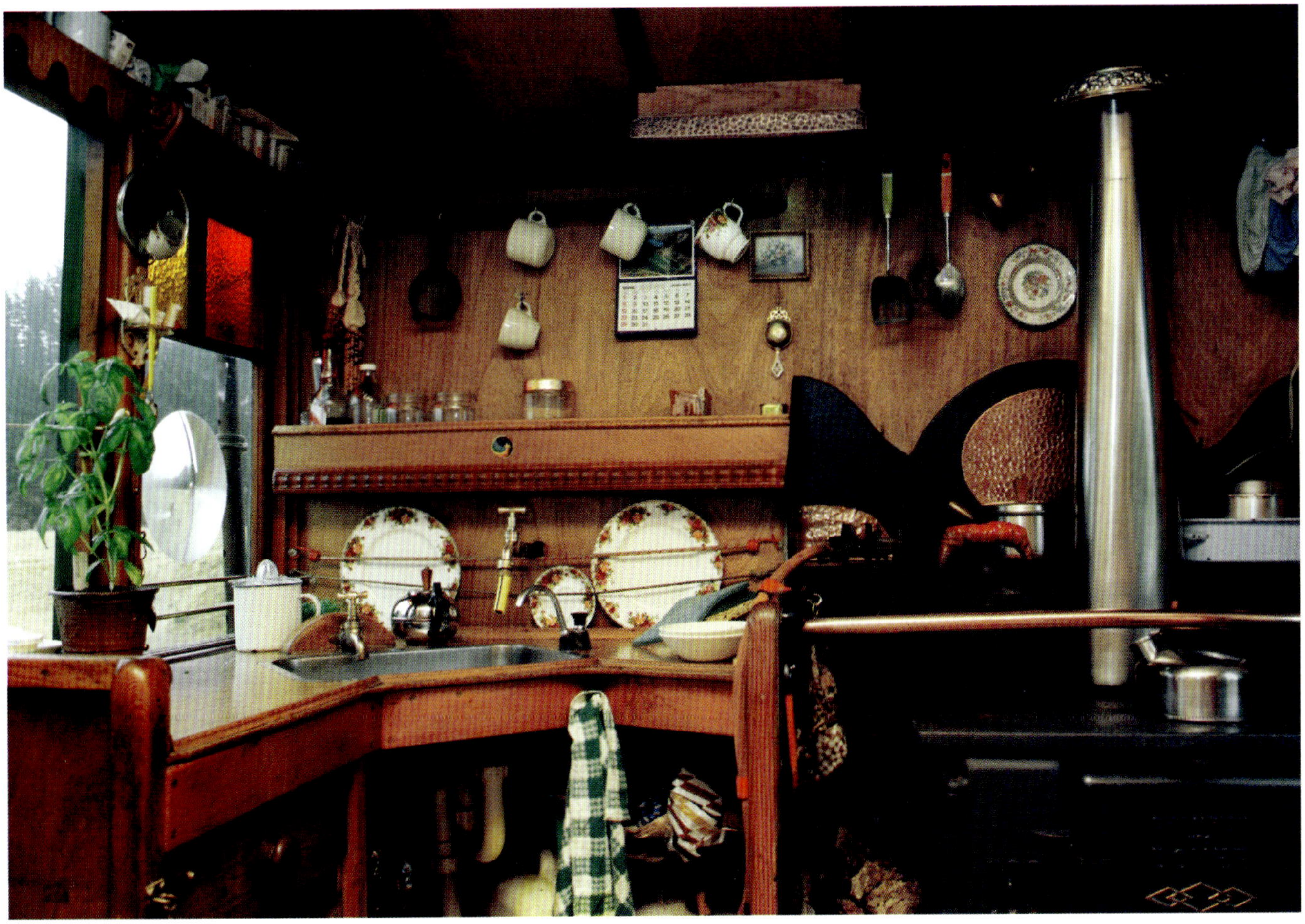

Gypsy Jeff's
'Stealth Truck',
Aratiatia, 1984

View from a window in David Sheridan's People's Circus bus, Aratiatia, 1984

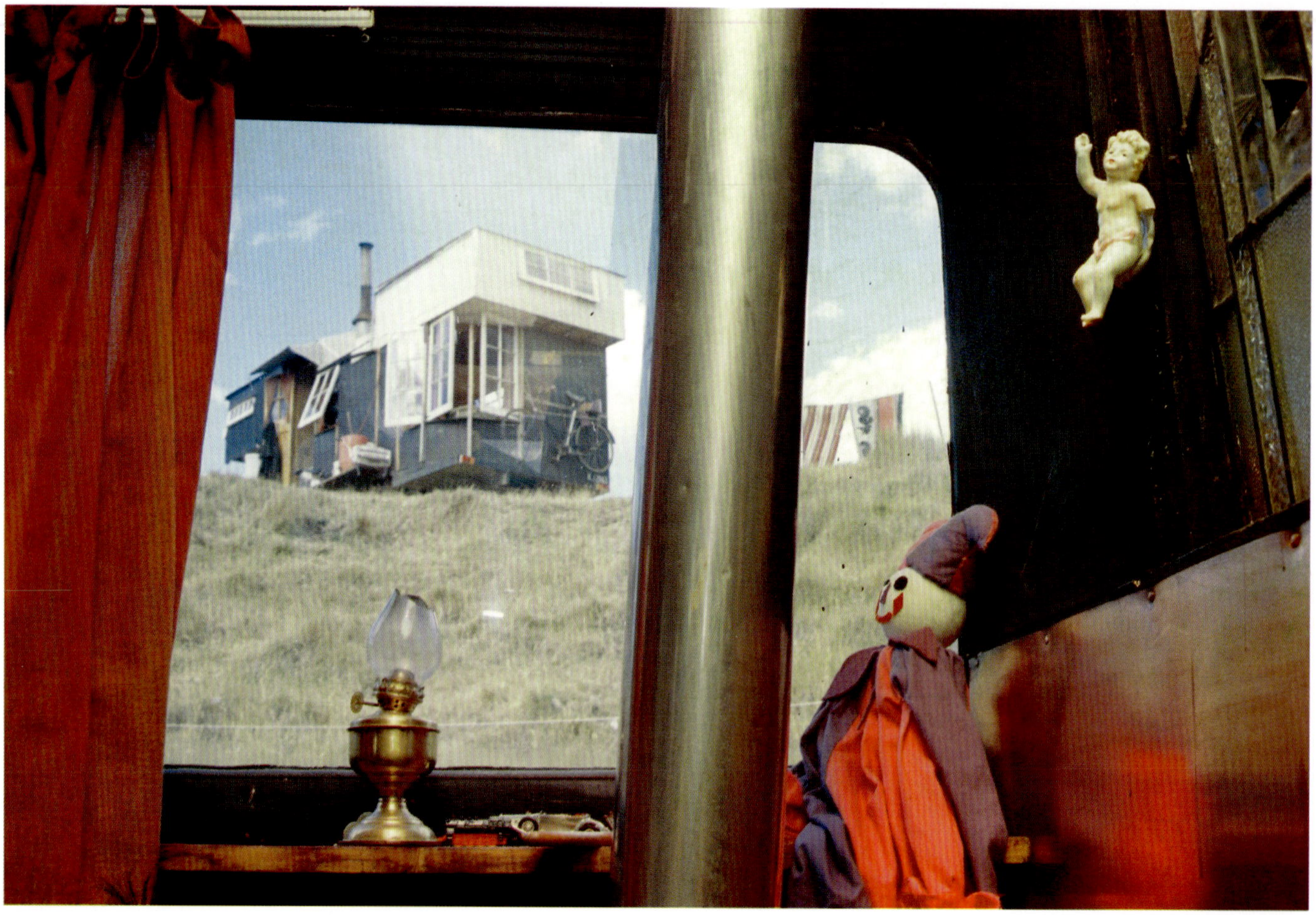

Interior, unidentified truck, Aratiatia, 1984

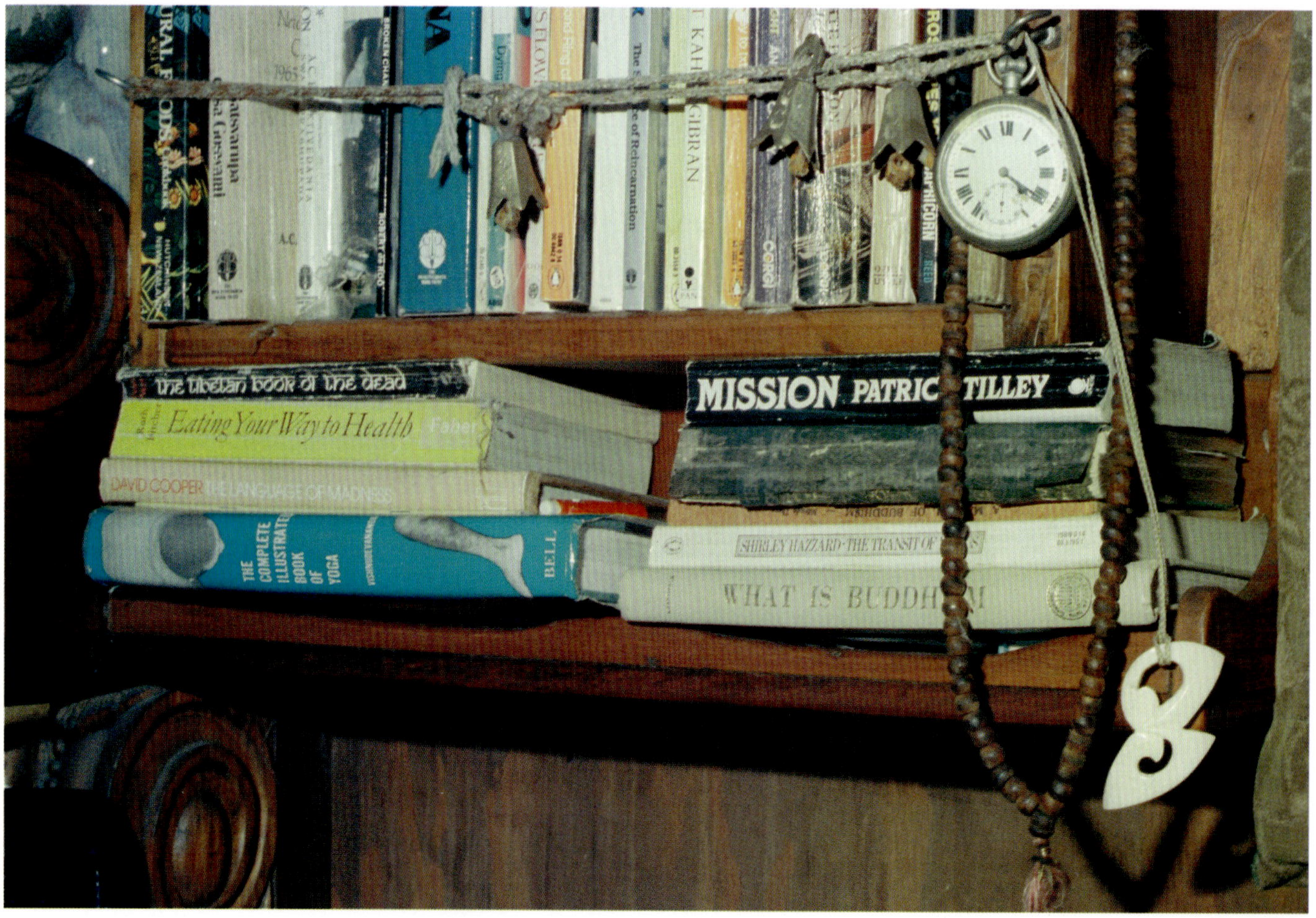

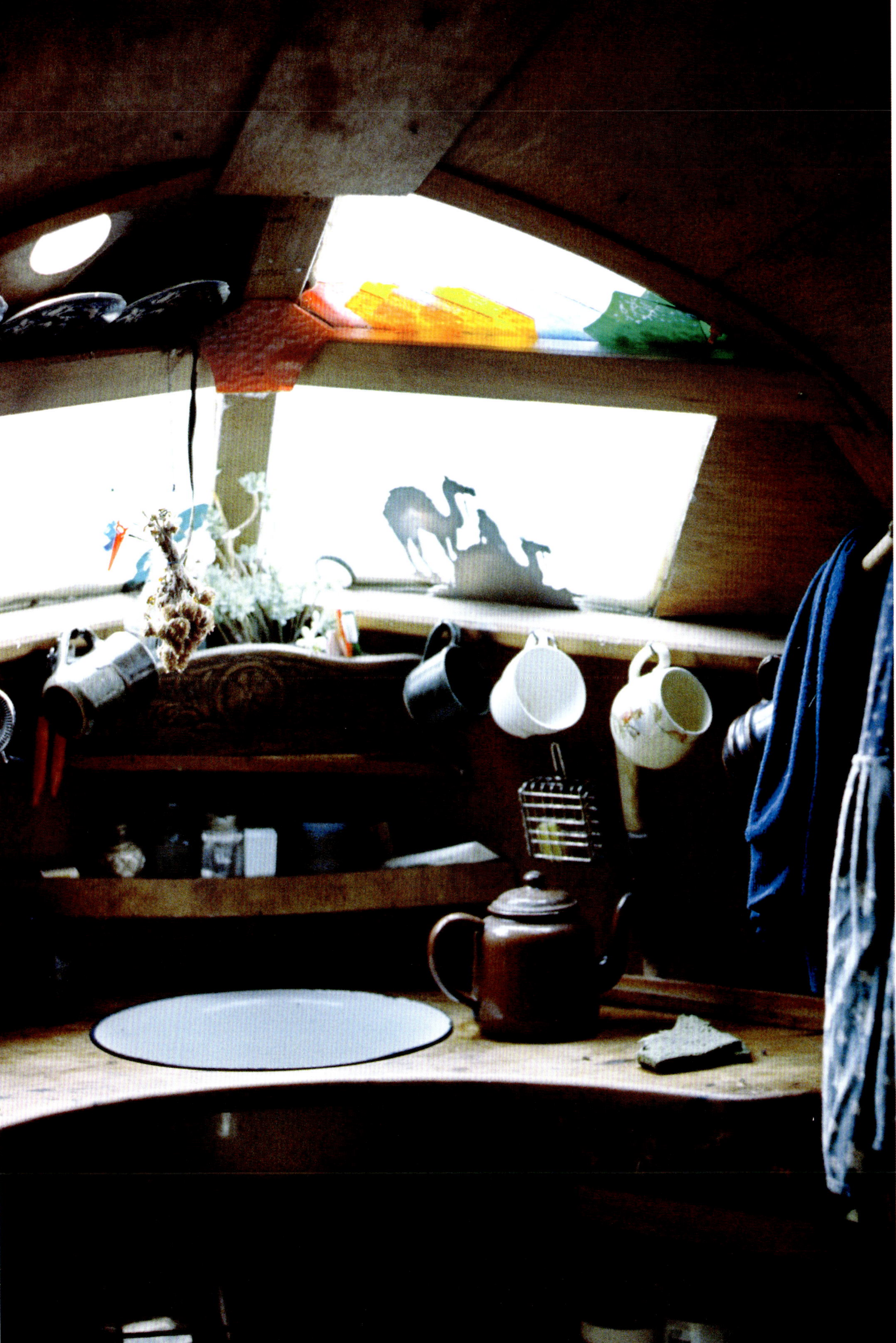

Inside Ewan's aeroplane caravan, Sweetwaters Music Festival, Ngāruawāhia, 1980

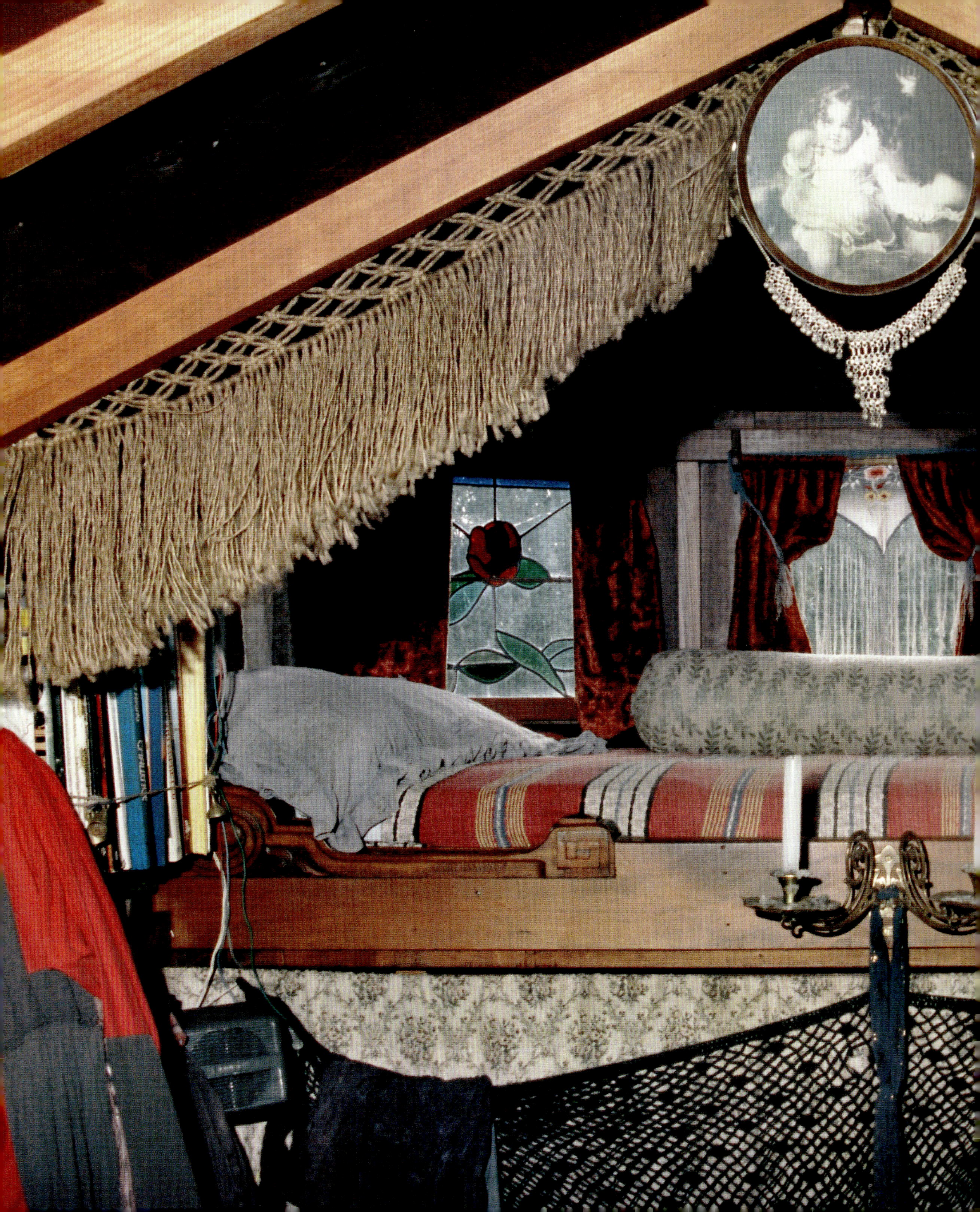

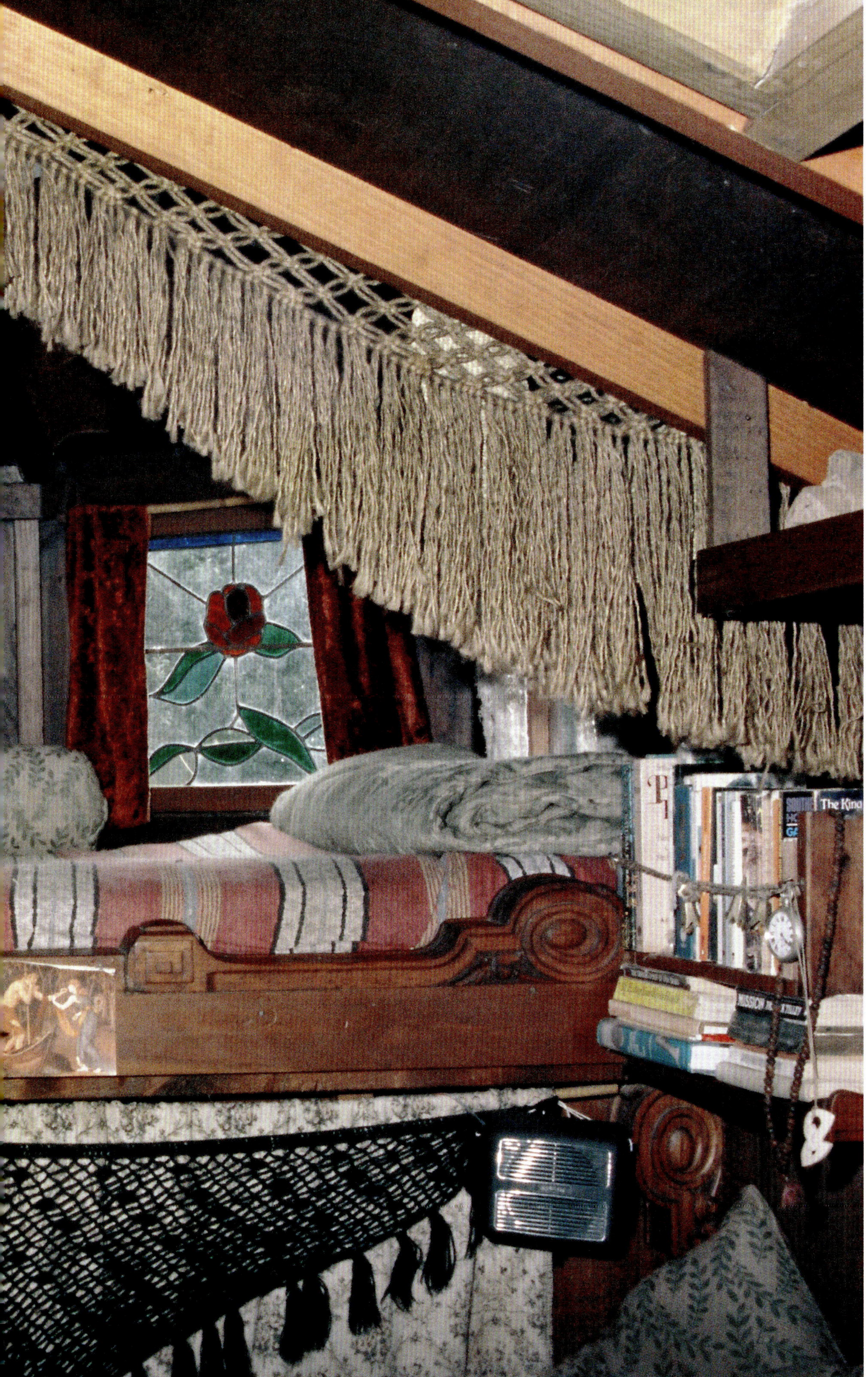

Interior, unidentified truck, Aratiatia, 1984

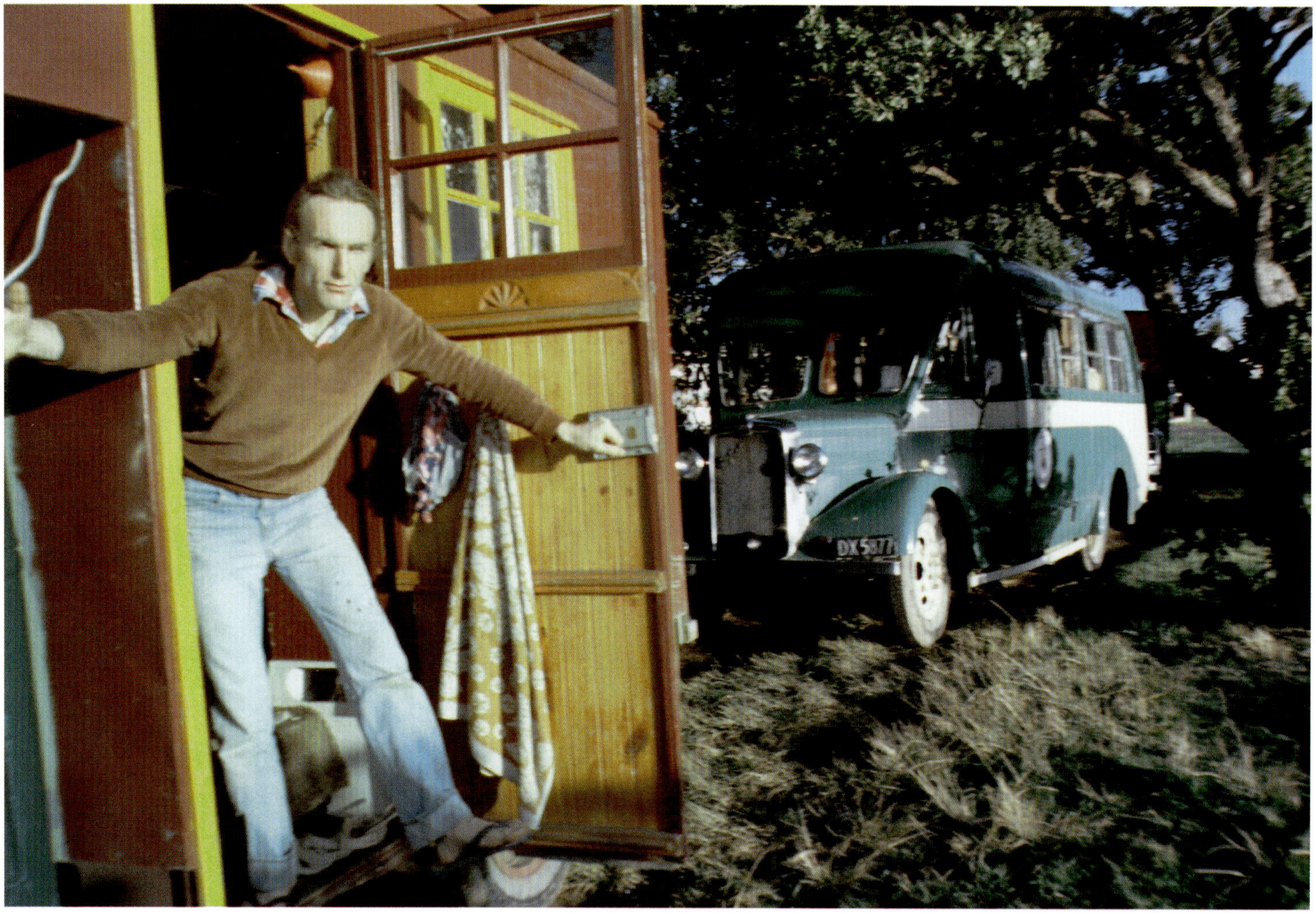

Jonathon Acorn in Walnut truck, Waikato tour, Raglan, 1981

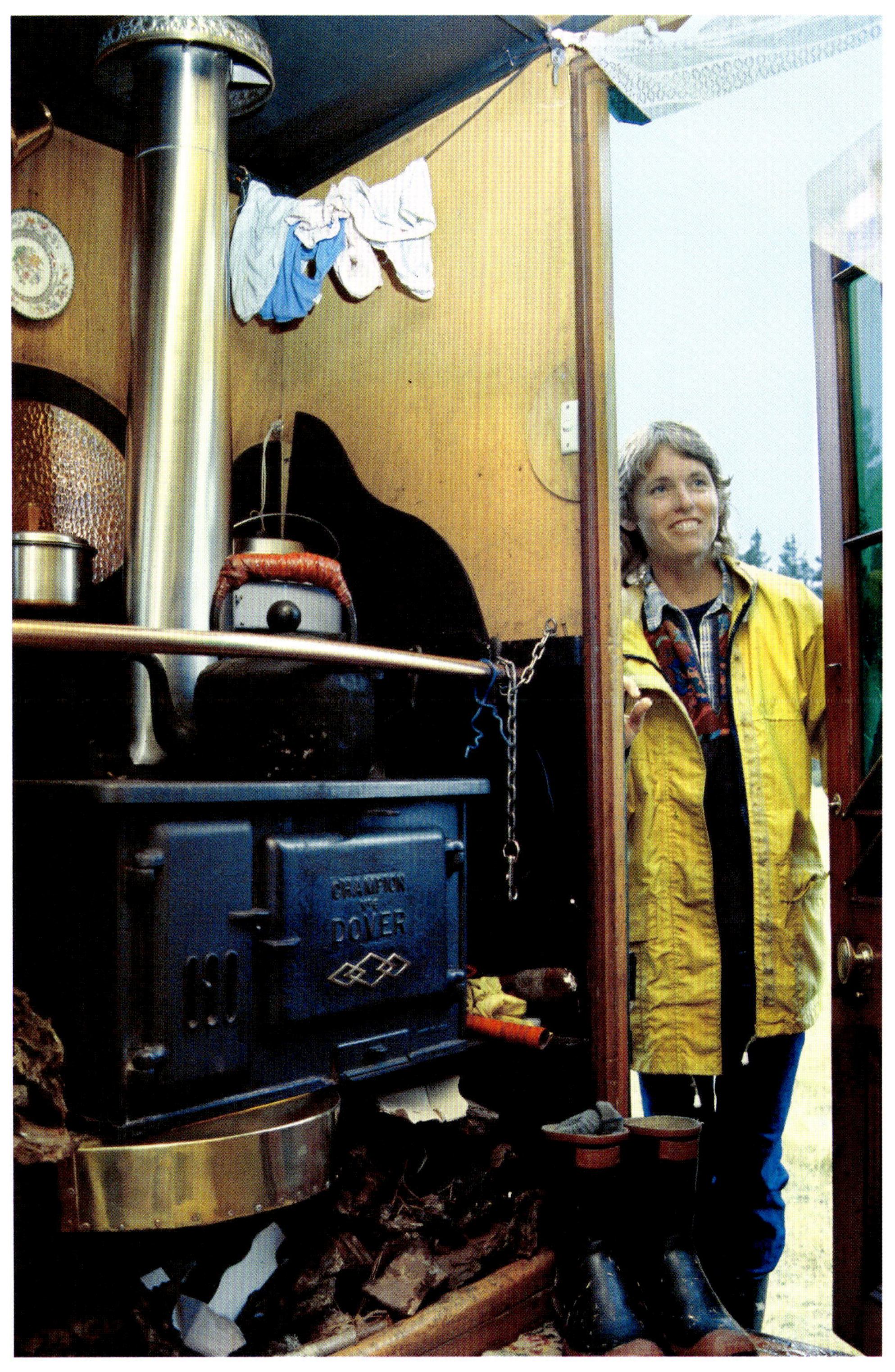

Marypat Ross visiting Gypsy Jeff's 'Stealth Truck', Aratiatia, 1984

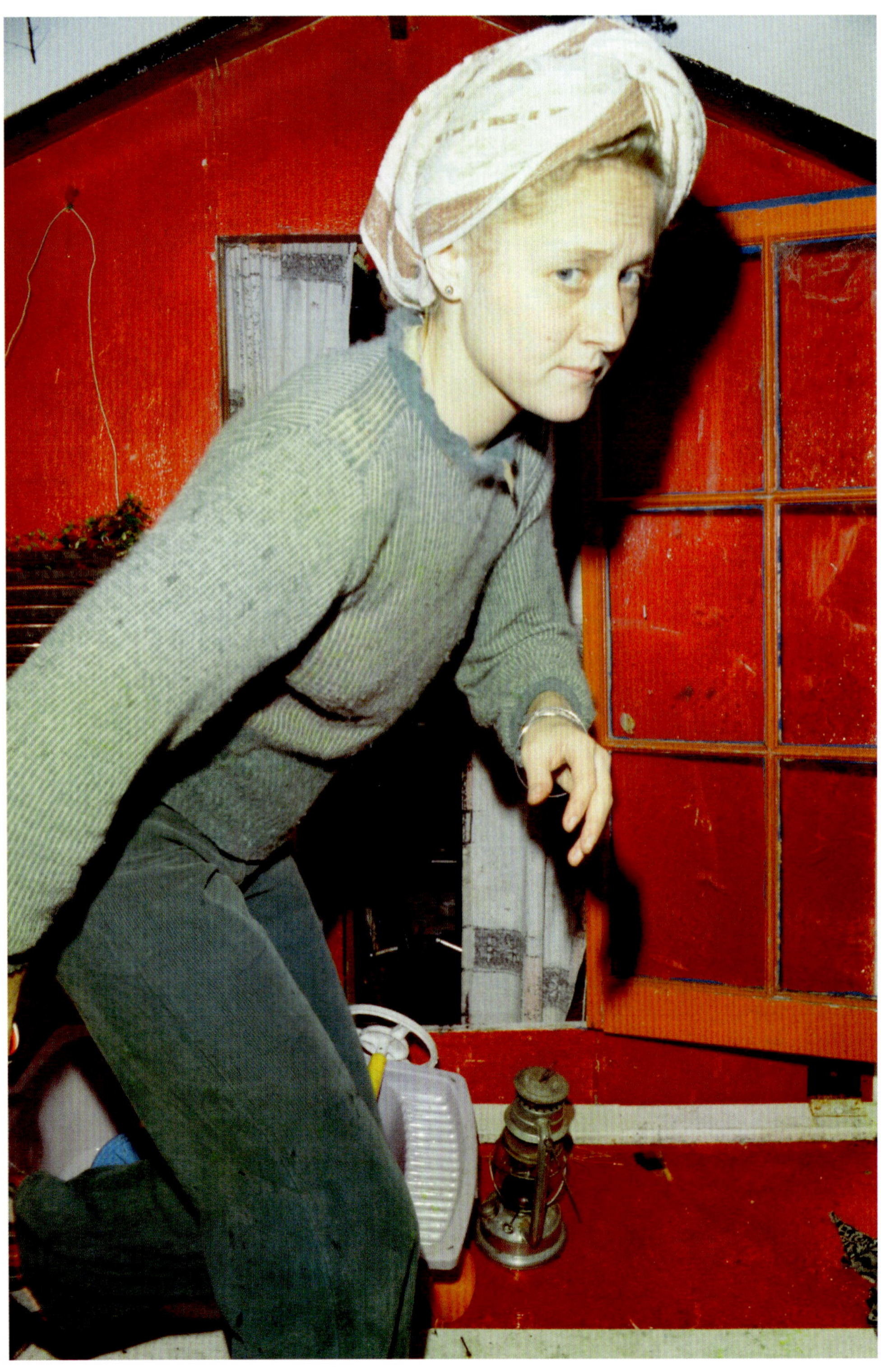

Hendrika Vaneveld inside Acorn's Puppet Theatre caravan, 1979

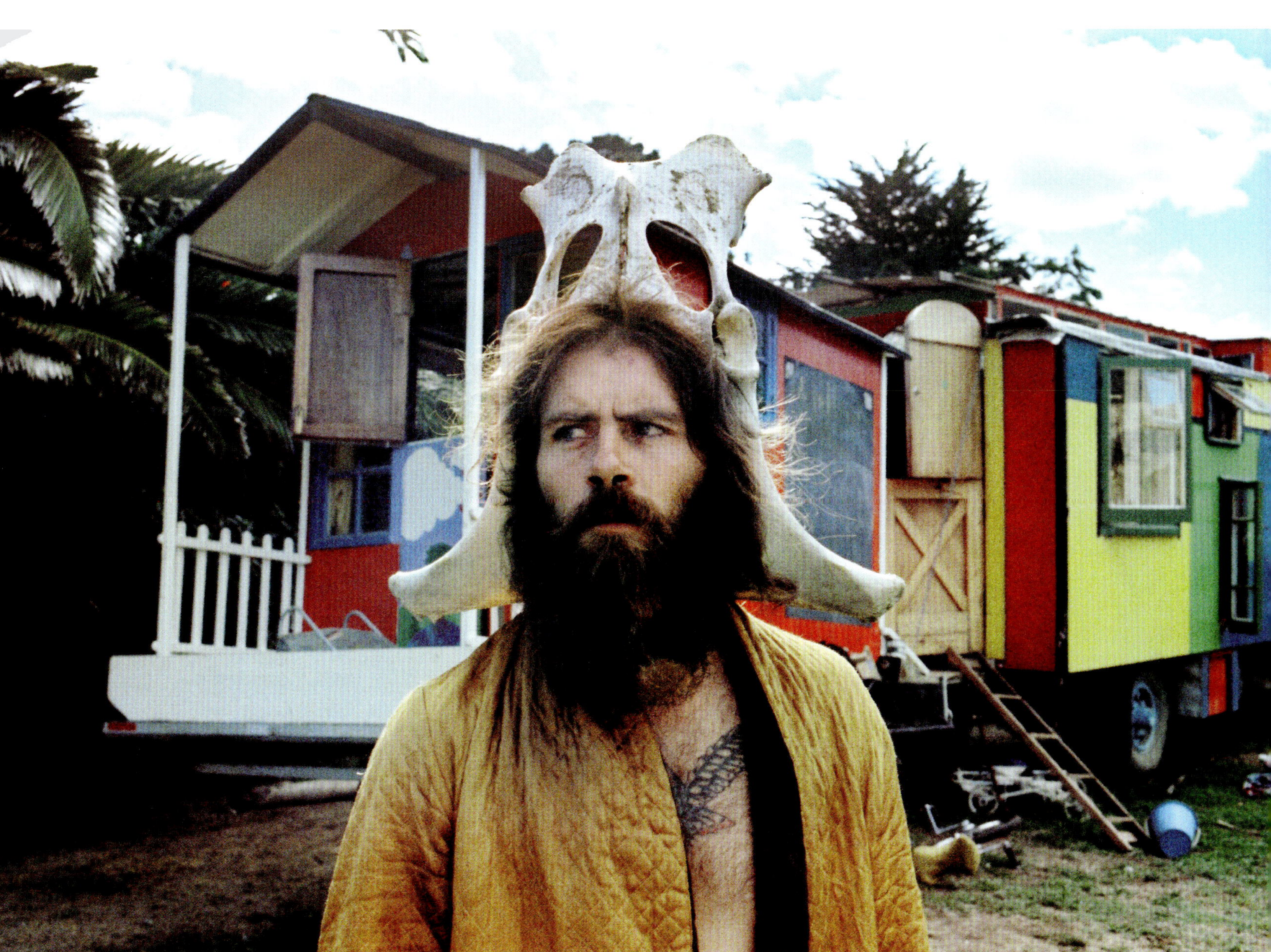

Podge with Acorn's Puppet Theatre truck and caravan, 1979

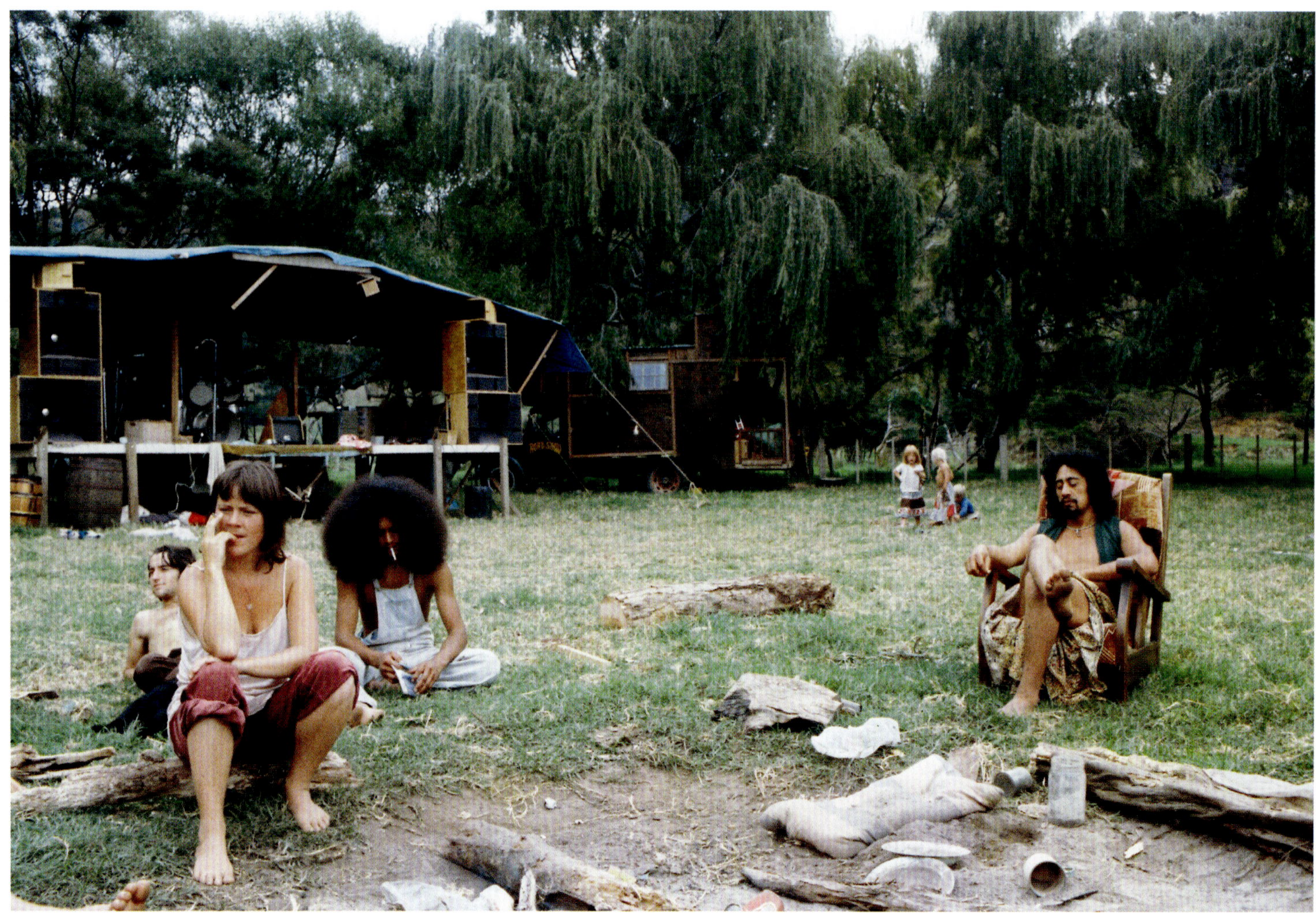

Stuart, Monica, Waka and Jimmy with the stage-truck, Moehau Community, Coromandel, 1979

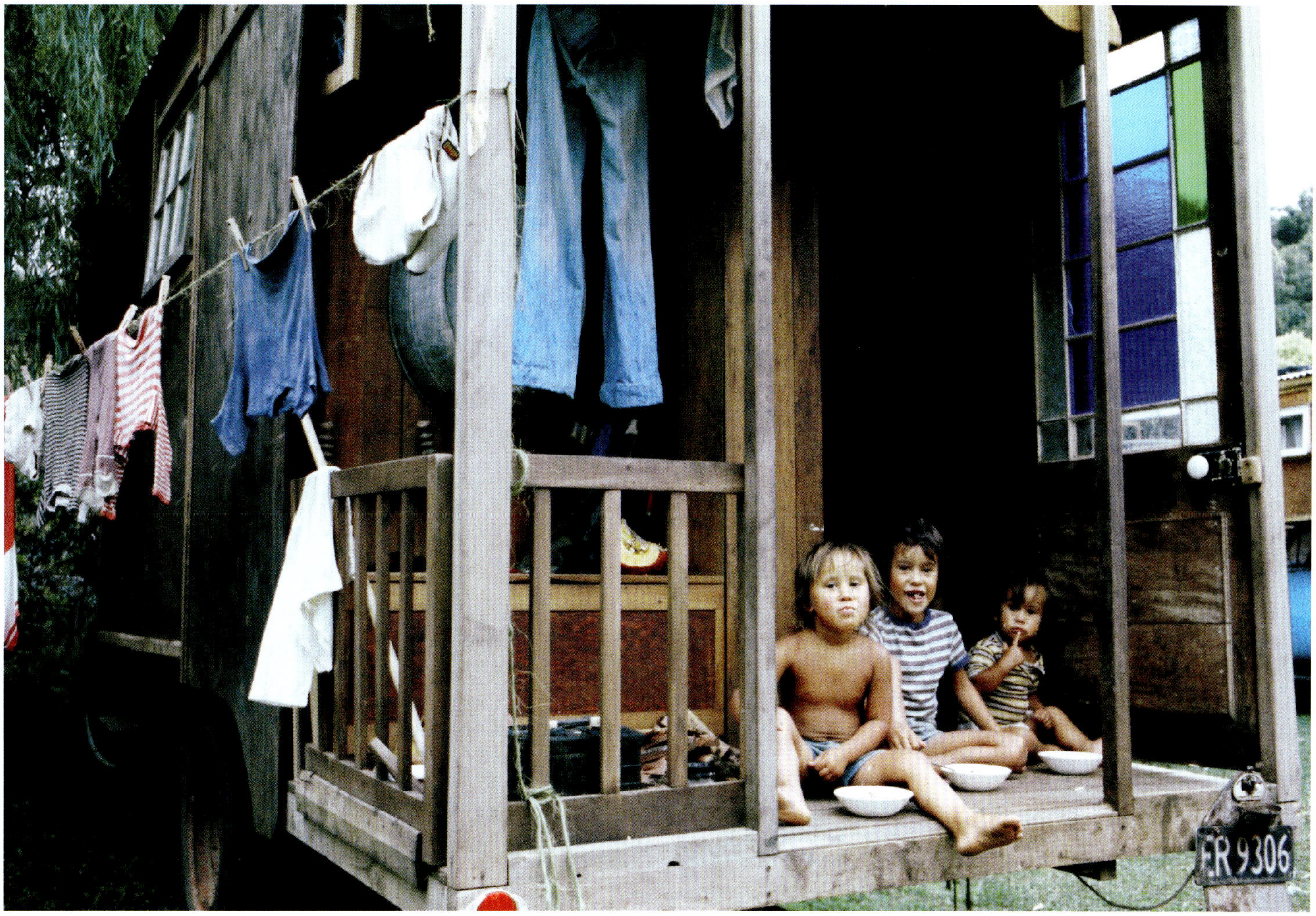

Jade, Jeremy and Wiremu at mealtime, Moehau, 1979

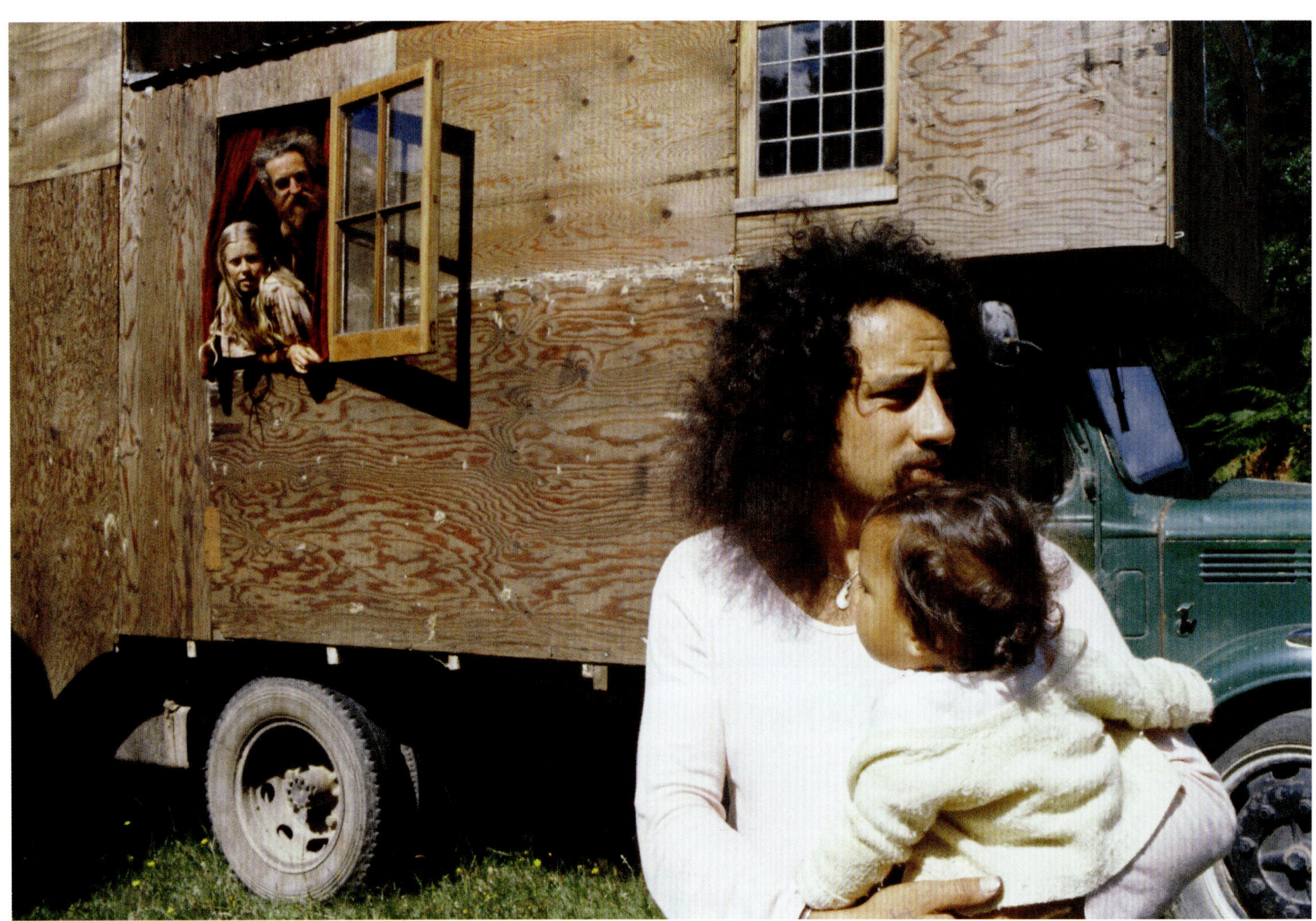

Yvonne Dixon and Mike Sweet framed by the window, with Jimmy Dixon holding Wiremu, 1979

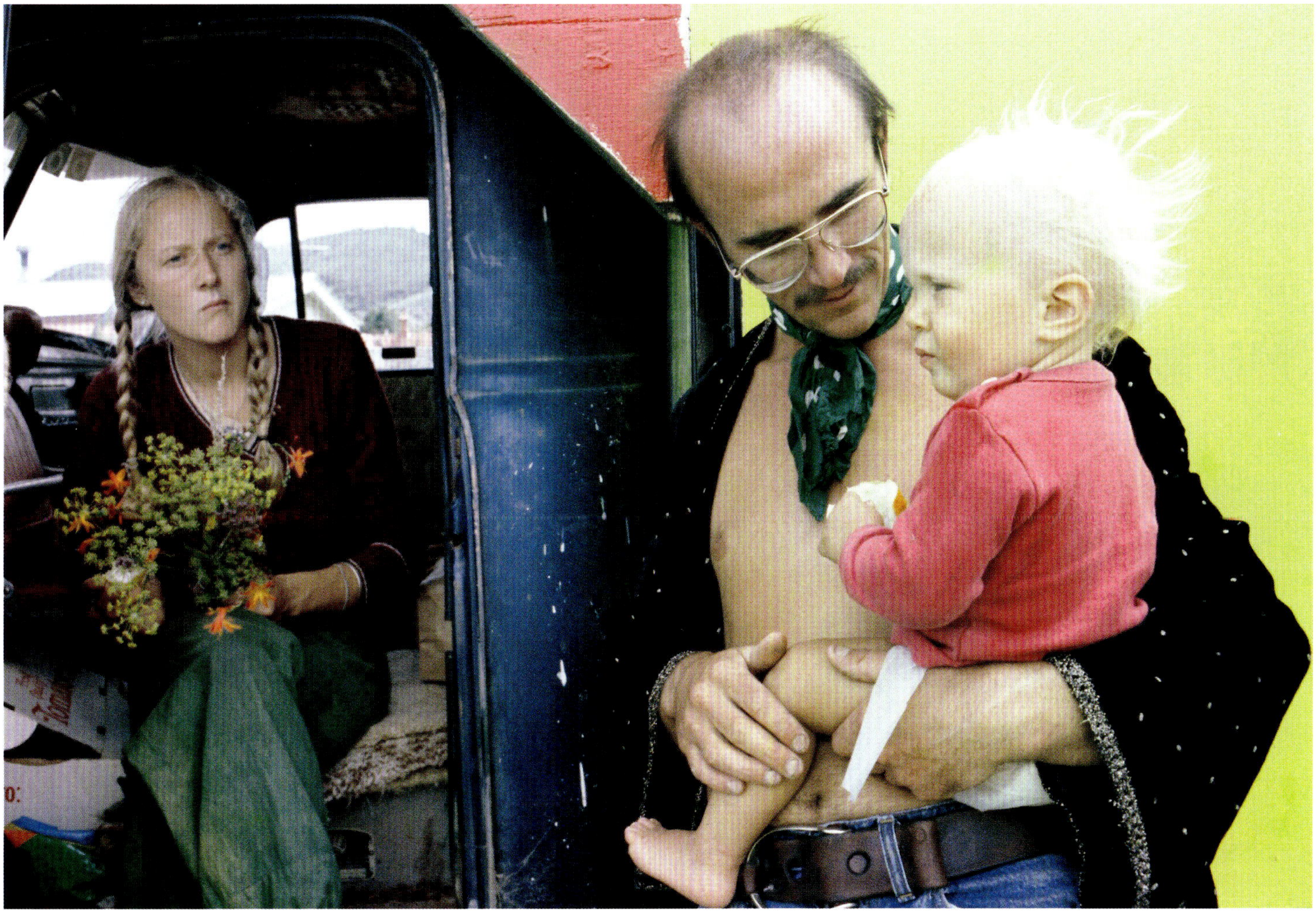

Hendrika Vaneveld and David Sheridan with Tahi, 1979

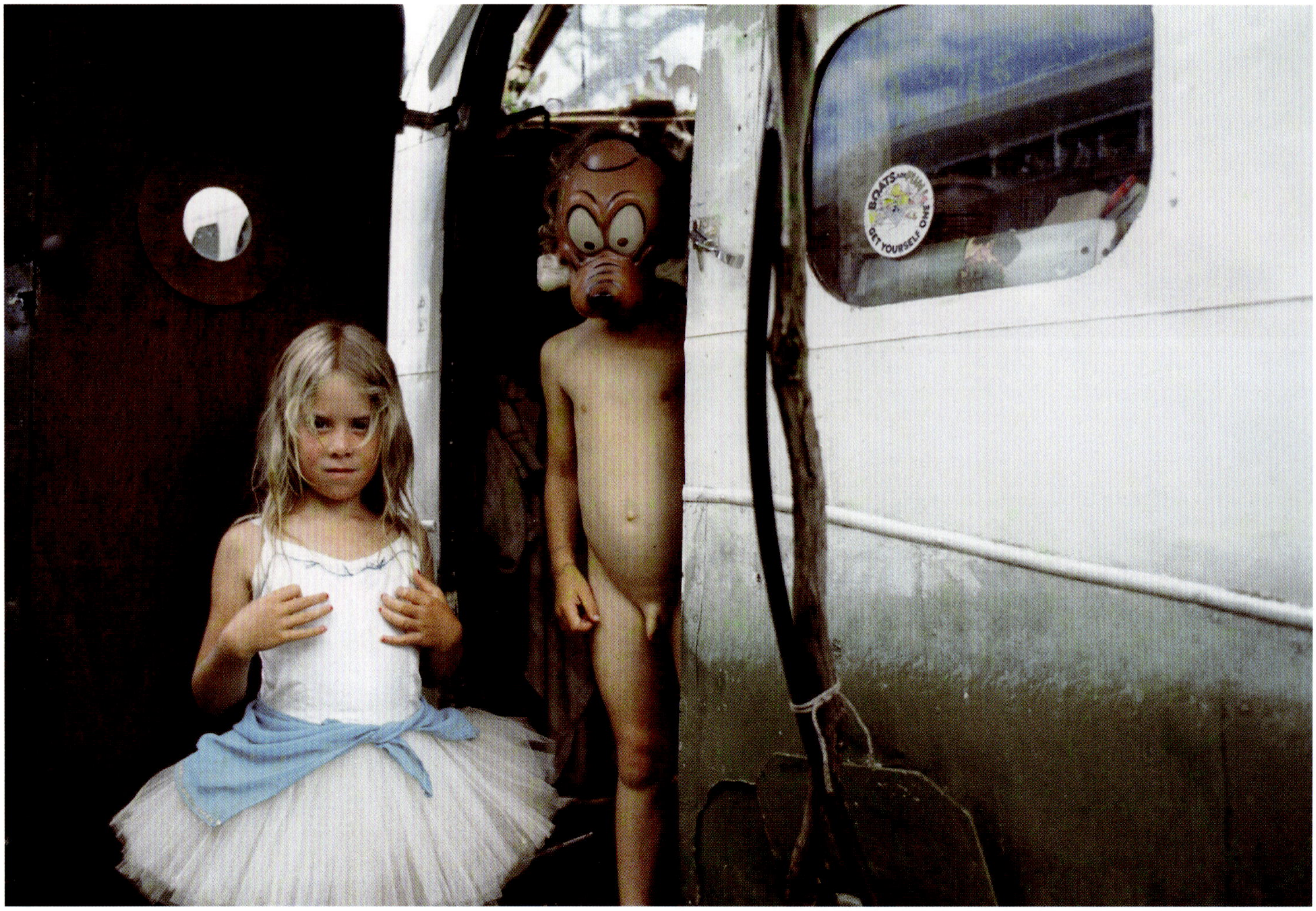

Children playing in the aeroplane caravan, Clifton Road, 1979

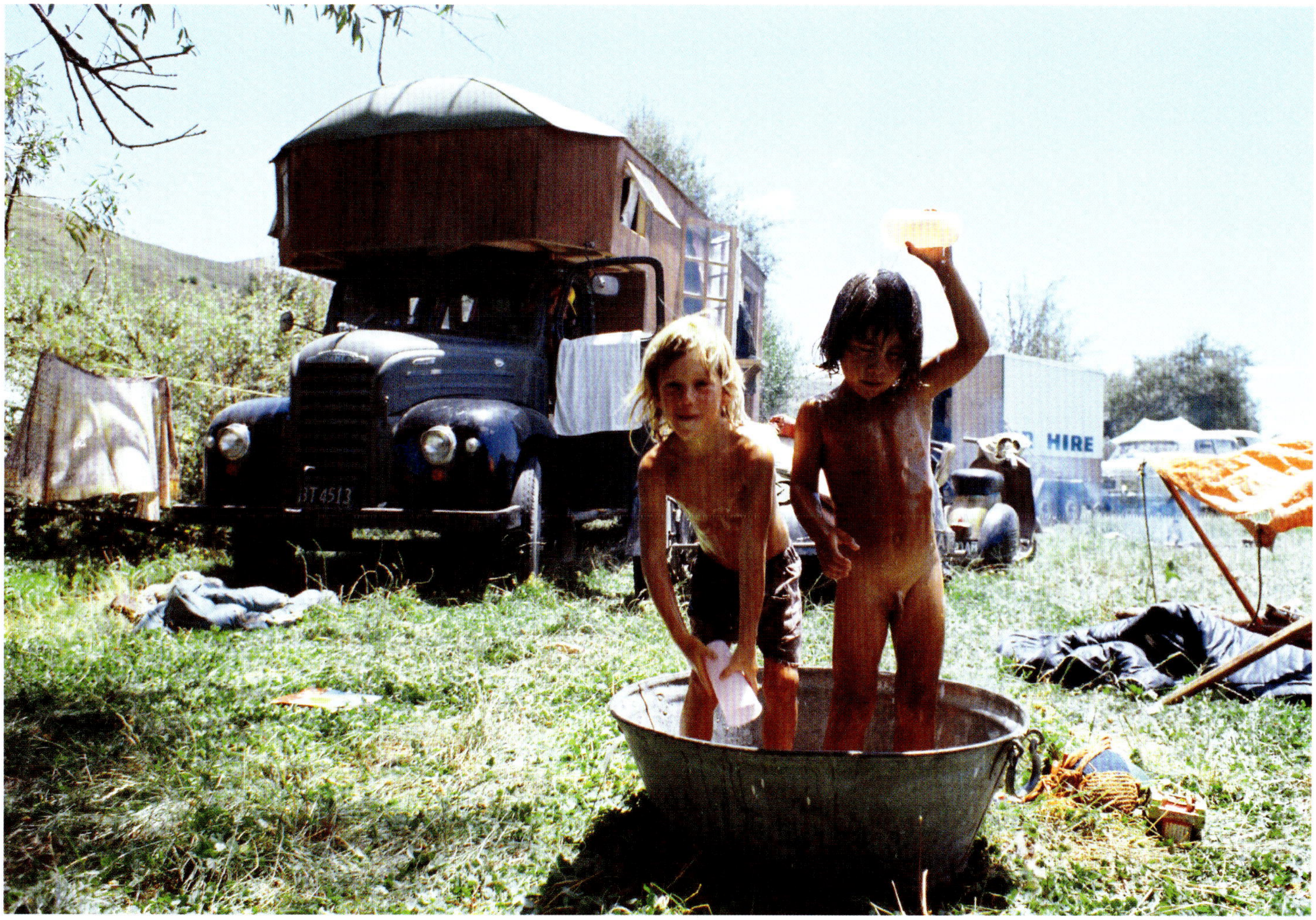

Children having a shower, Sweetwaters, 1982

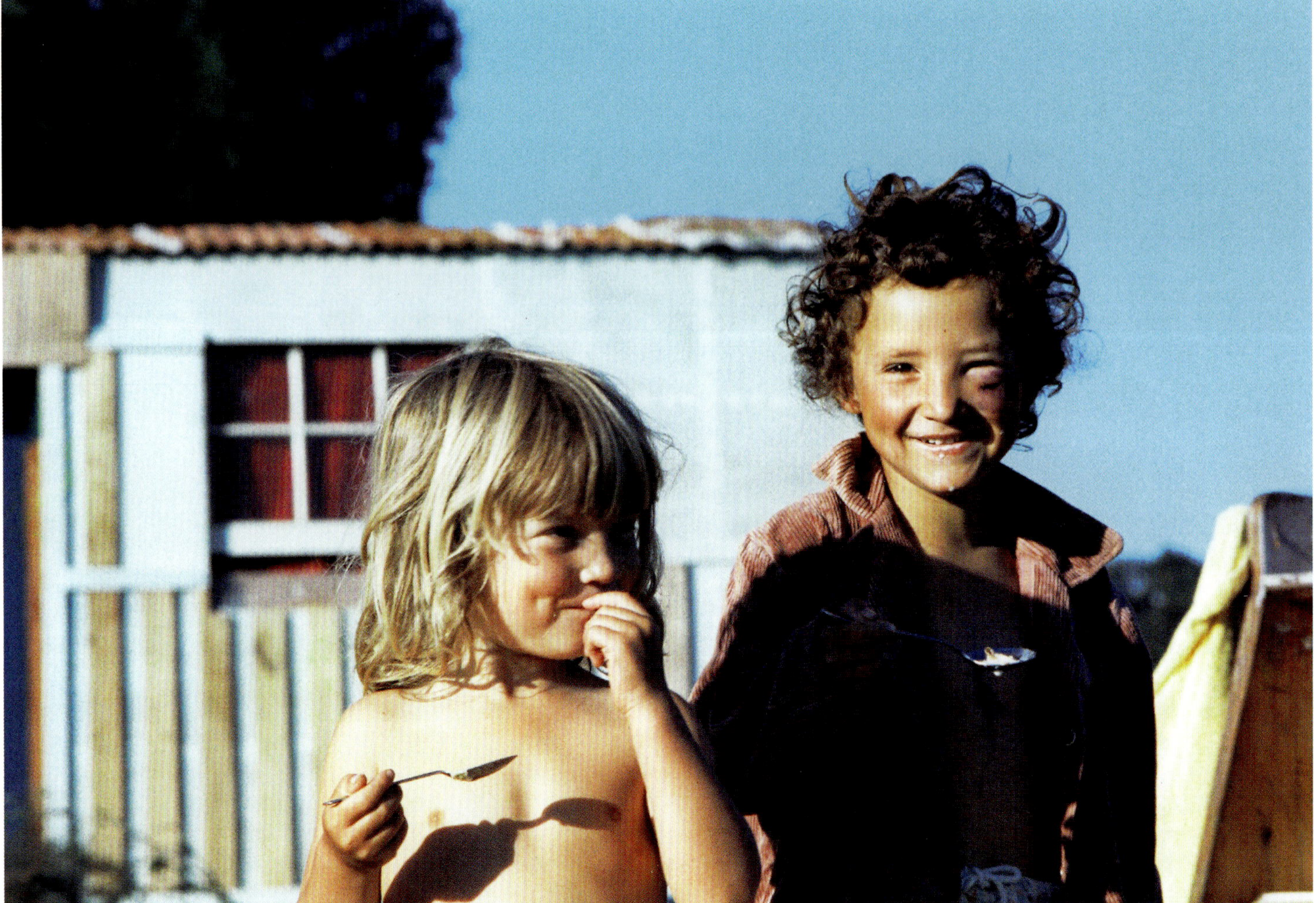

Children from the Kipling Avenue communal house at Sweetwaters, 1980

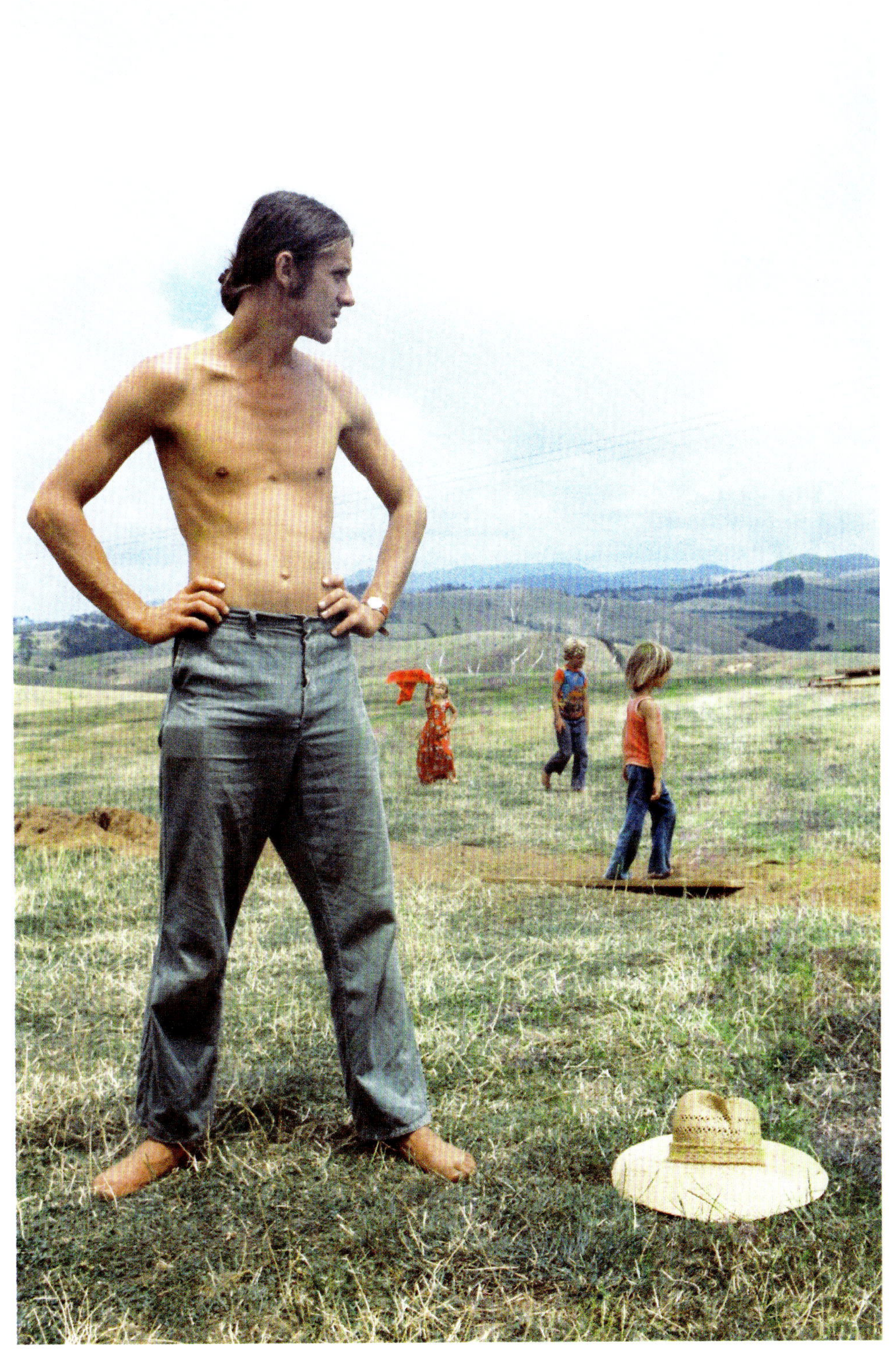

John Tucker, Nambassa, 1979

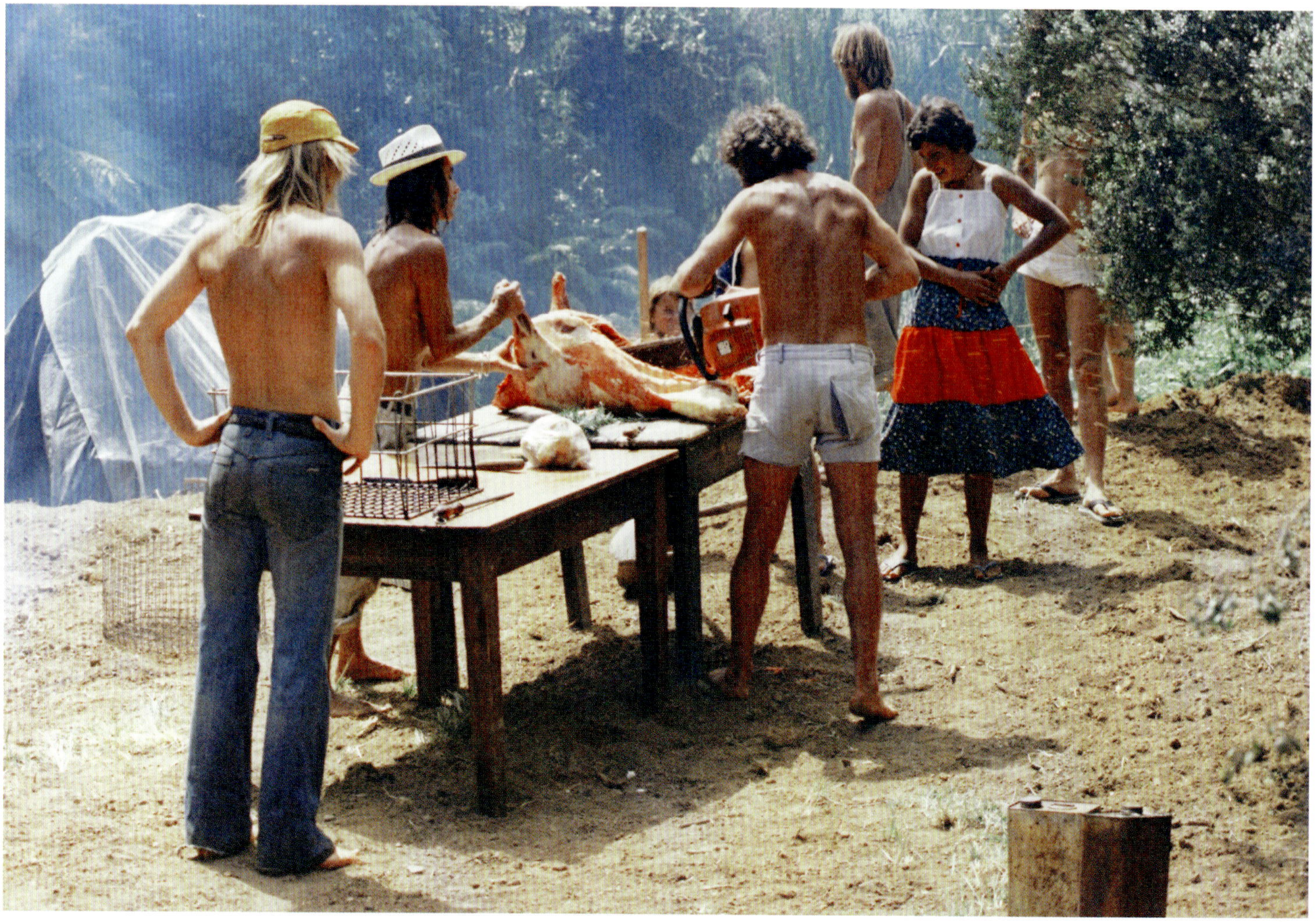

Preparing for a hangi, Fern Flat, Northland, 1978/79

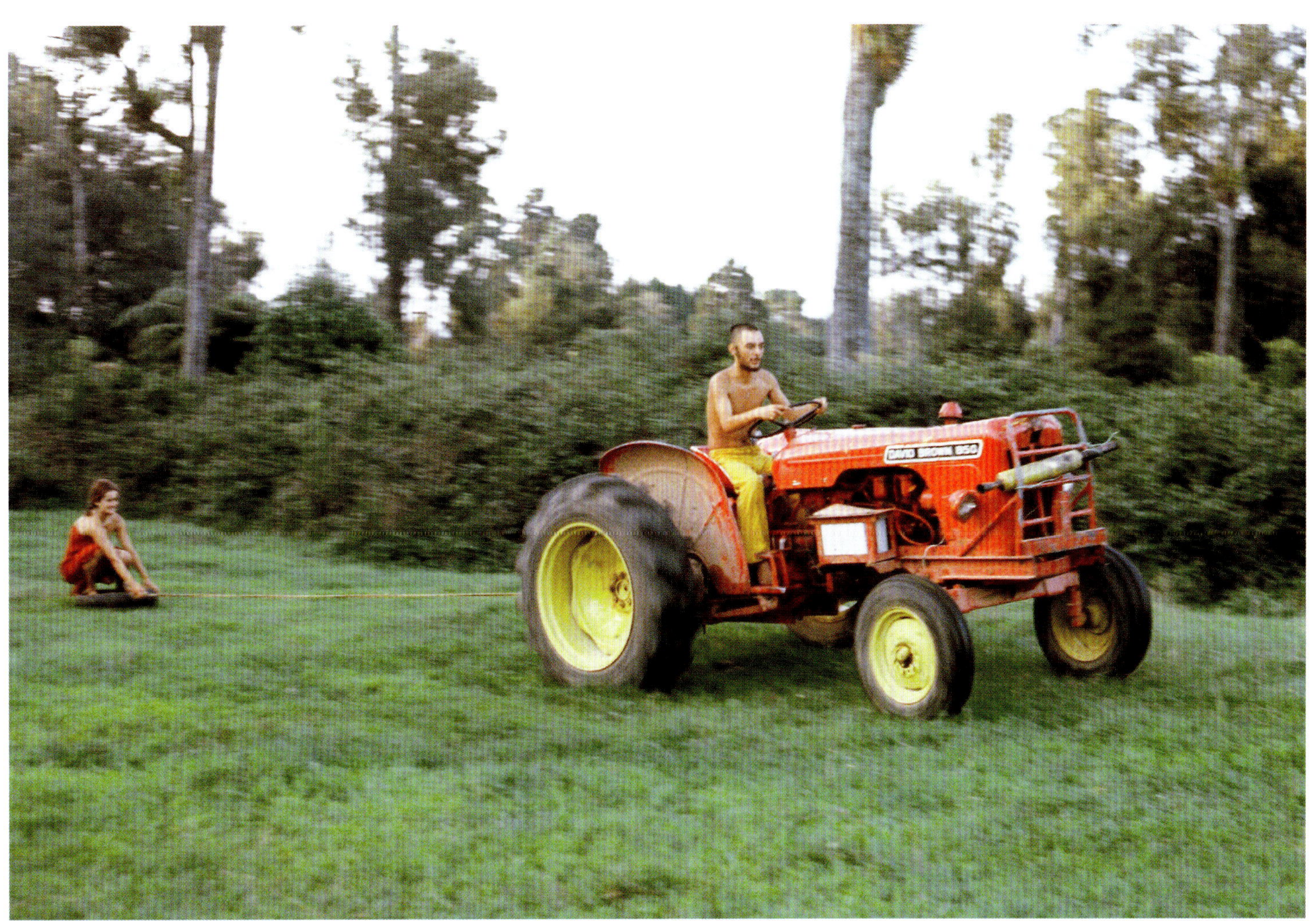

Lennox driving, Judarne riding, Katikati, 1981

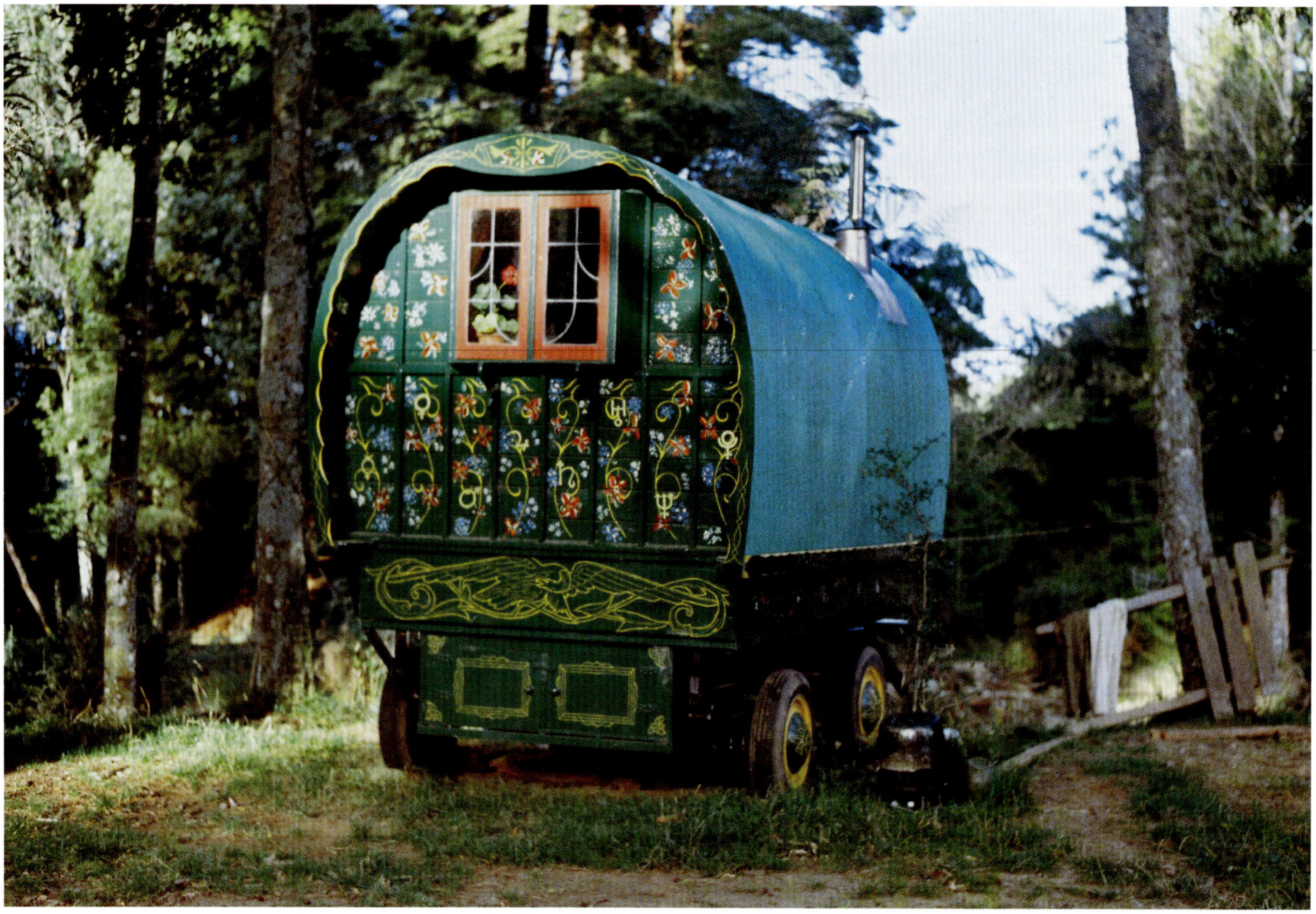

Simon's wagon, Nambassa, 1981

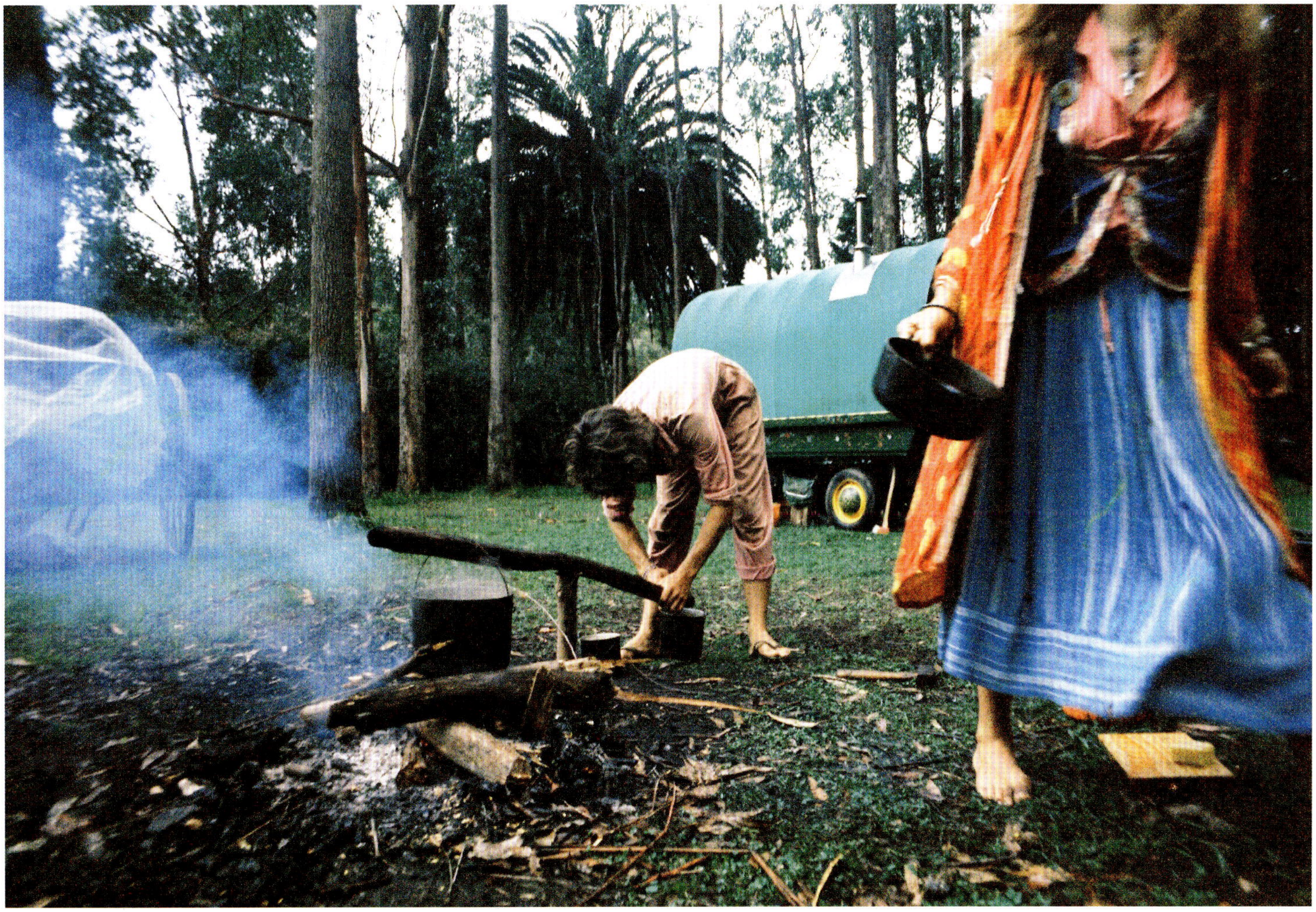

Journey back to Katikati, after Nambassa, 1981

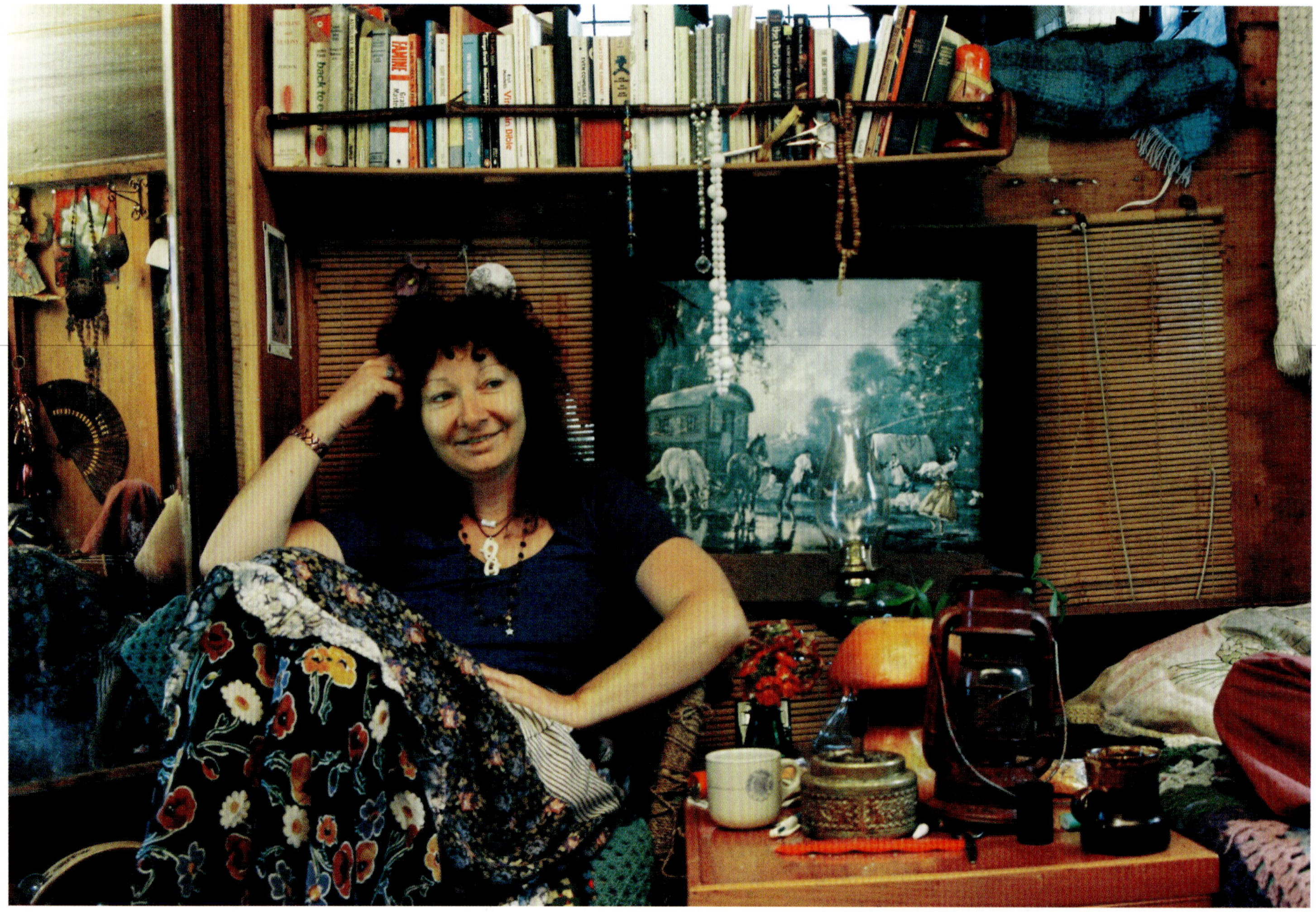

Jade in her truck, Aratiatia, 1984

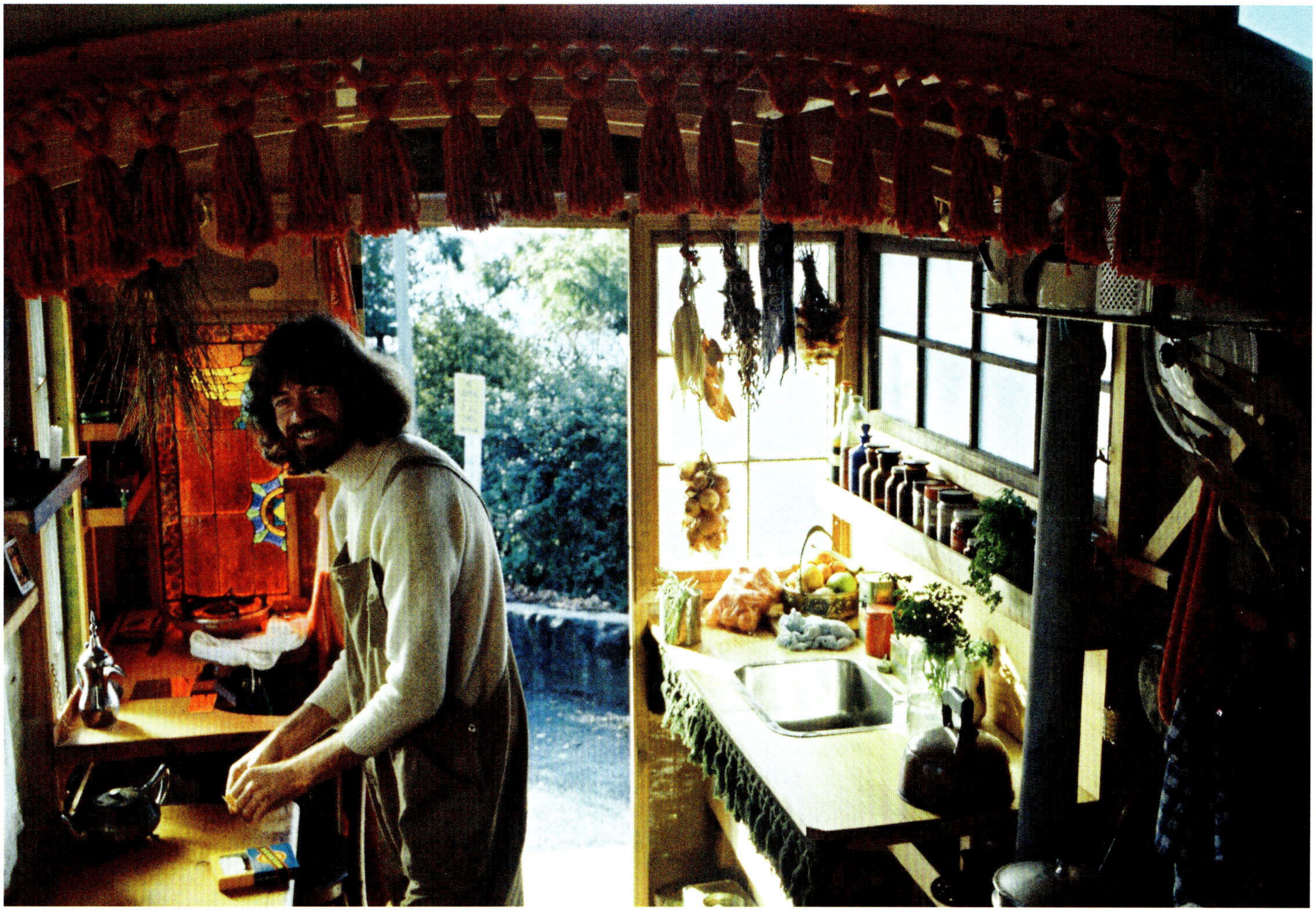

Unidentified trucker, c.1980

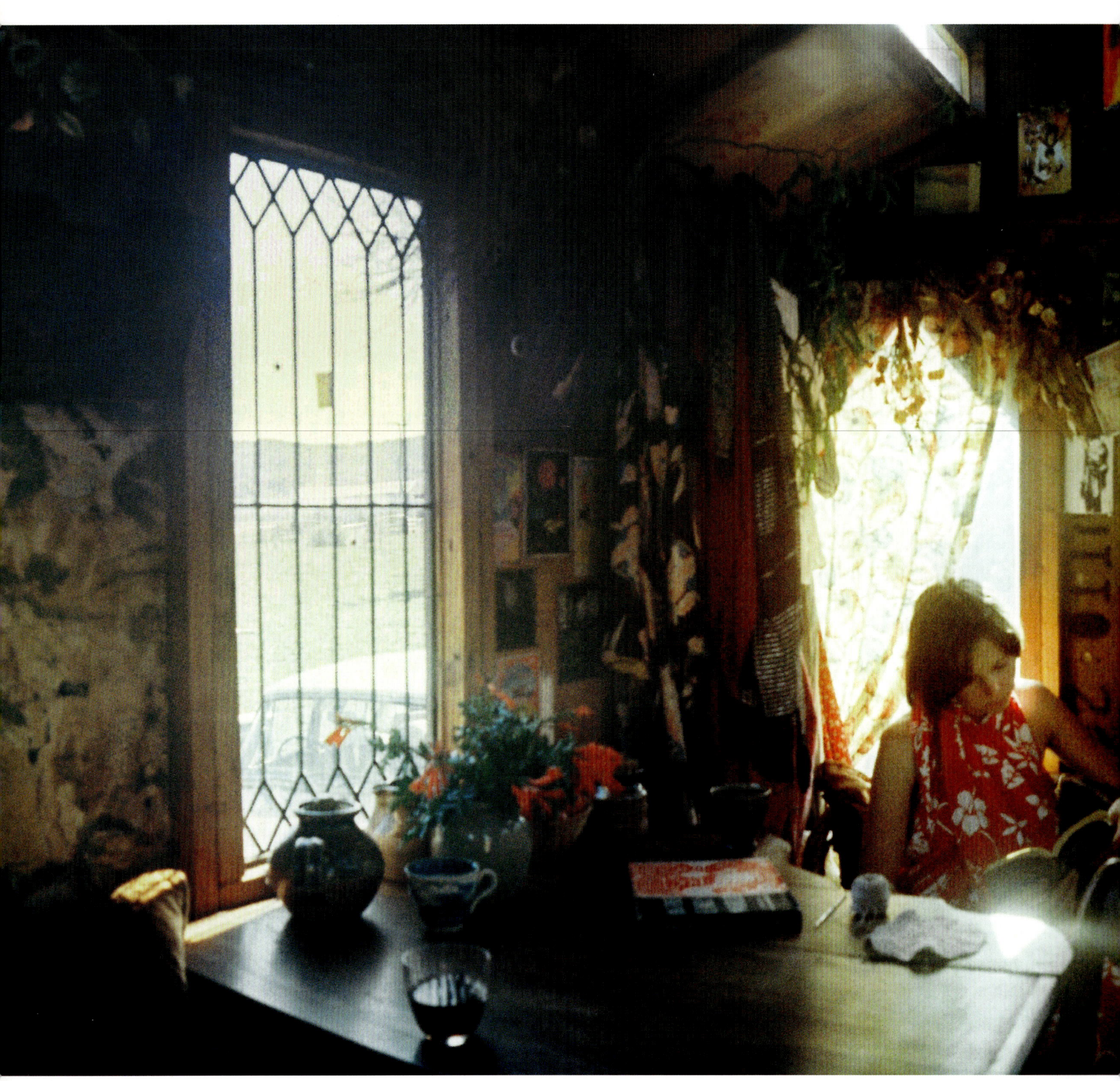

Marie in her truck, Nambassa, 1979

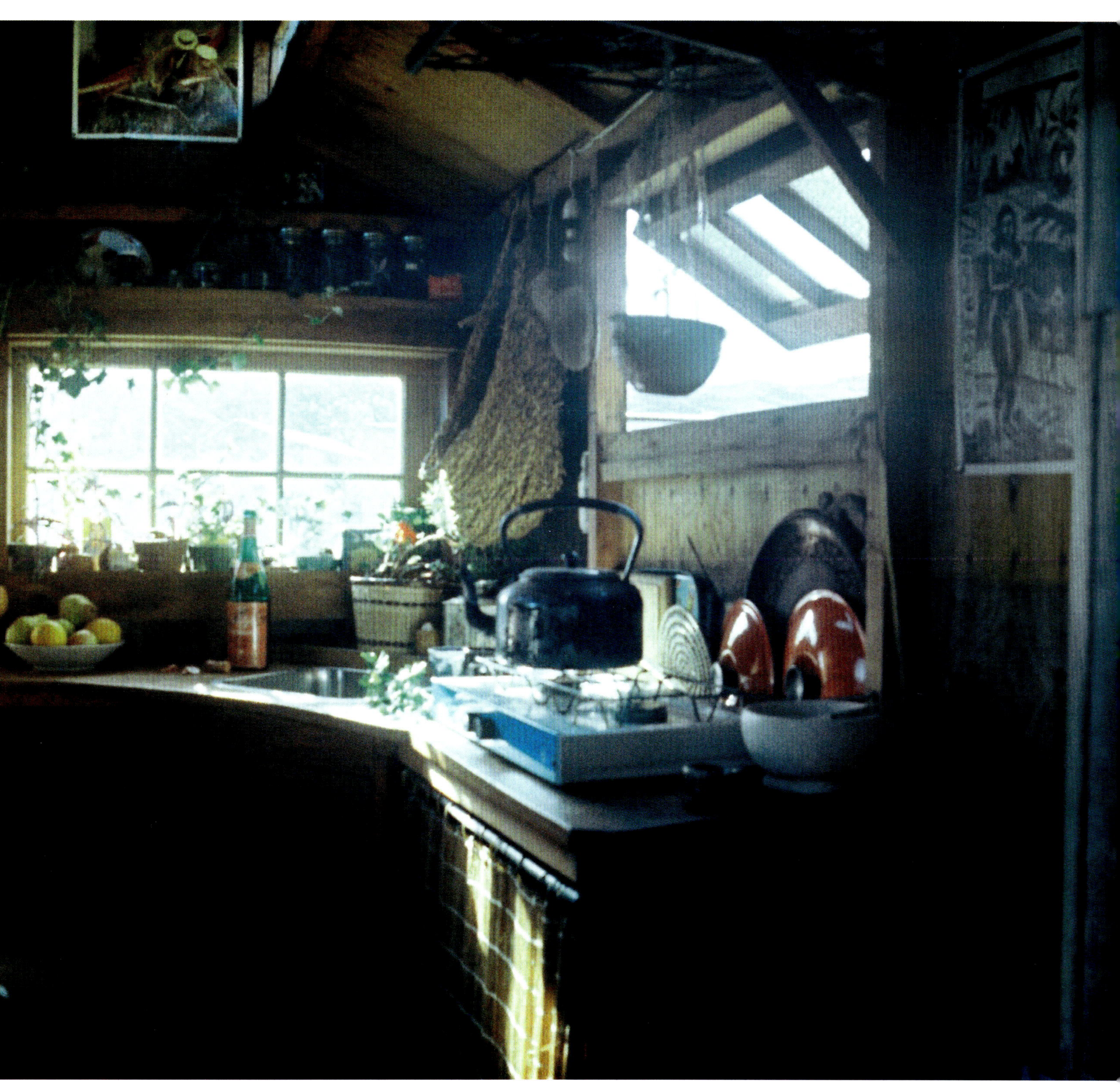

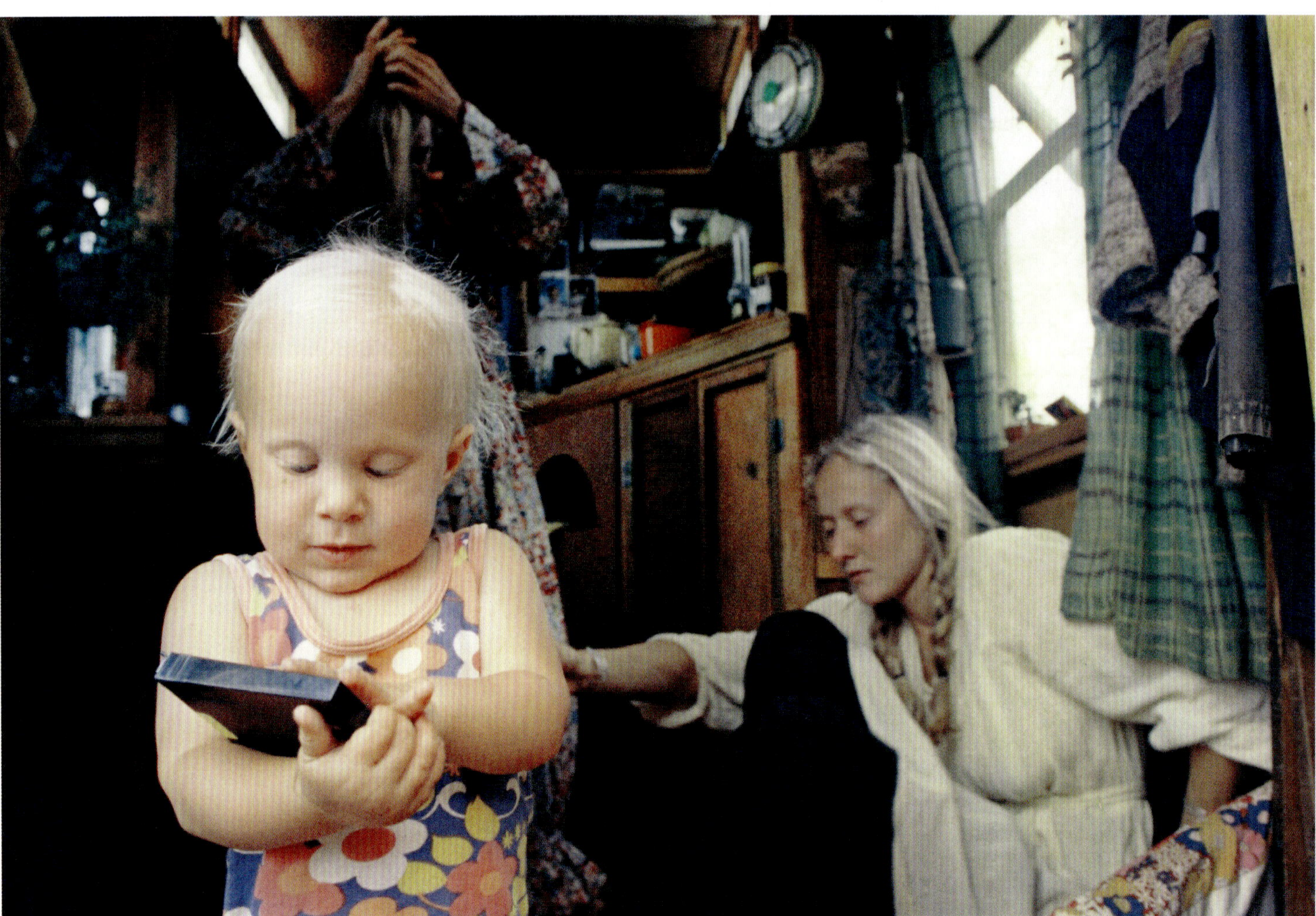

Jonathon, Hendrika and Tahi, Northland tour, 1979

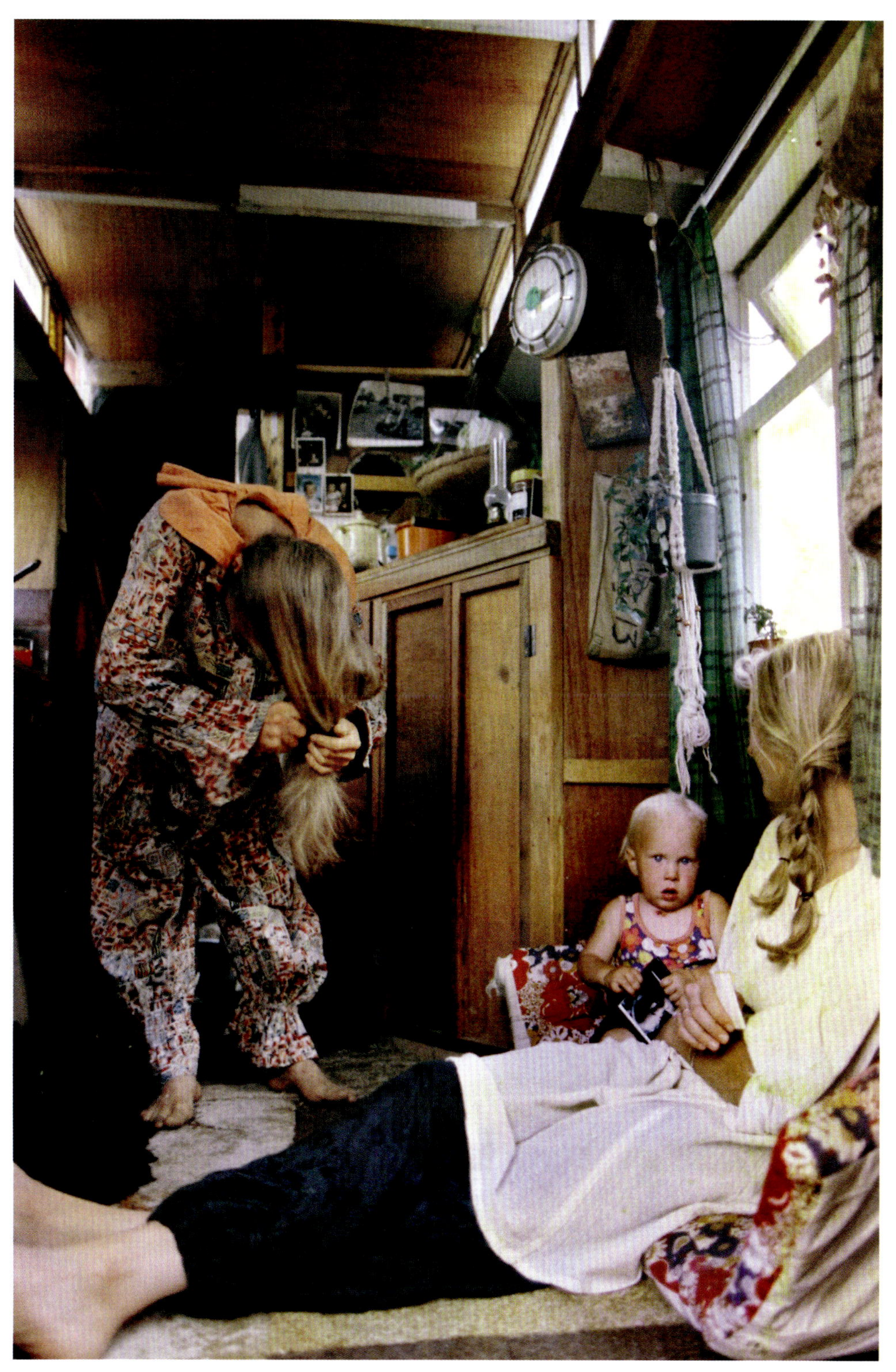

Jonathon, Hendrika and Tahi, Northland tour, 1979

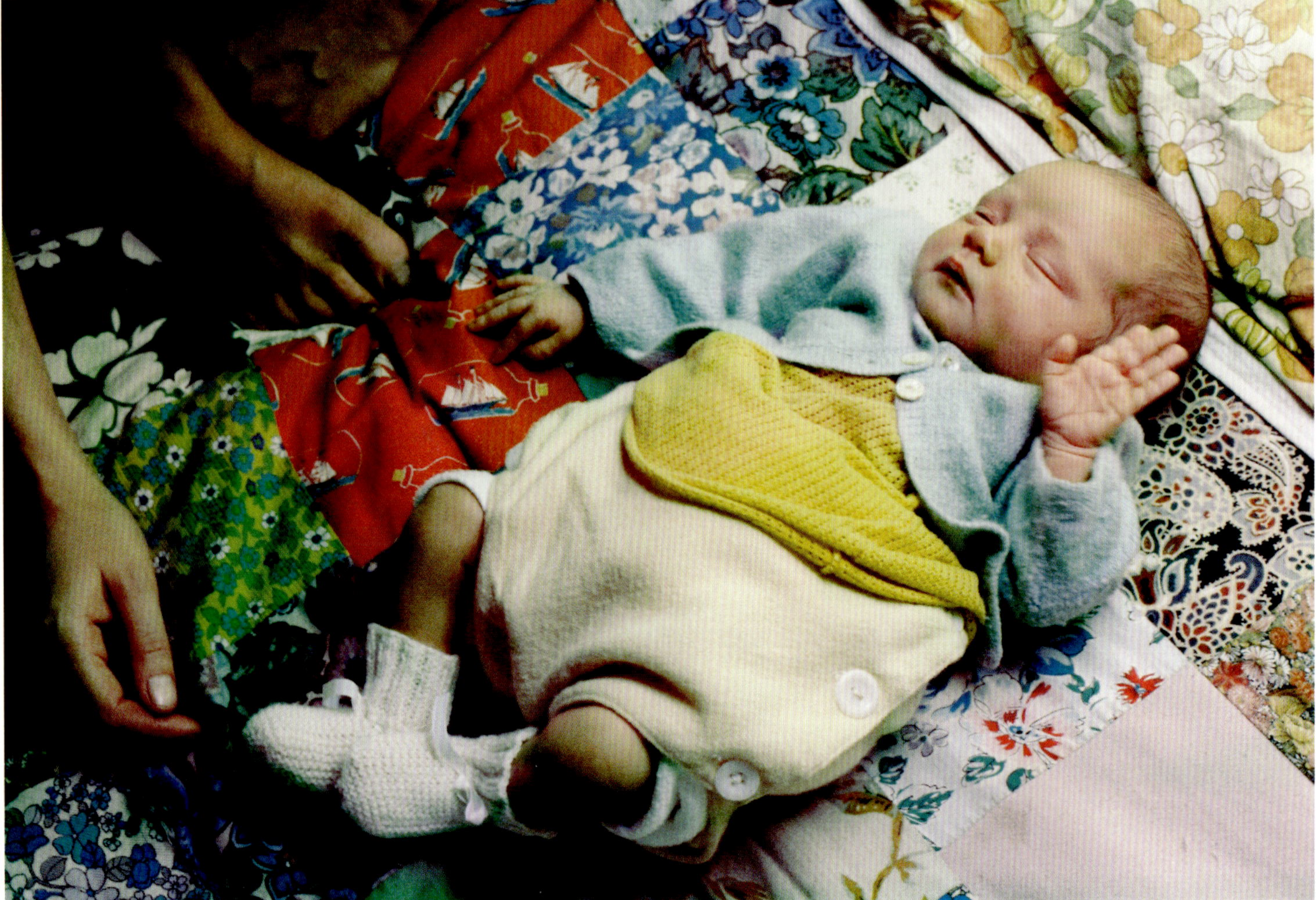

Aya, house truck baby, 1981

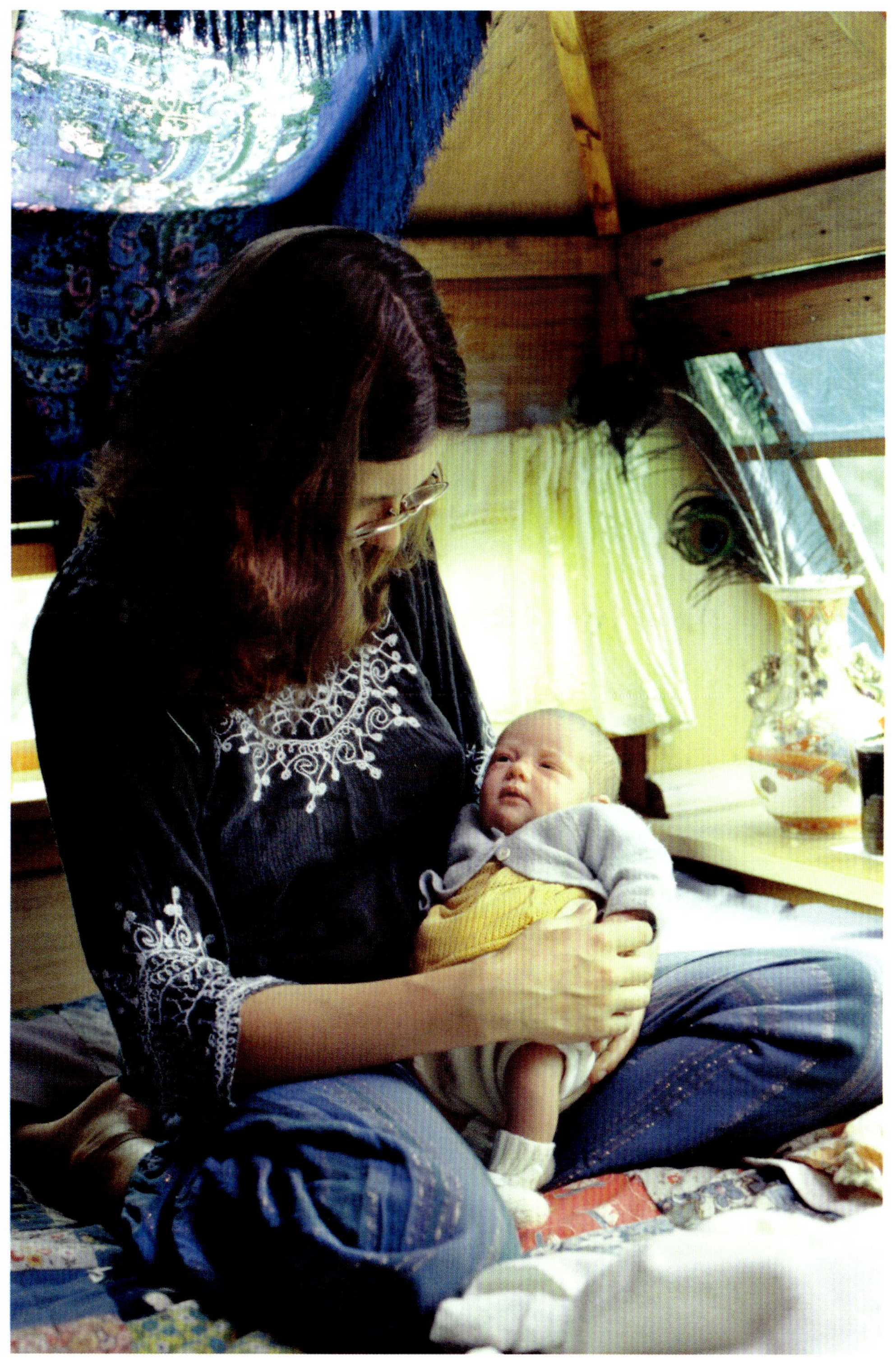

Helen and Aya, 1981

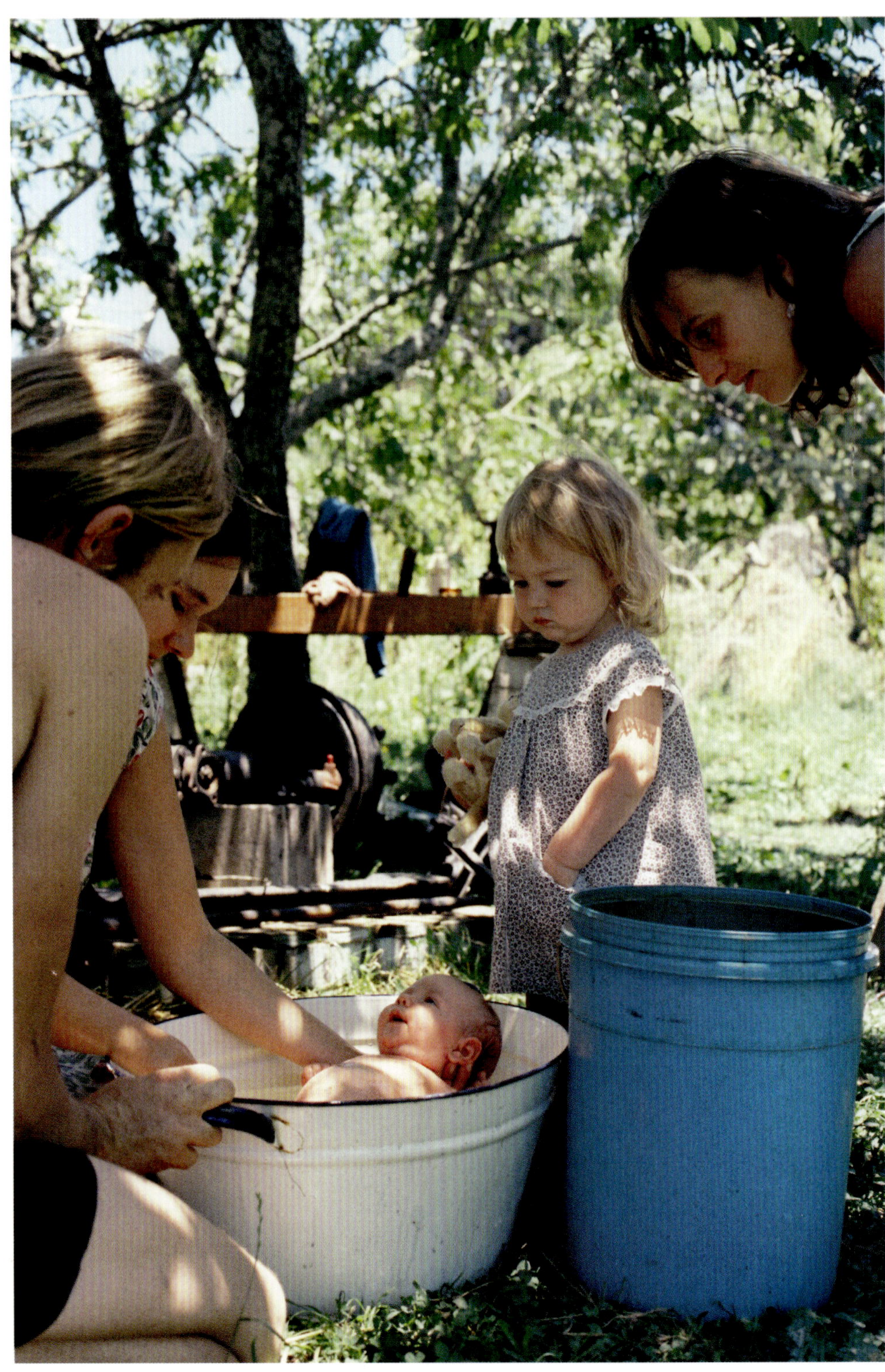

Ewan and Helen with baby Aya, Michala and Sue, Taupaki orchard, Auckland, 1981

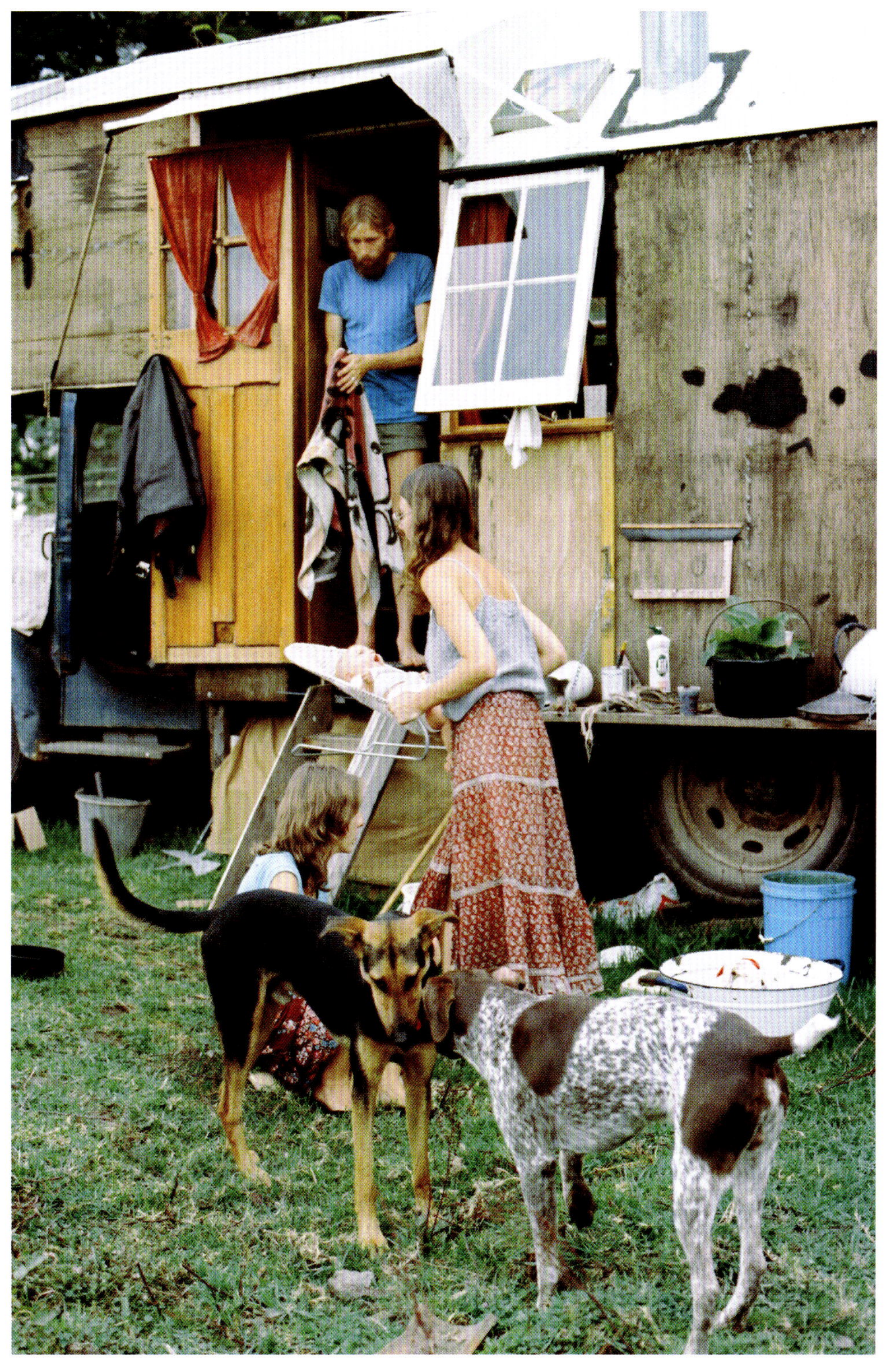

Sue, Ewan, and Helen with baby Benjamin, Brookby, Auckland, 1983

Ewan with baby Benjamin who was born in the house truck, and his sister Aya, at Brookby, Auckland, 1983

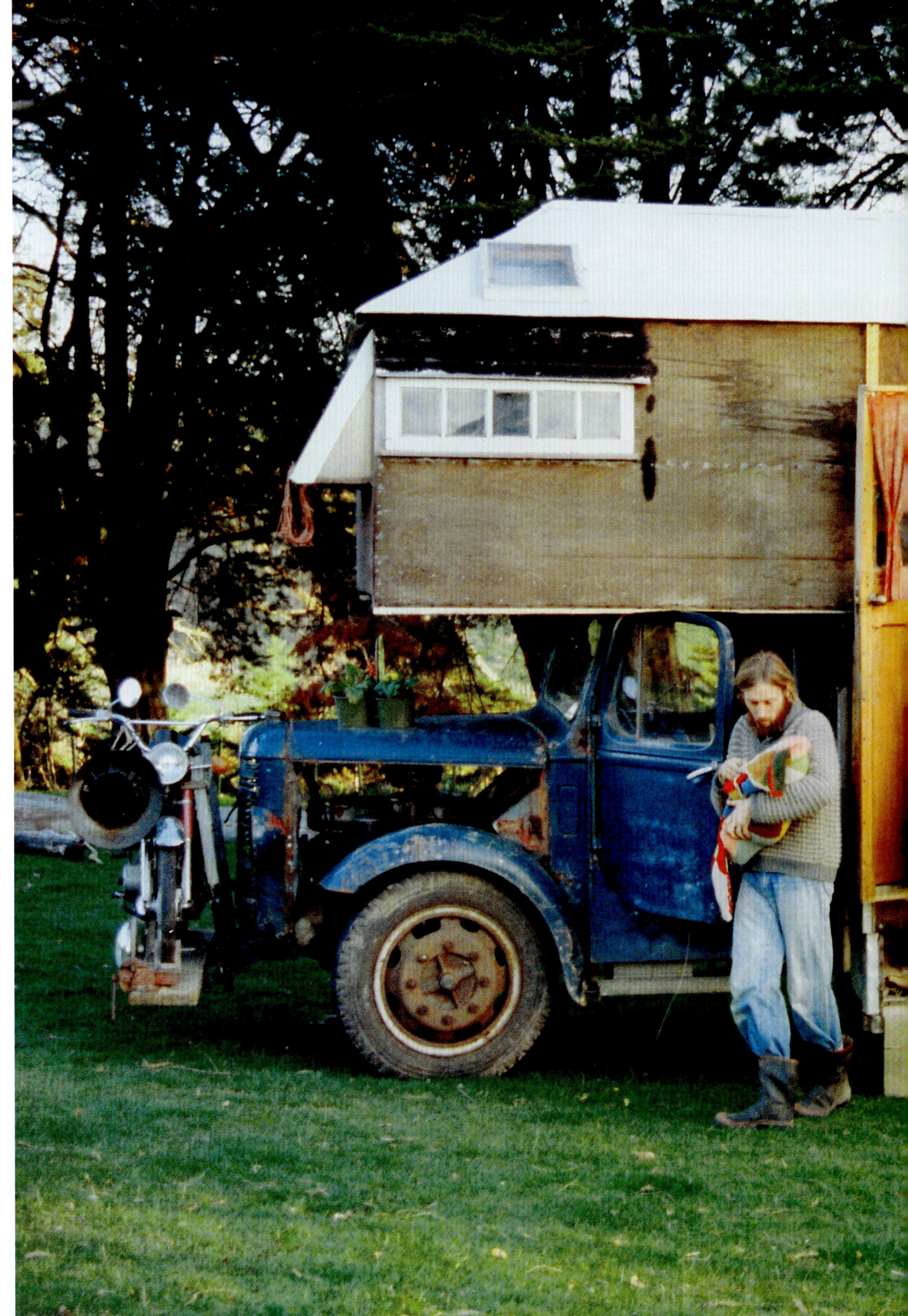

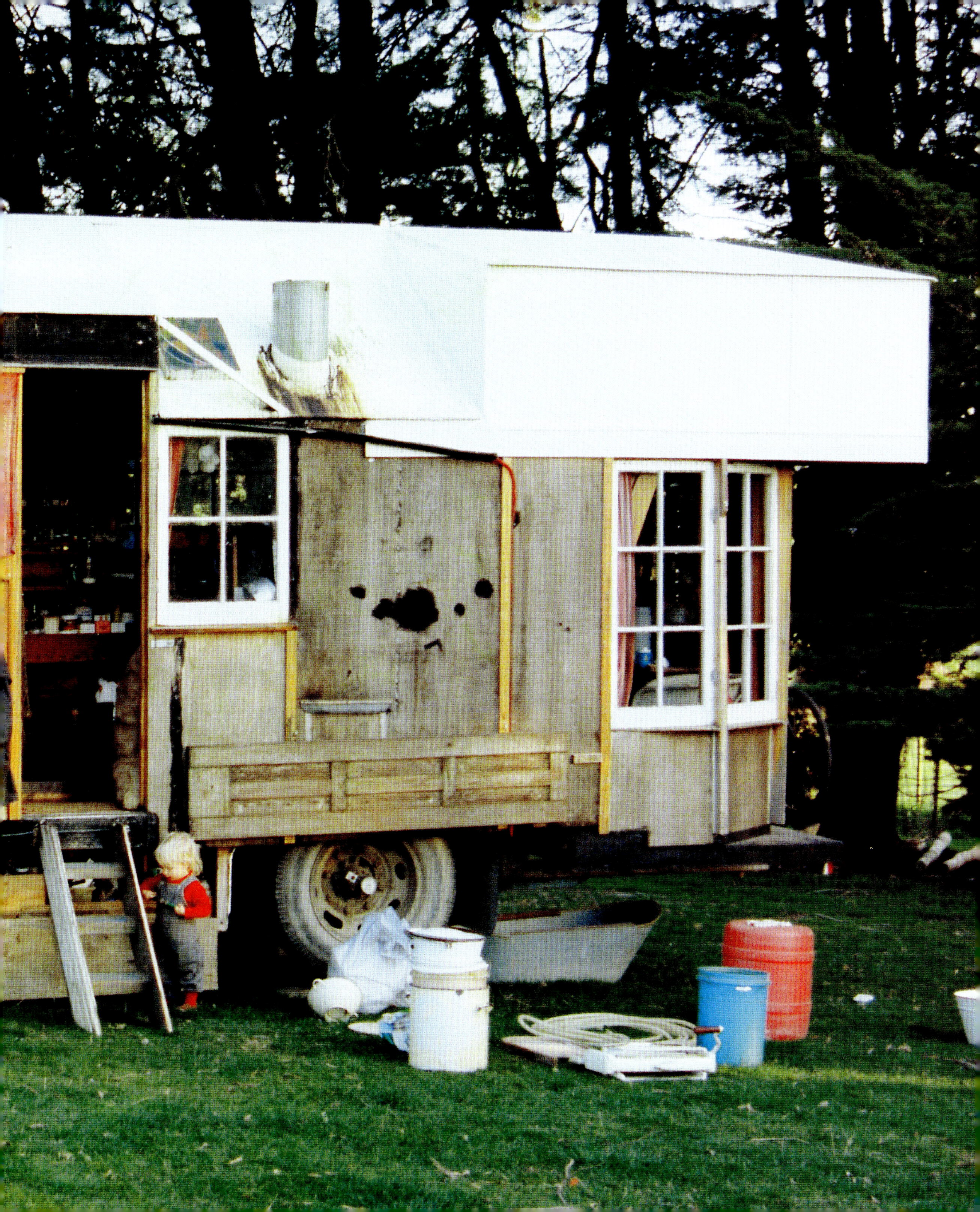

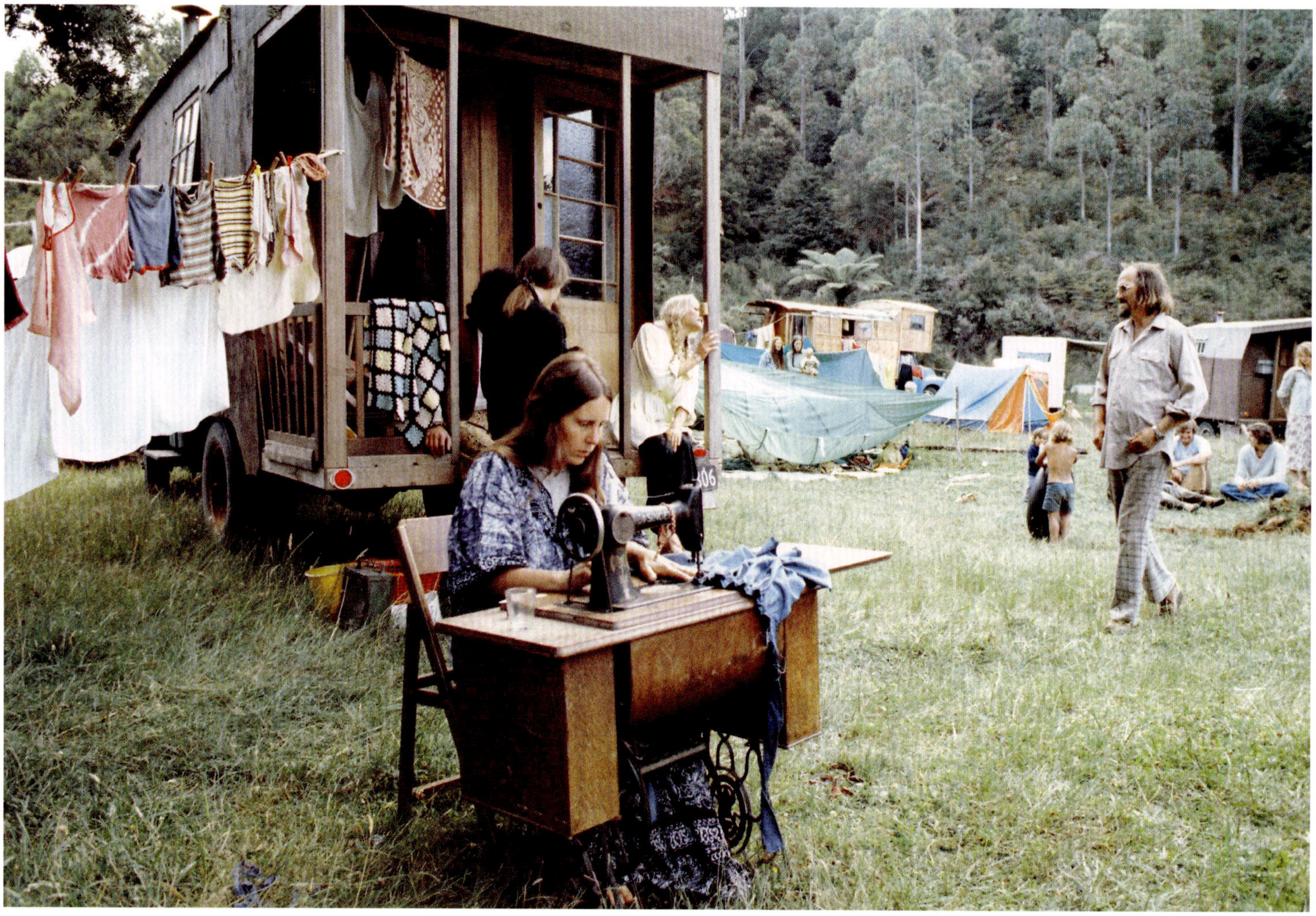

Fern Flat camp, Far North, 1978/79

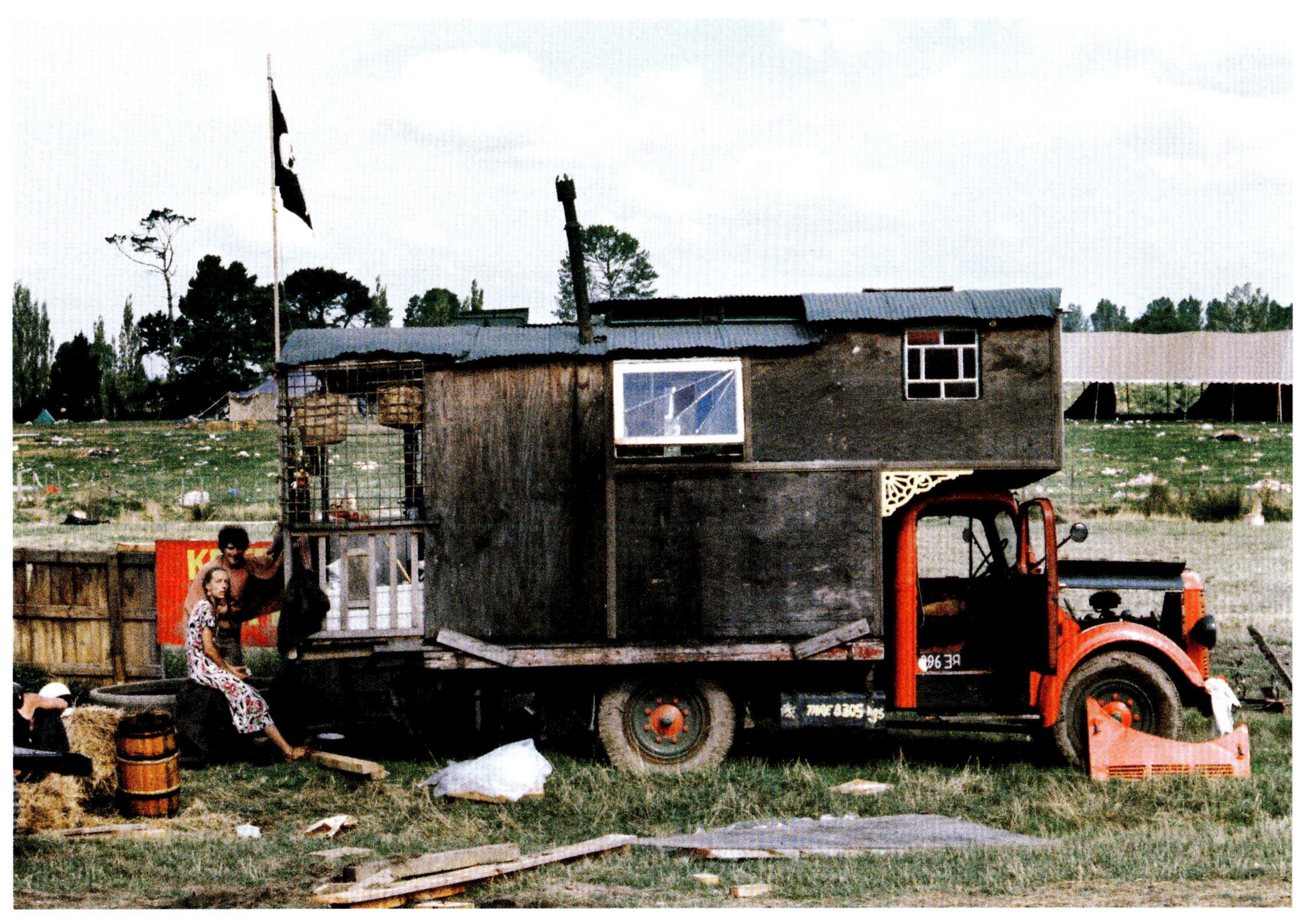

Robyn's house truck after 1982 Sweetwaters Festival; the first house truck built by John Tucker, 1982

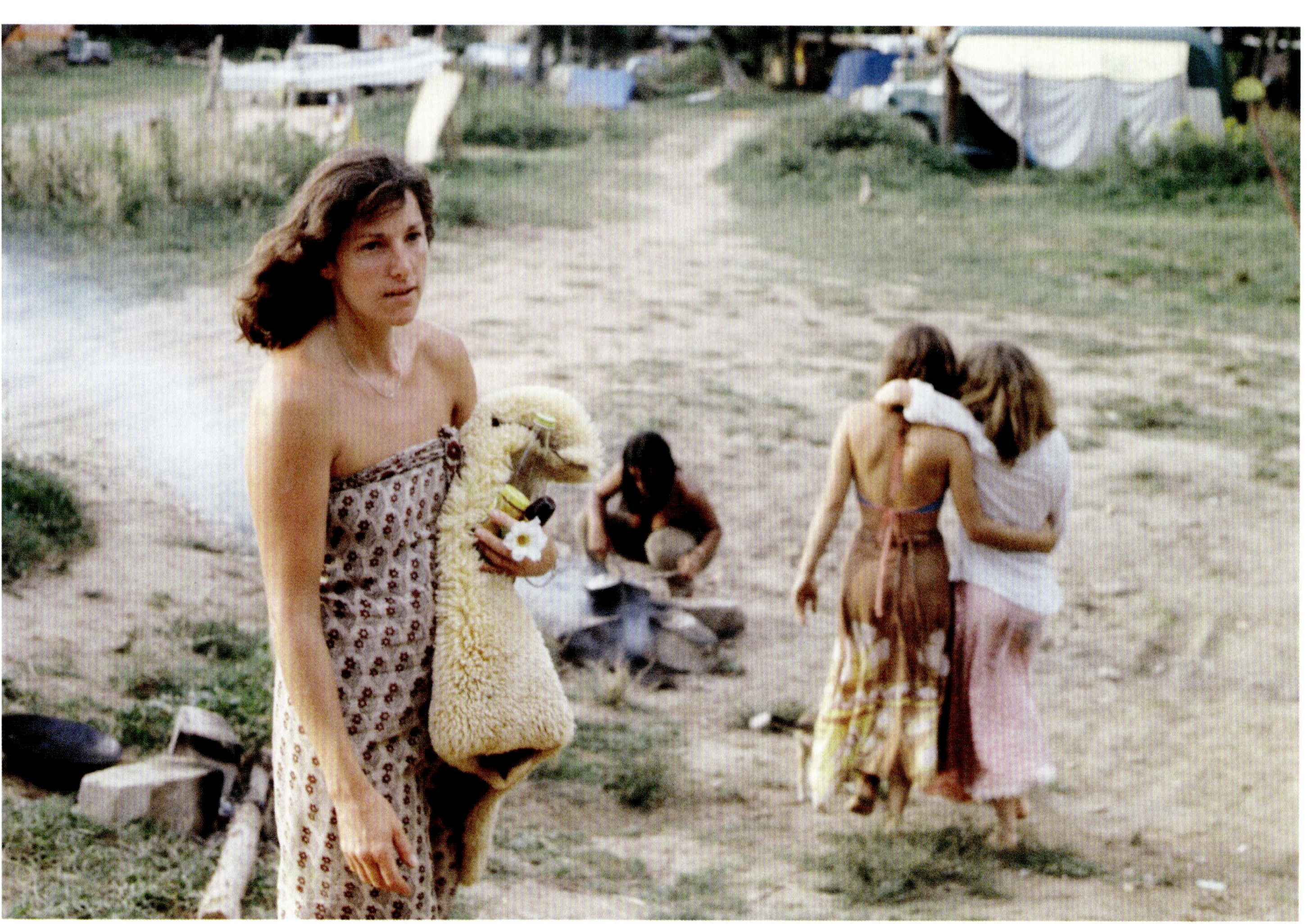

Nambassa, 1981

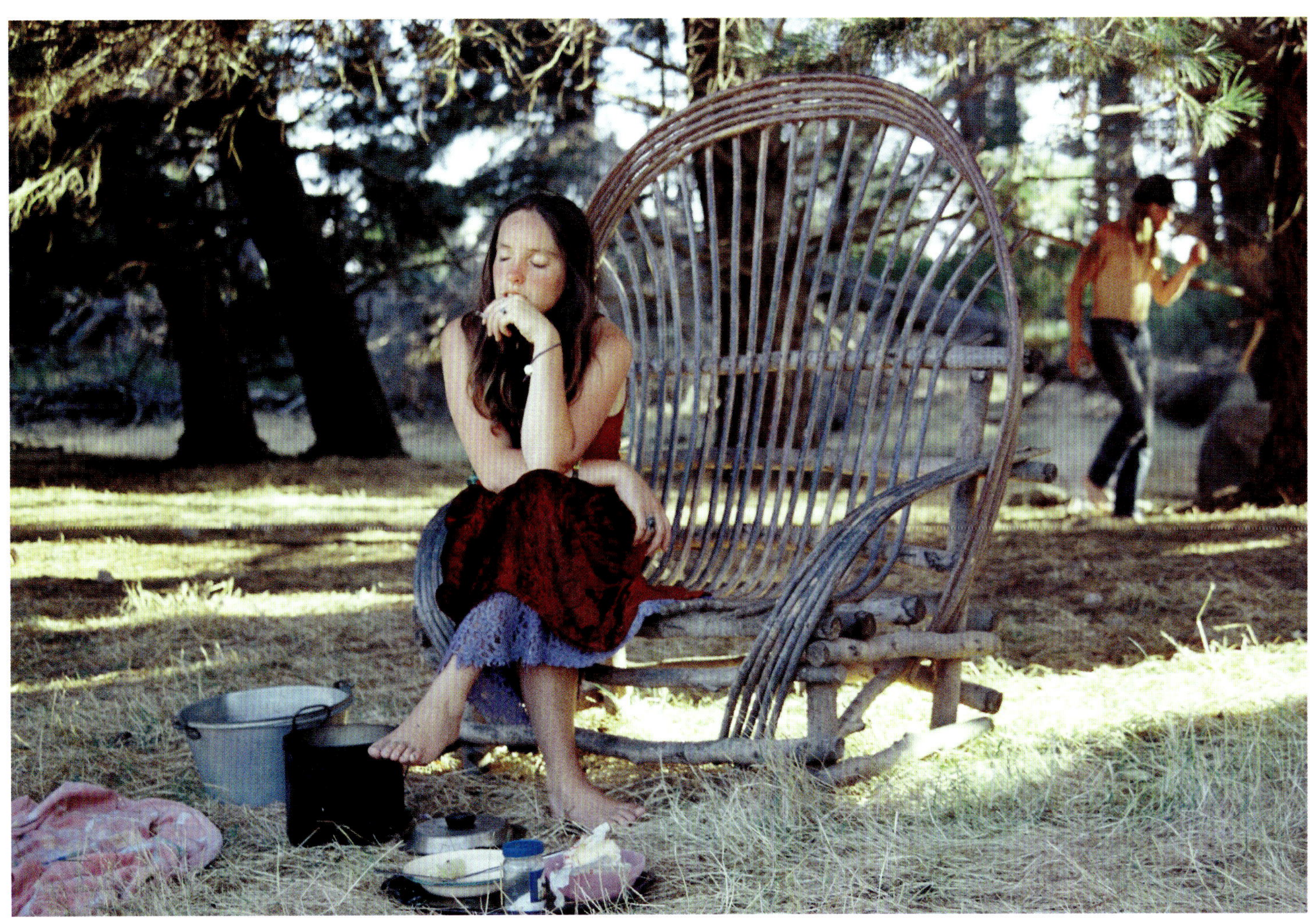

Aratiatia, 1984

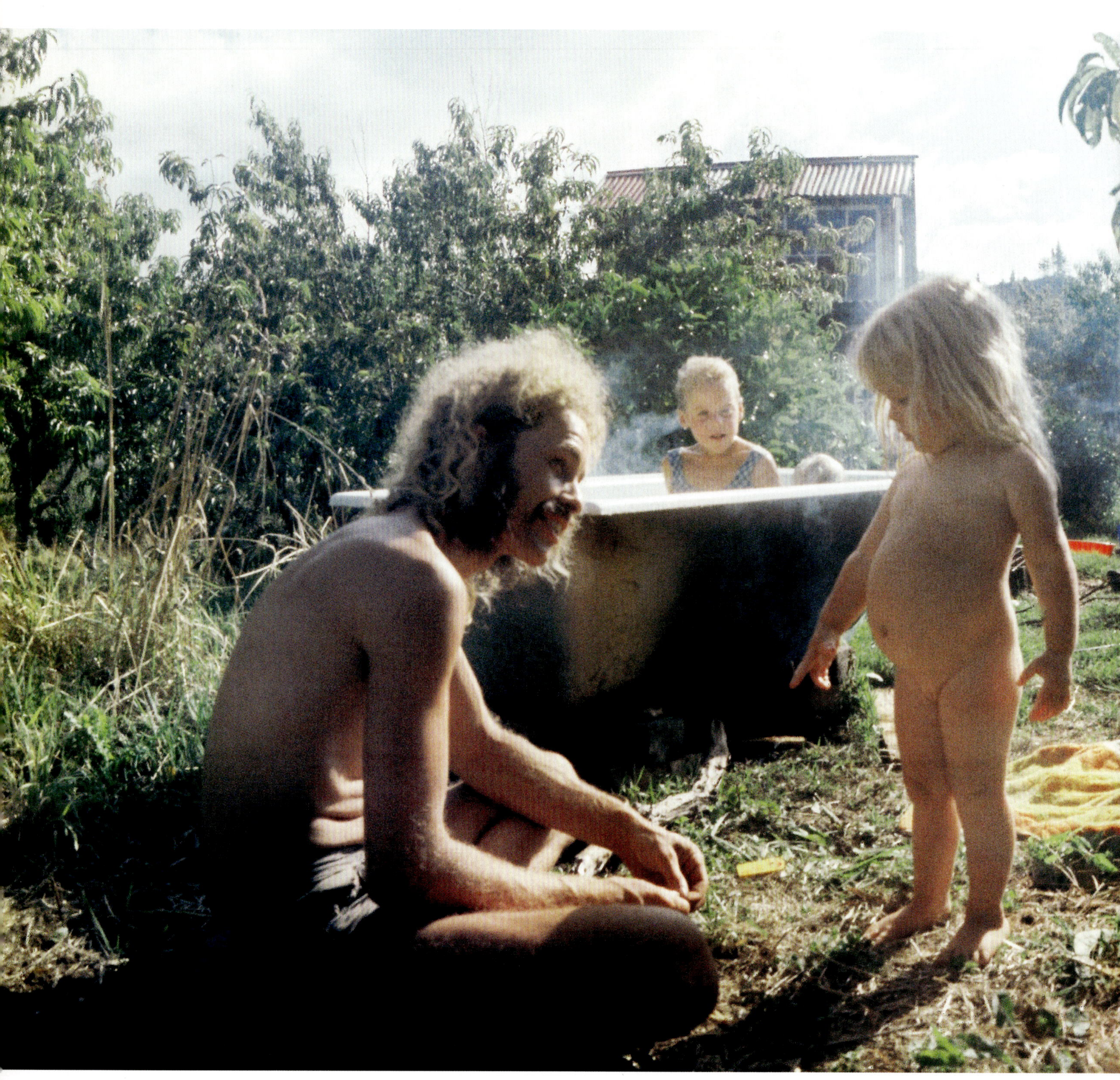

Phil Toms and Leone from the band Tribrations, with John Tucker and children at the Wilderland community, Coromandel, 1979

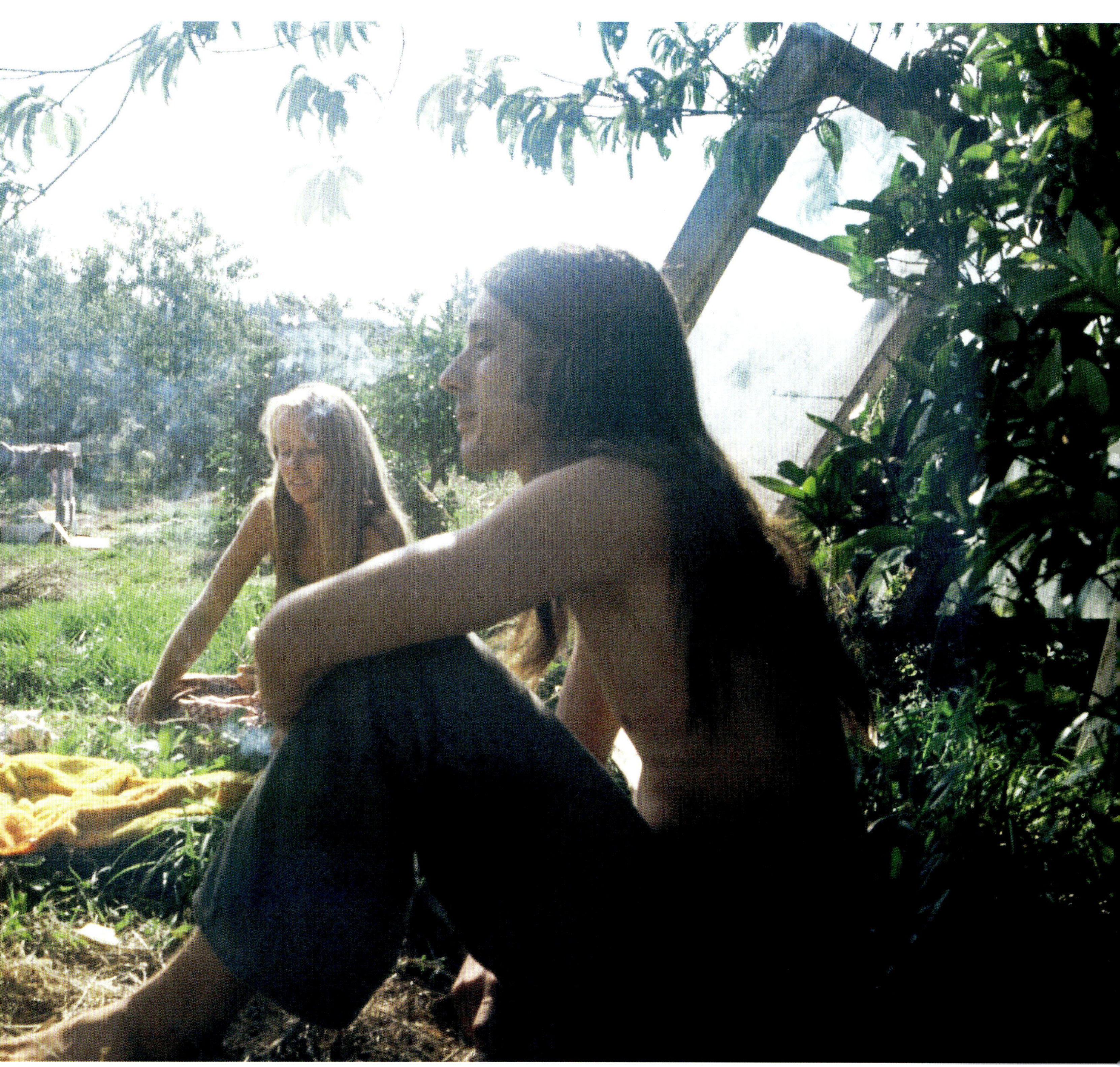

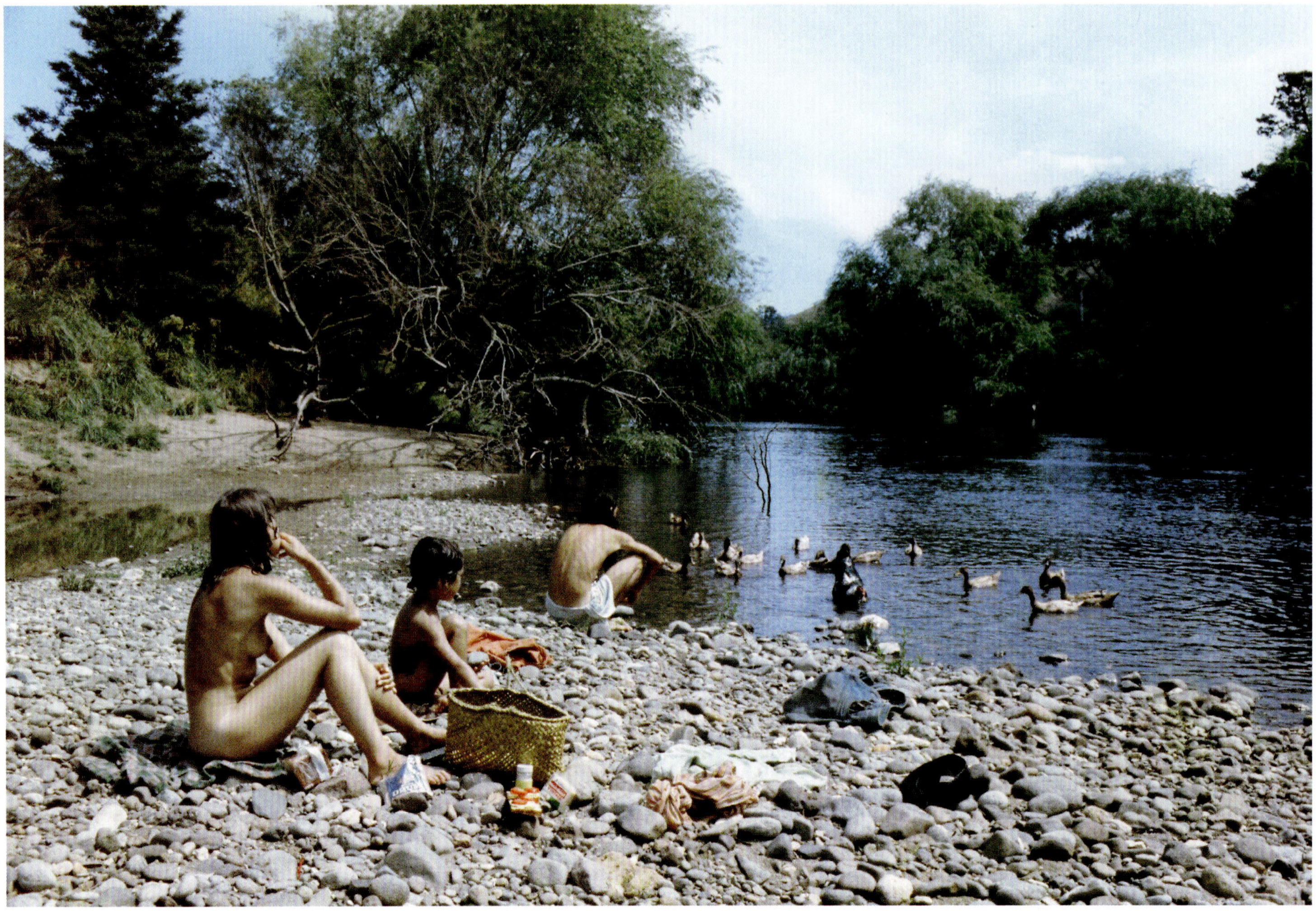

Hikutaia River, Maratoto, Coromandel, 1978

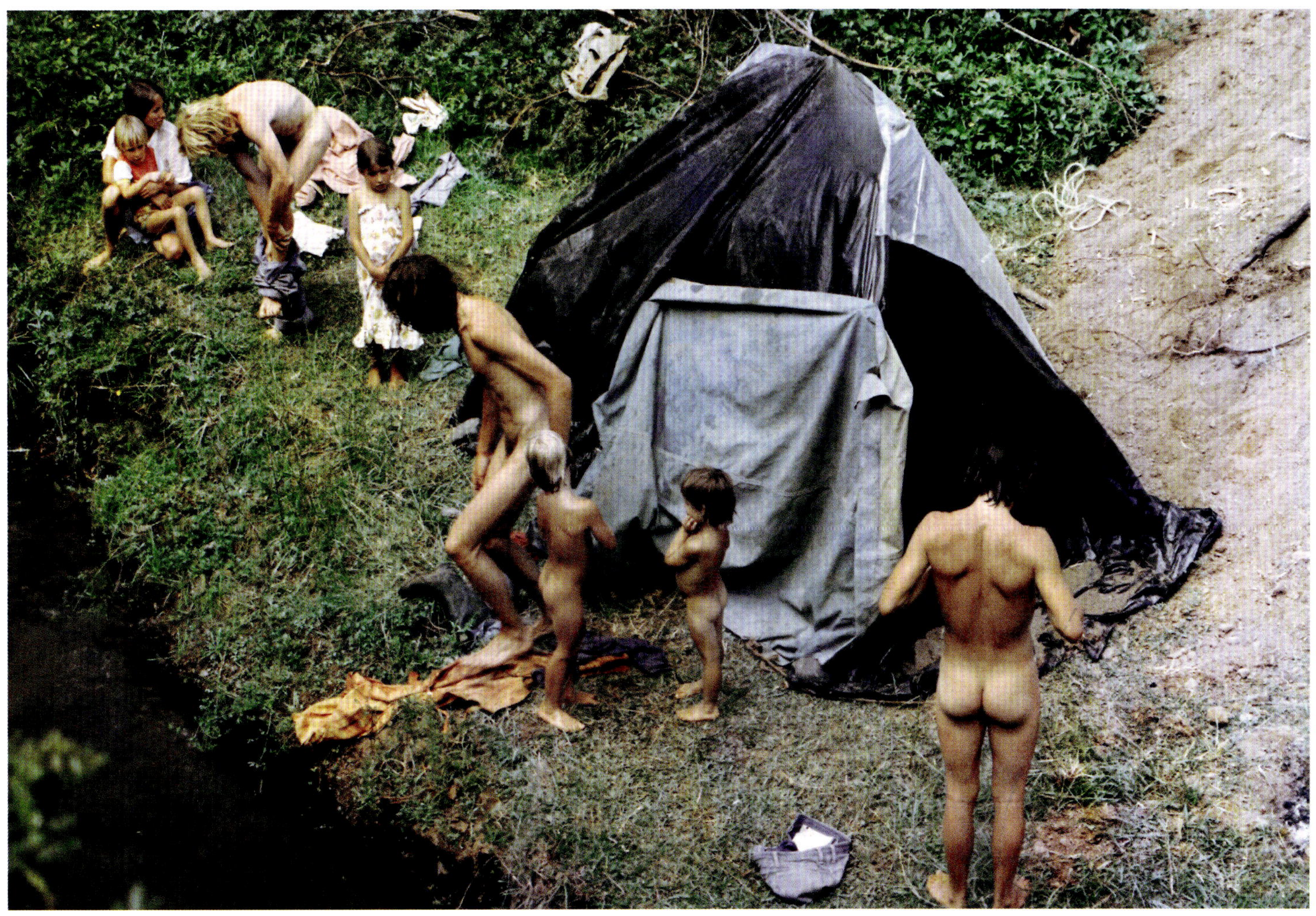

Sweat/steam bath at camp in Fern Flat, 1978/79

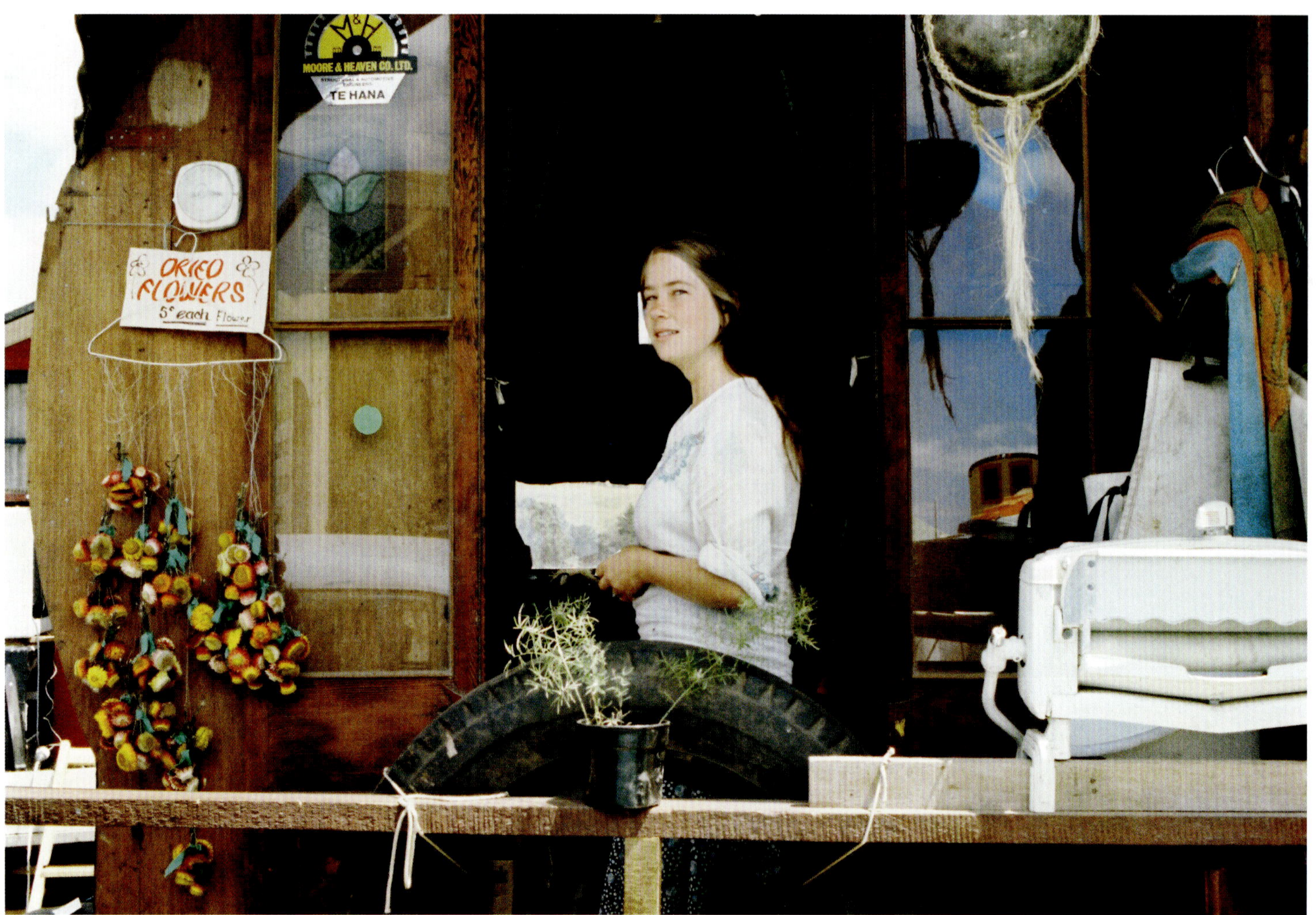

Diana in her and Podge's truck, Nambassa, 1979

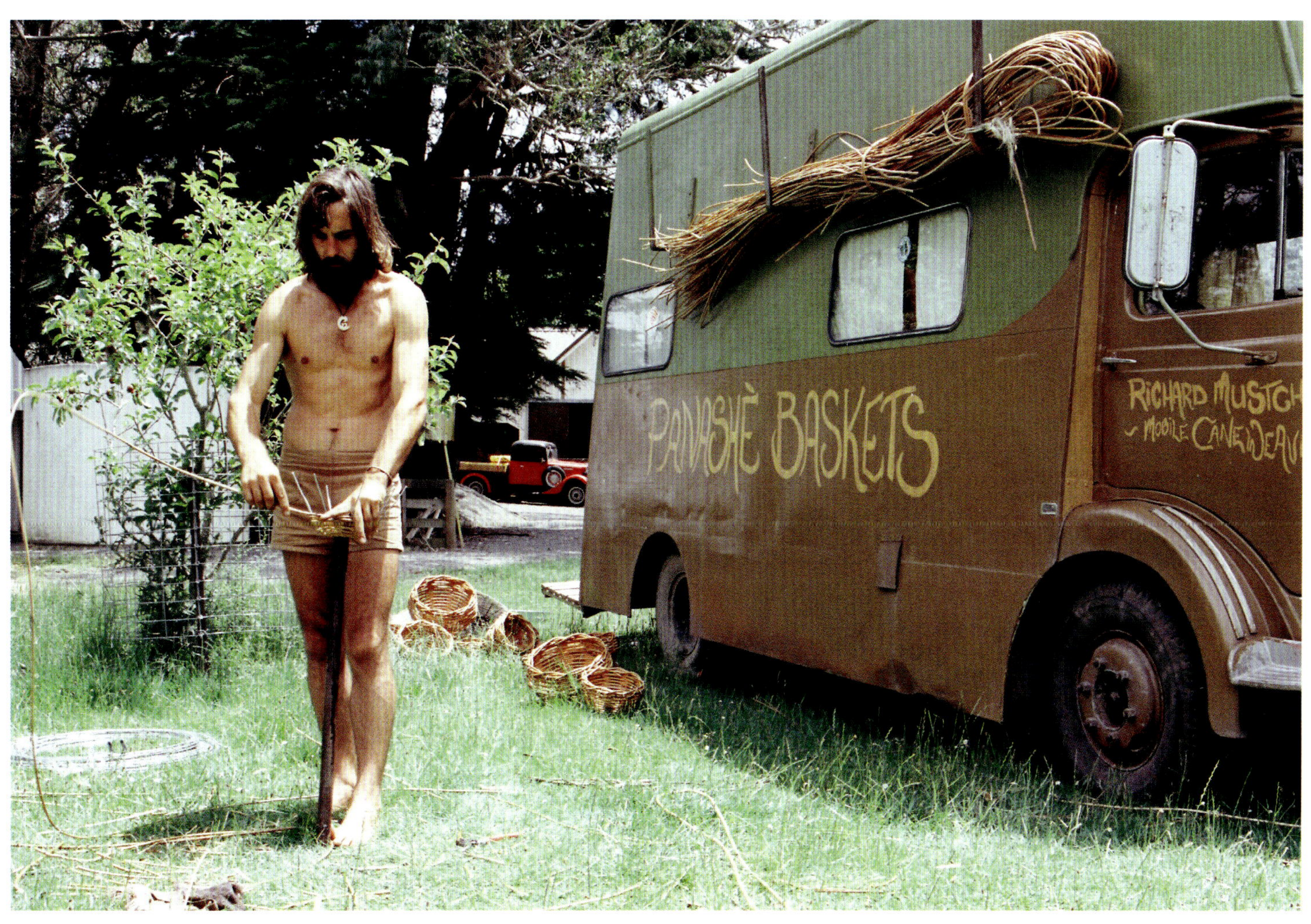

Richard, cane weaver, at Brookby, 1983

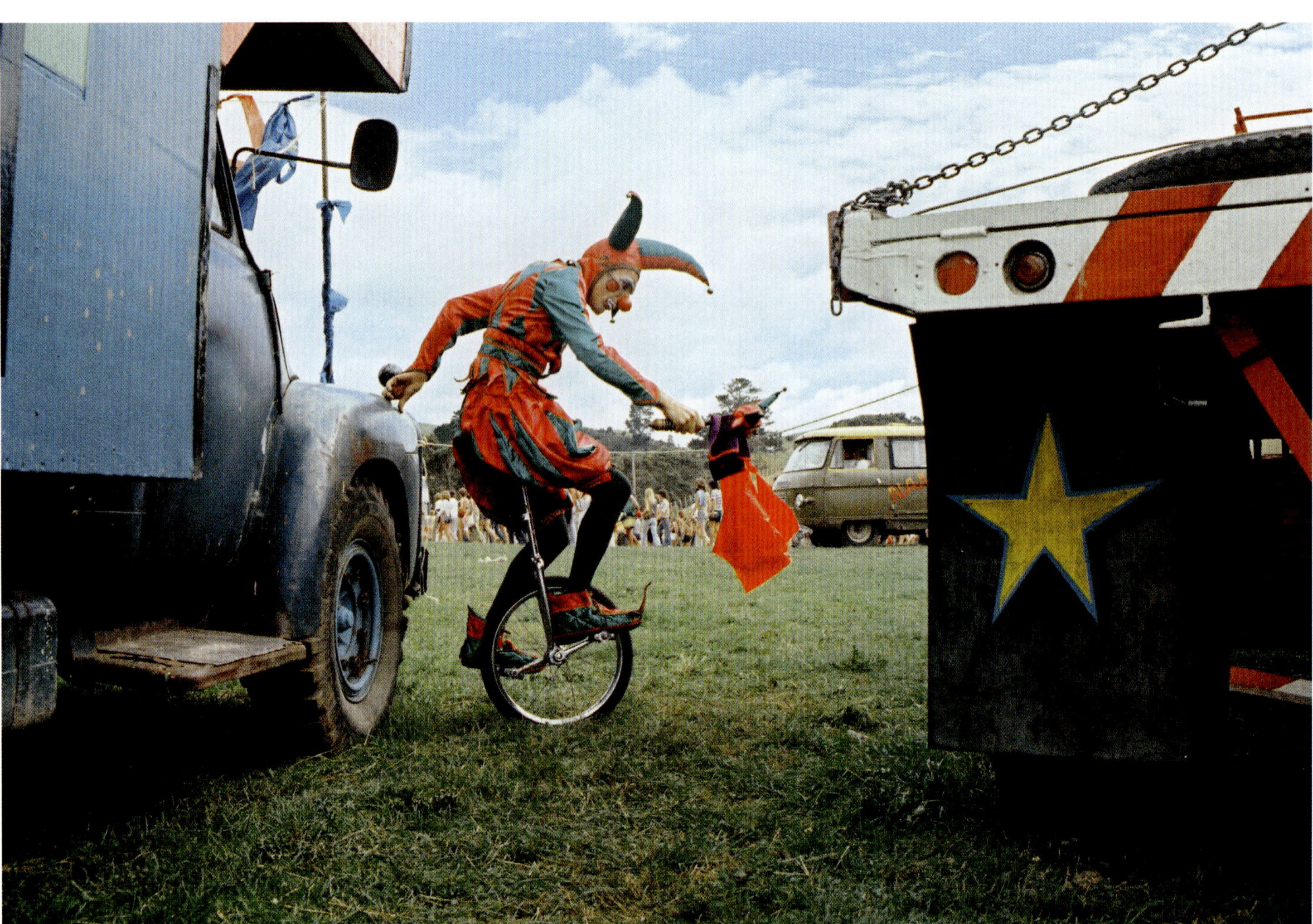

Jonathon Acorn, Nambassa, 1979

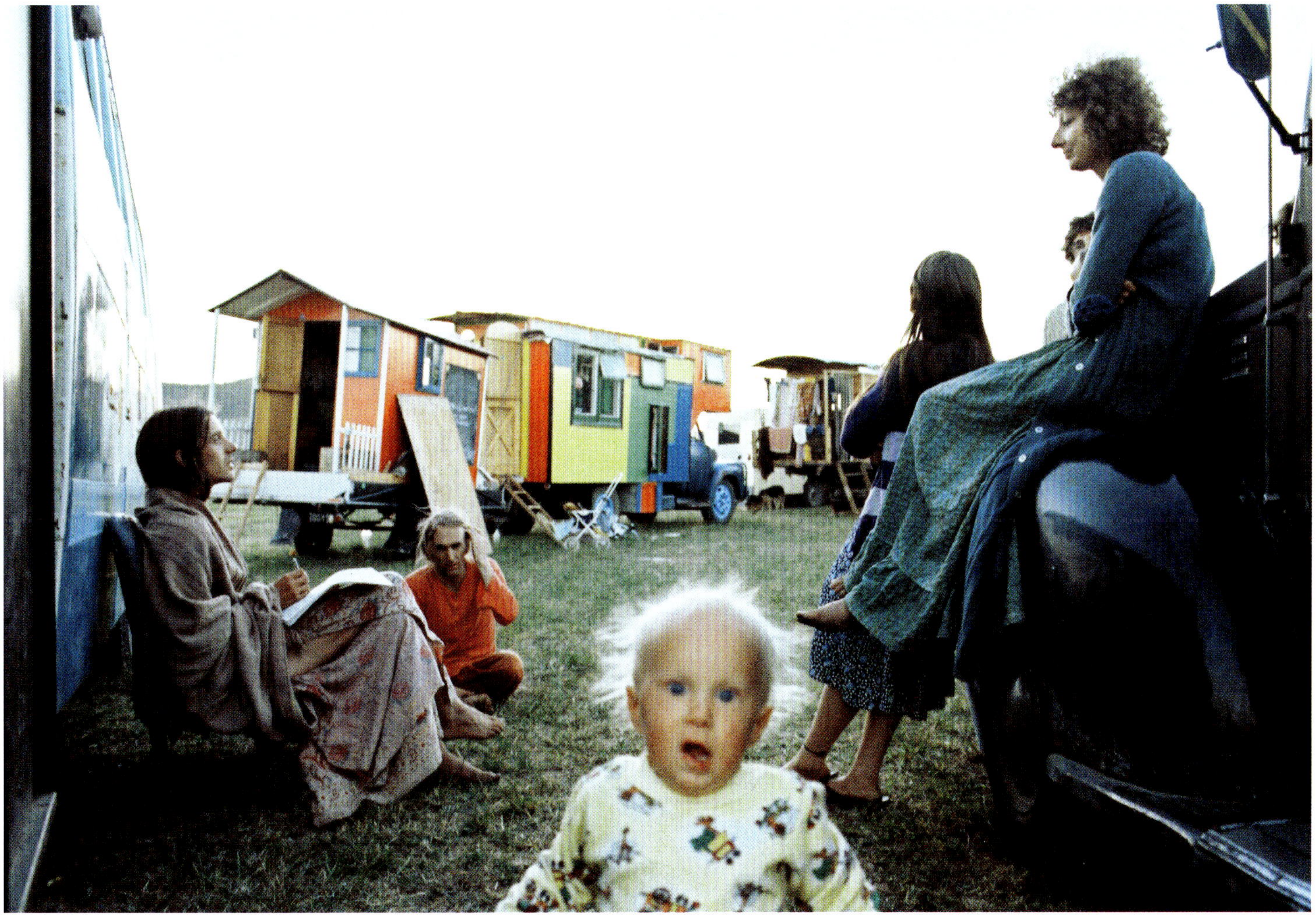

John Tucker, Jonathon Acorn, Tahi and Jade at Nambassa, 1979

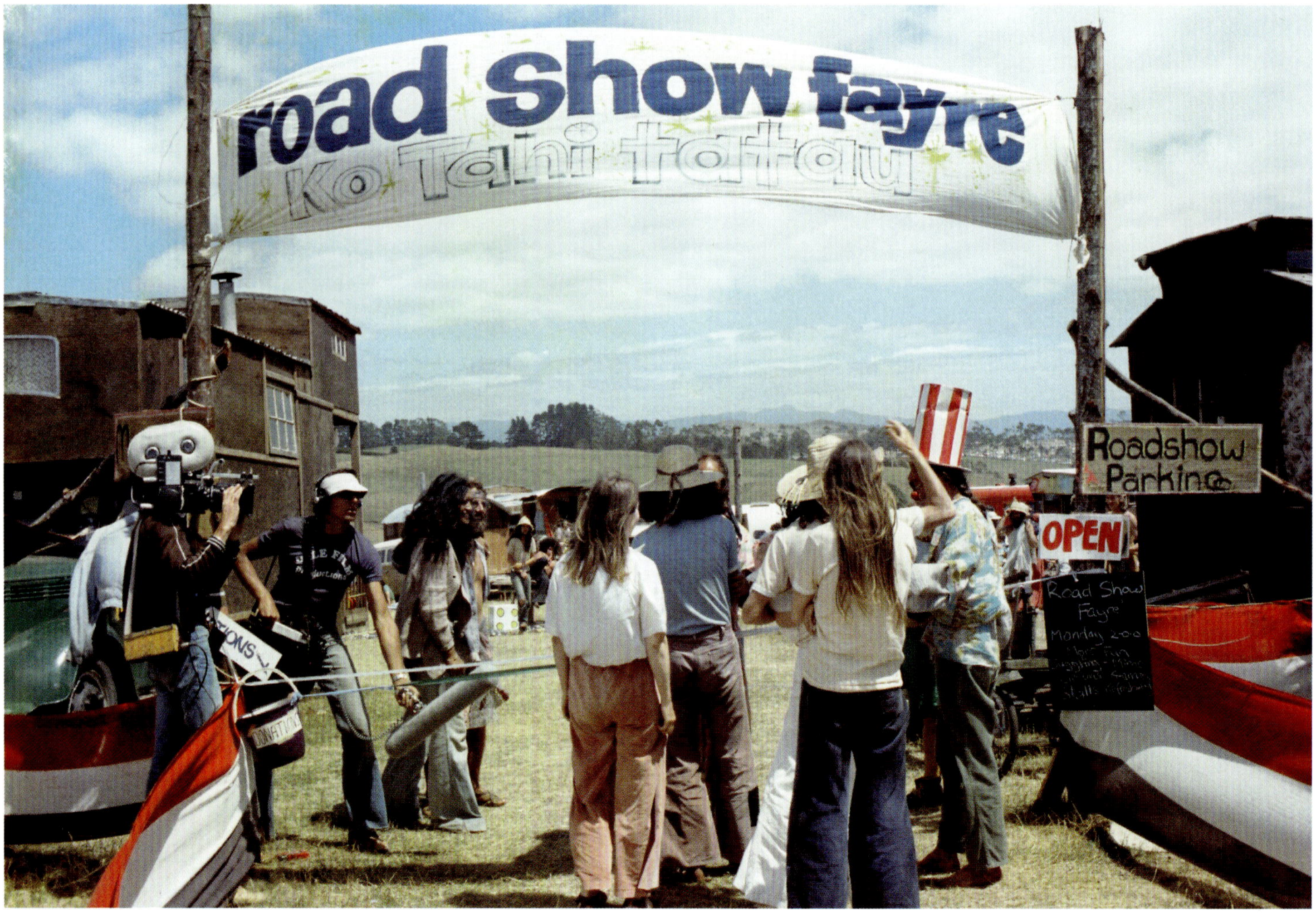

Road Show Fayre compound, with media attention, Nambassa, 1979

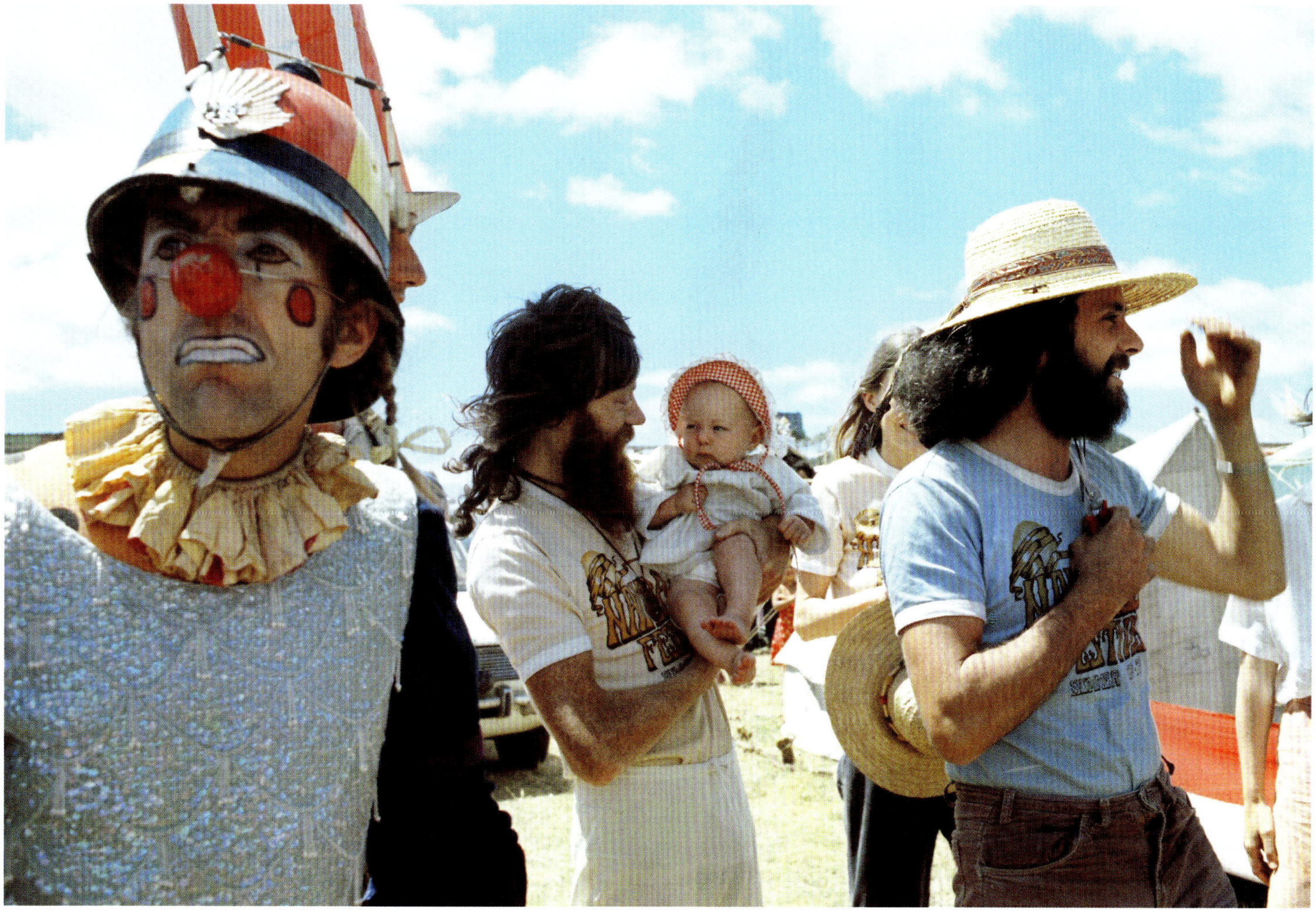

Jonathon Acorn, Nambassa directors Fred Alder, Cynthia Alder (obscured) and Peter Terry at the entrance to the Road Show Fayre compound, Nambassa, 1979

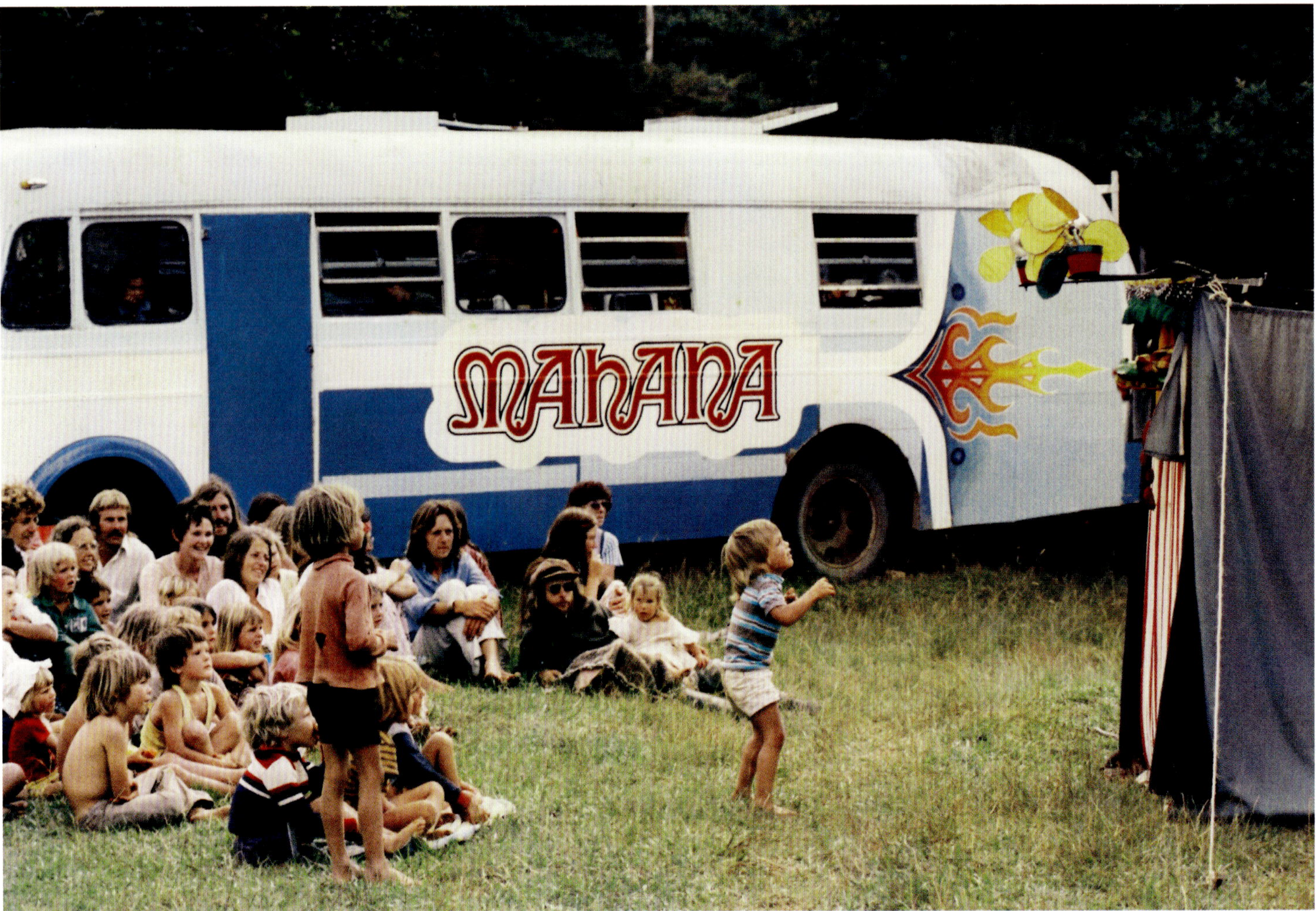

Acorn's Puppet Theatre and Mahana bus, The Original Travelling Road Show, Peria mini festival, Northland, 1978/79

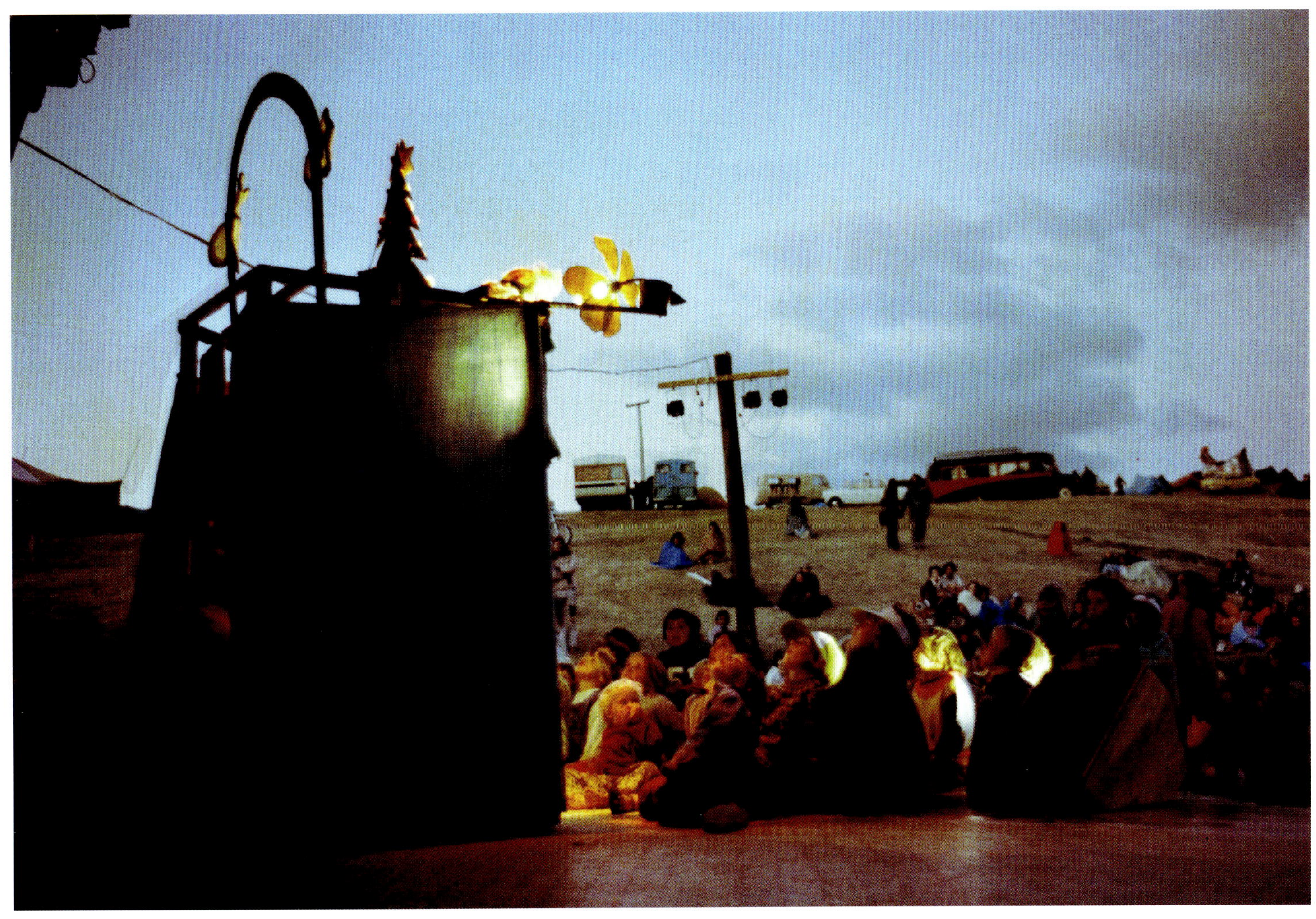

Acorn's Puppet Theatre, Aerial Railway Stage, Nambassa, 1979

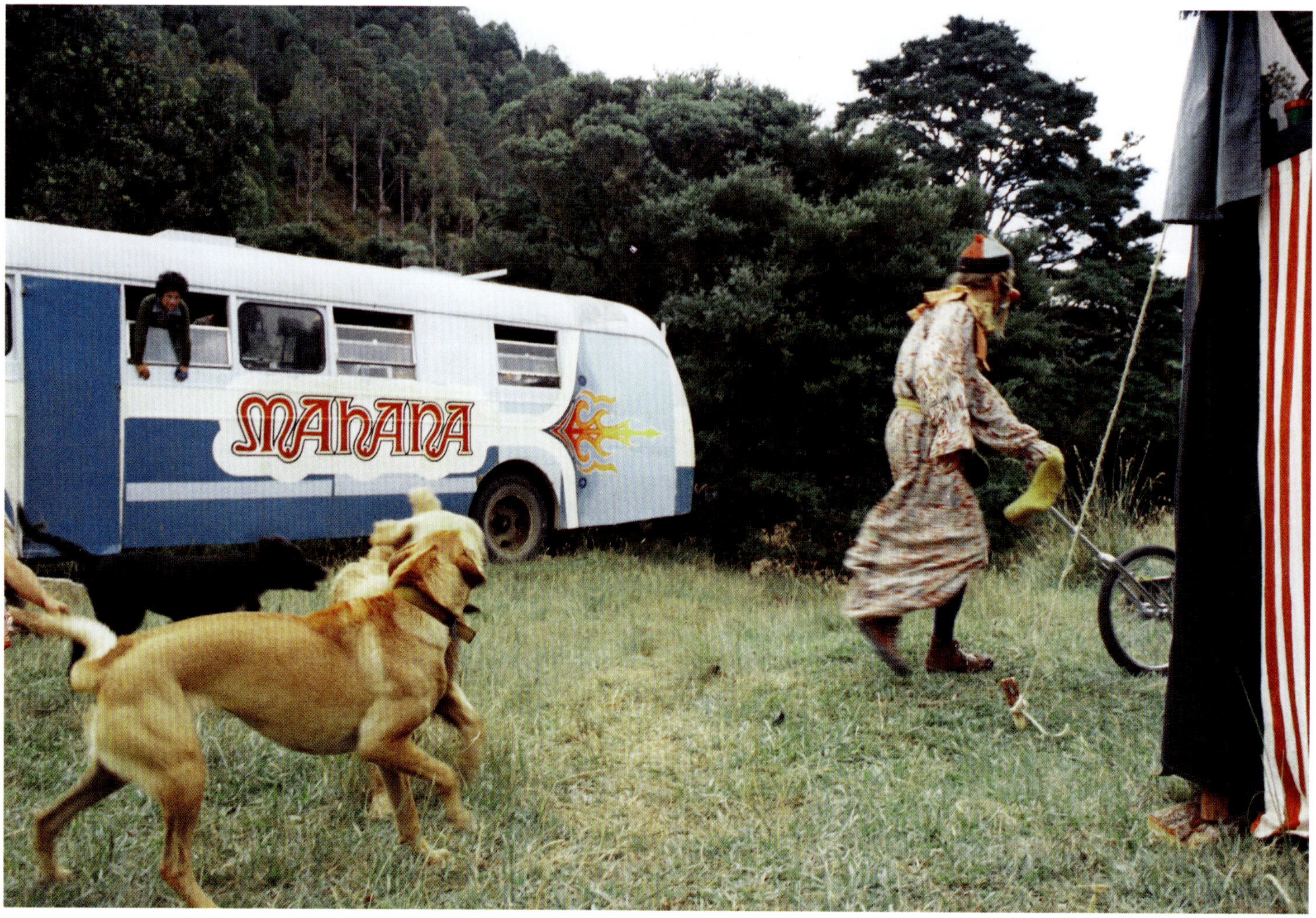

Jonathon Acorn and Mahana bus, The Original Travelling Road Show, Peria mini festival, Northland, 1978/79

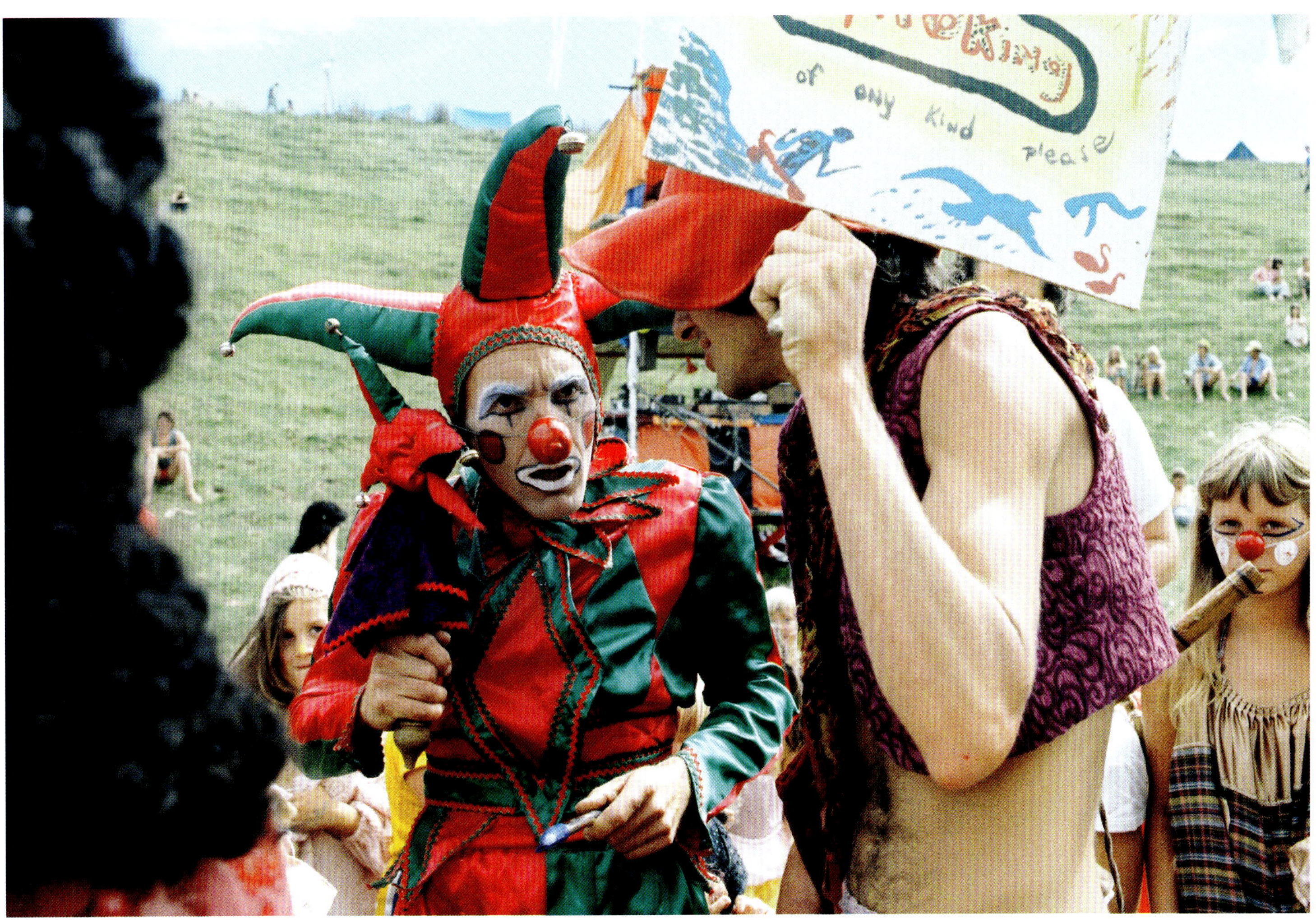

Jonathon Acorn and Brian Tracey (with a sign warning about the dangers of fire), Sweetwaters, 1982

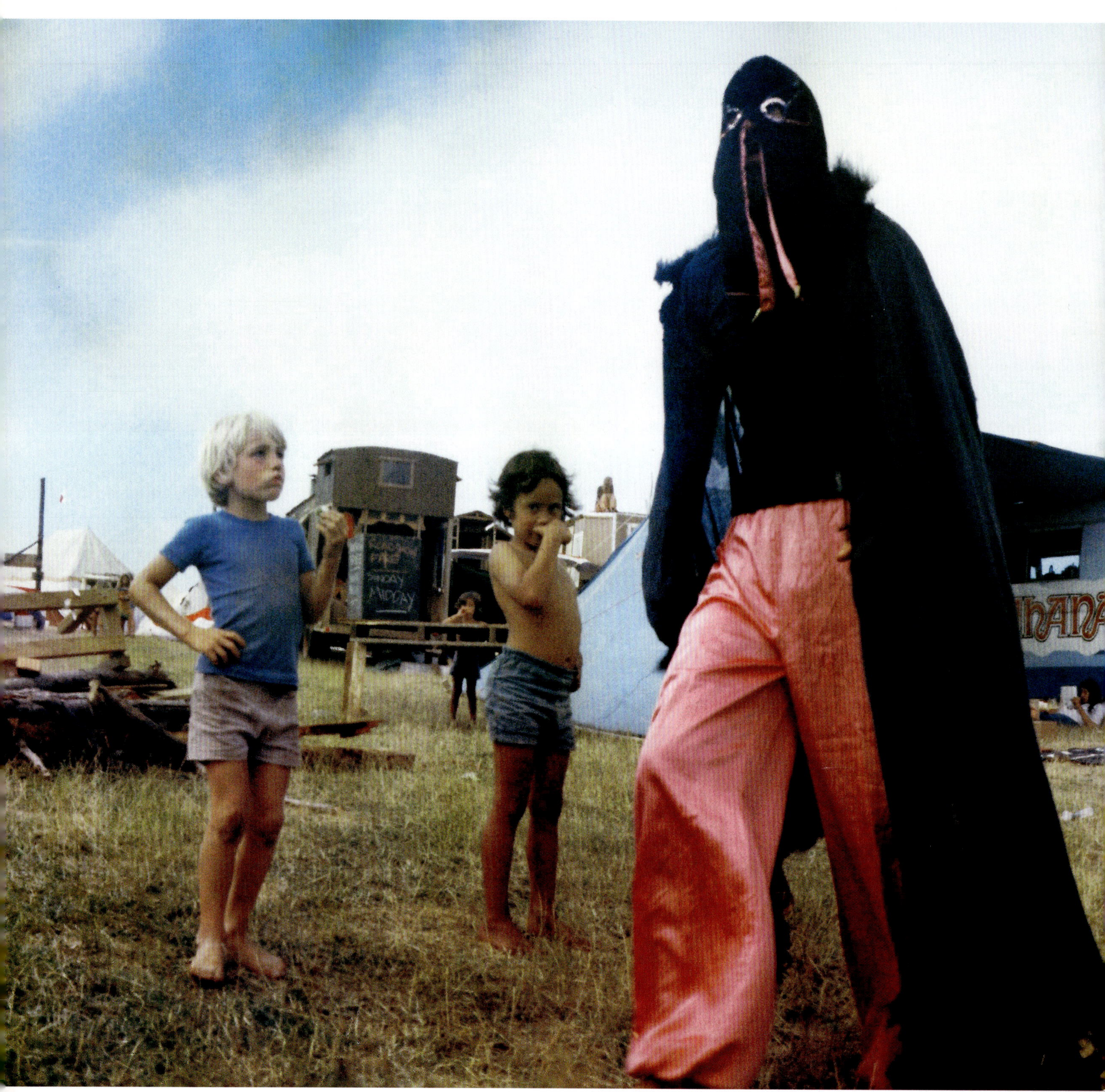

The Black Marauder attempts to steal the fun, Road Show Fayre, Nambassa, 1979

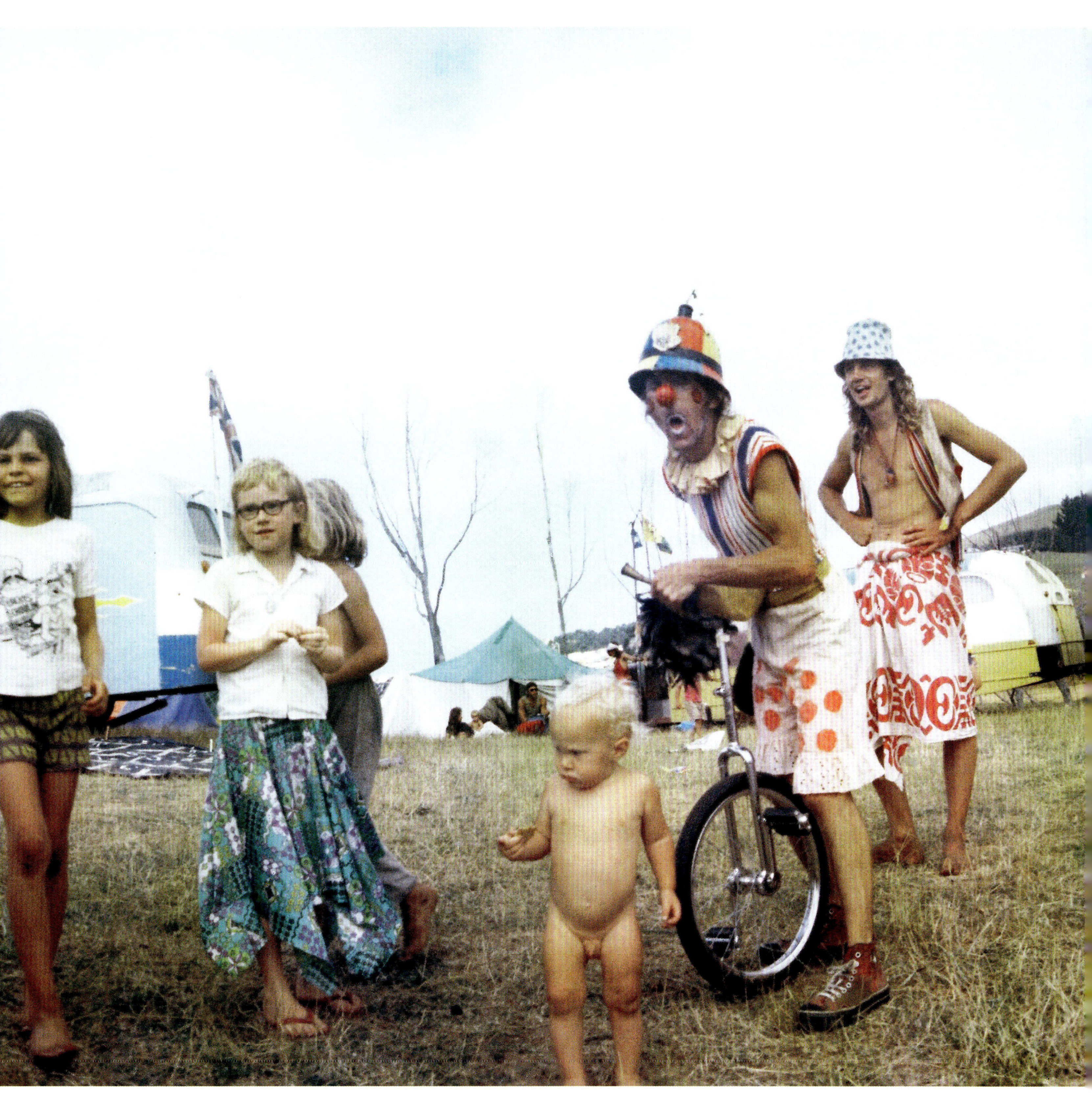

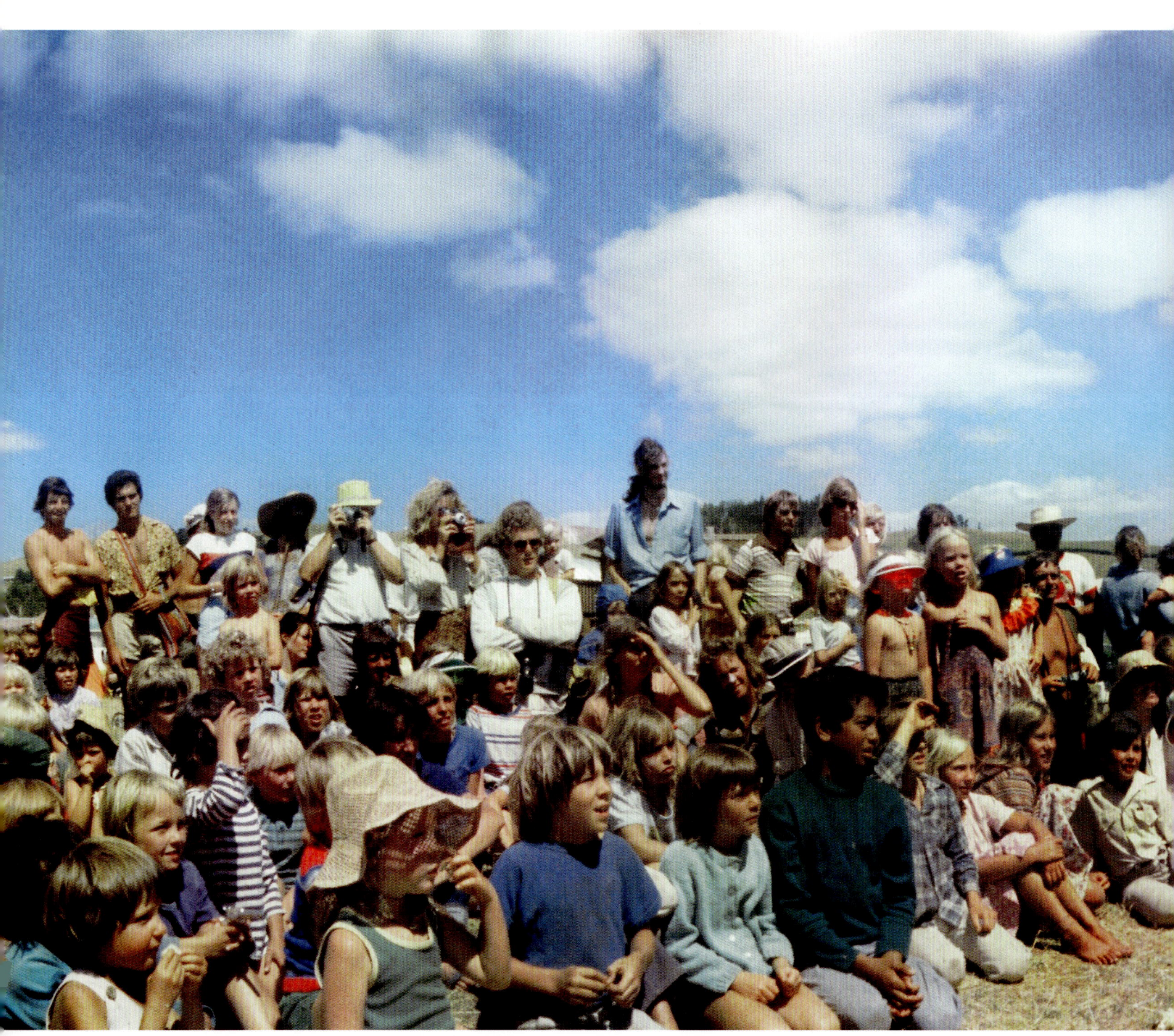

David Sheridan's fire eating act, Road Show Fayre, Nambassa, 1979

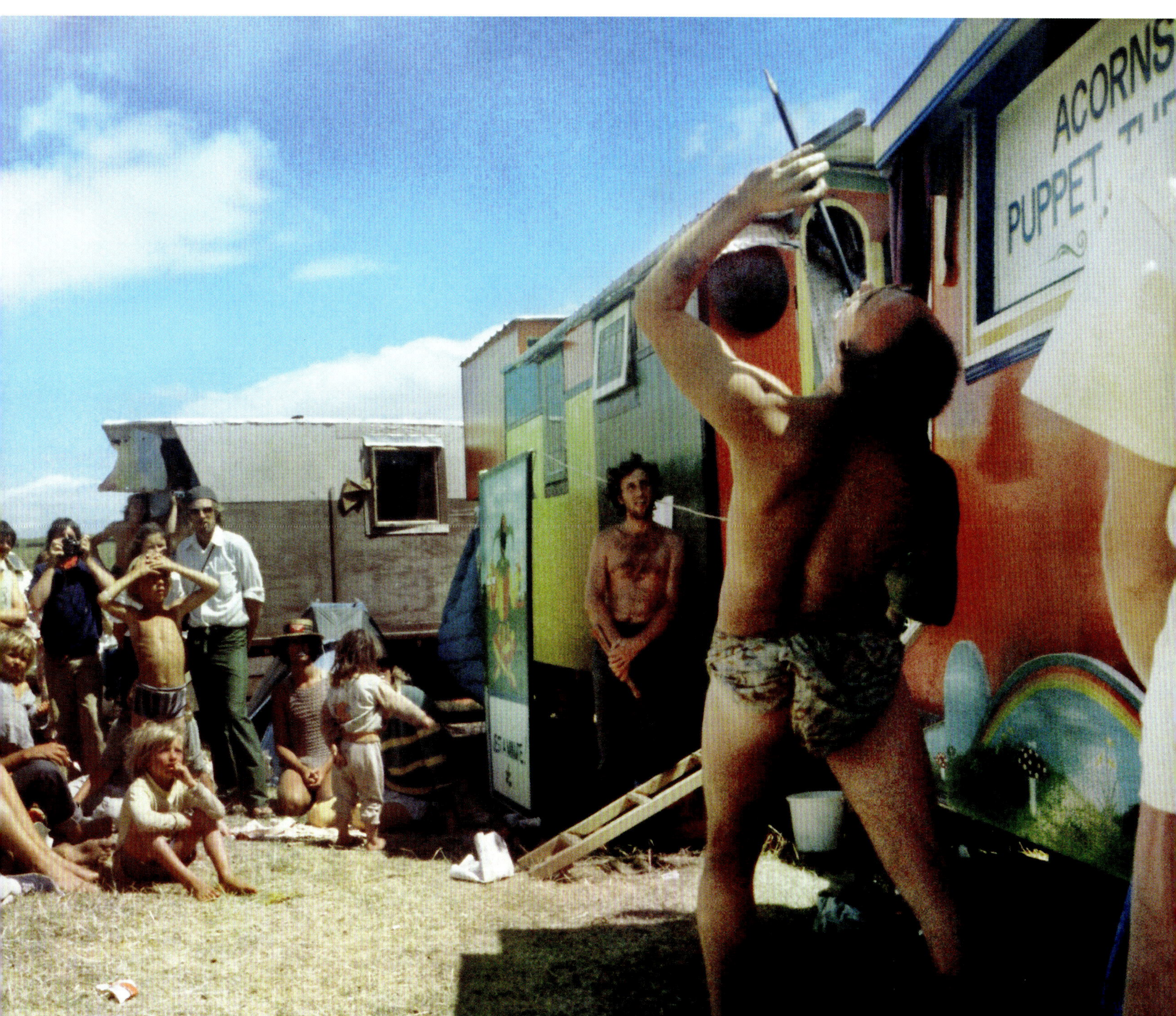
ACORNS
PUPPET

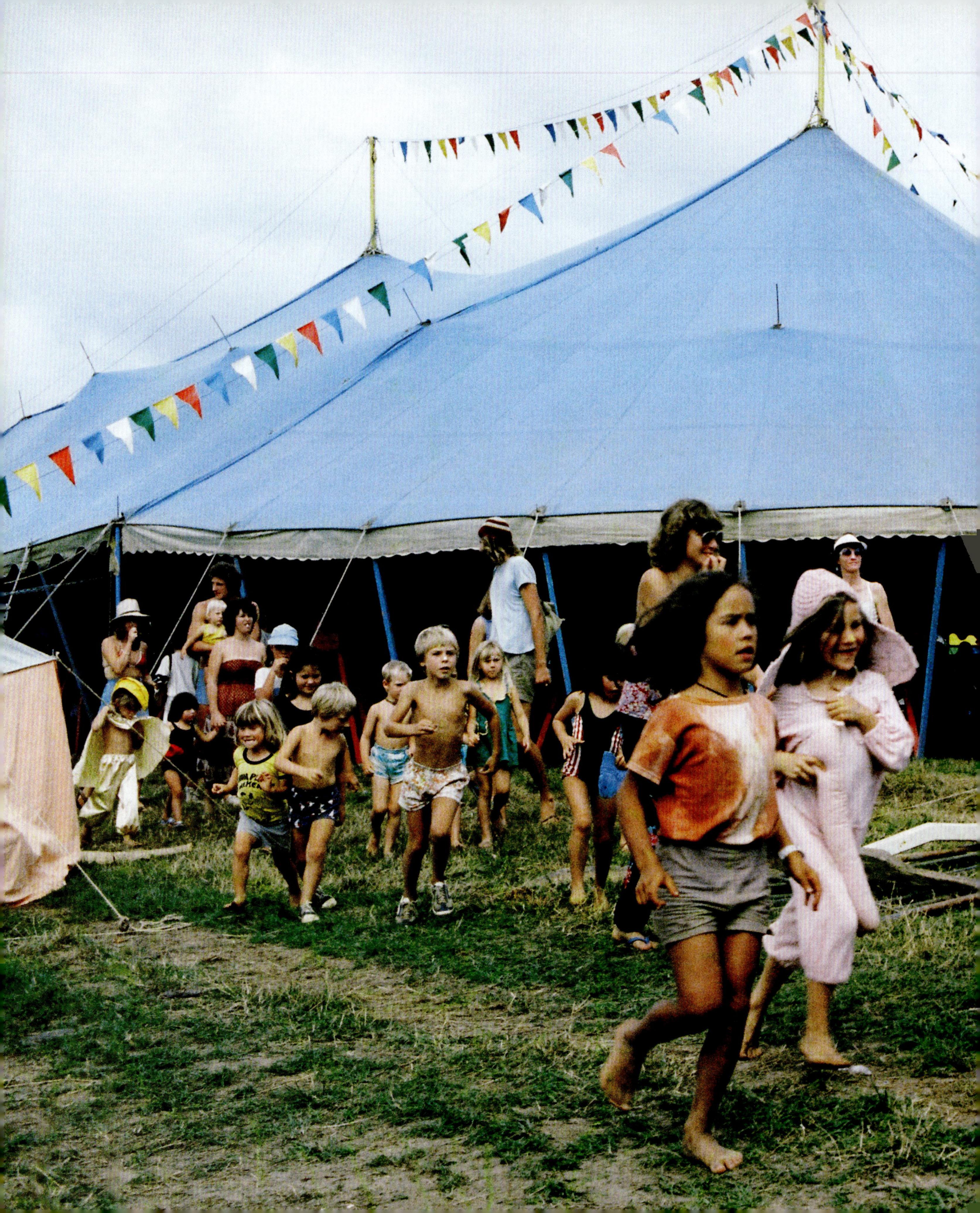

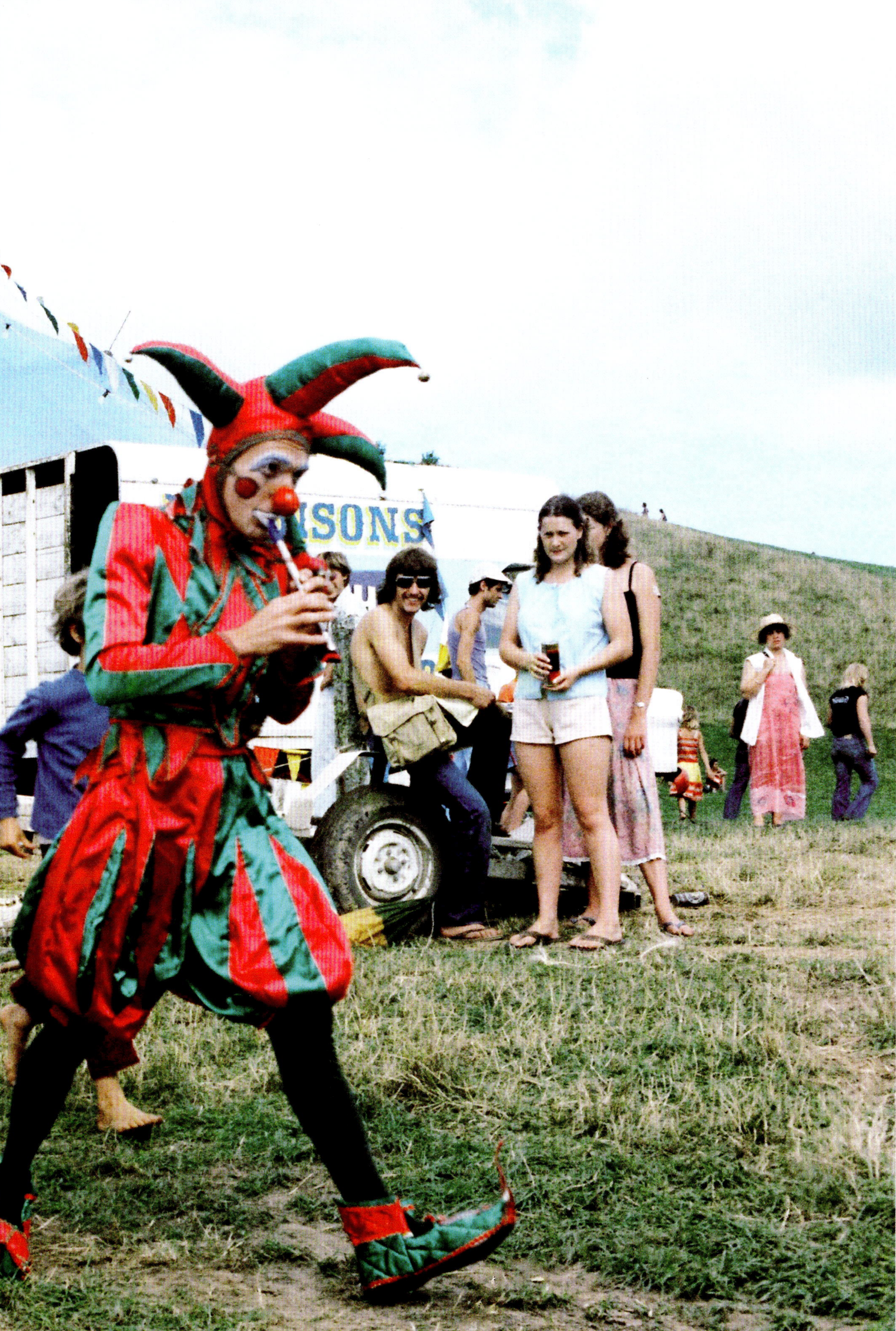

Big Top tent and
Jonathon Acorn
as the Pied Piper,
Sweetwaters, 1982

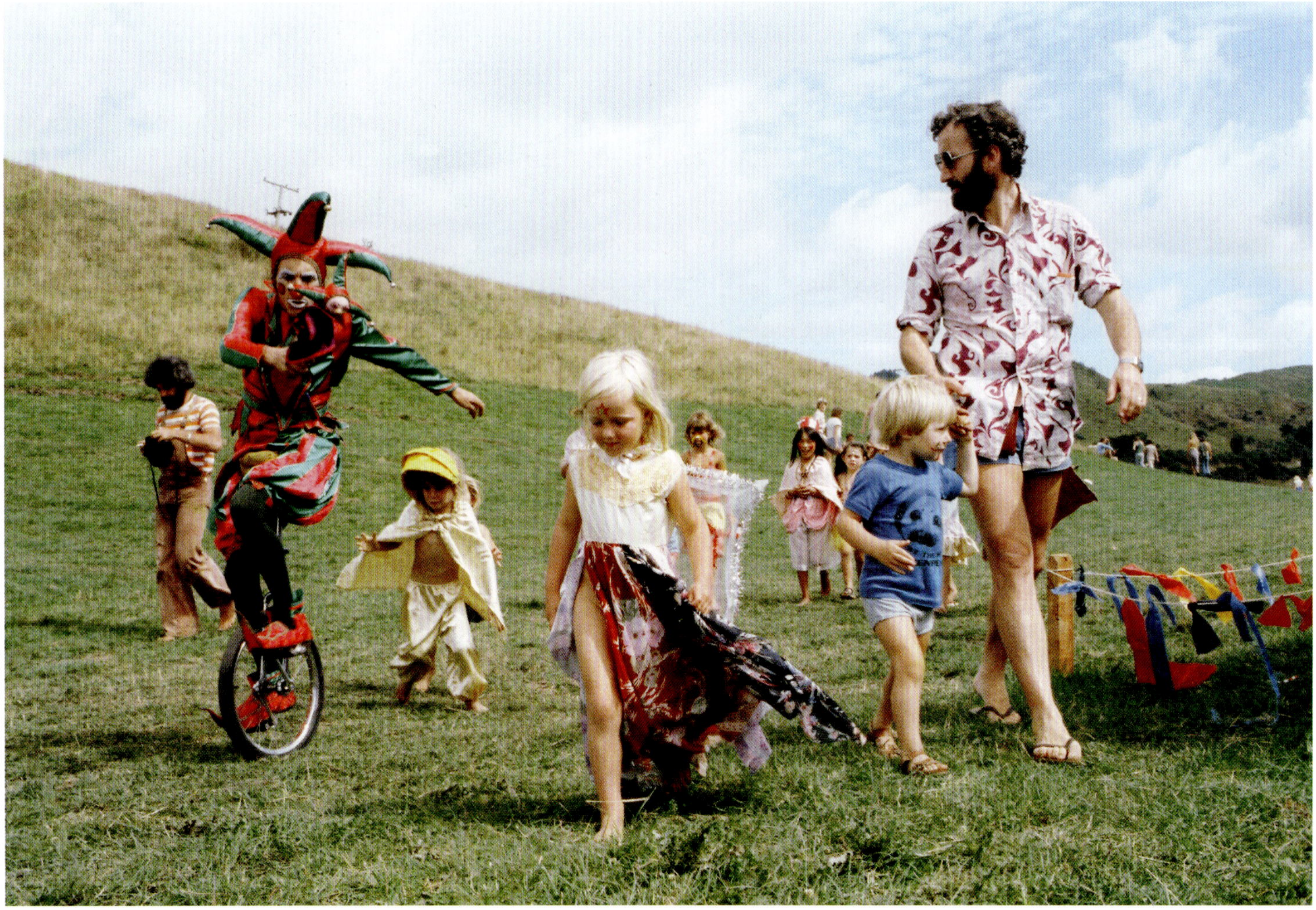

Jonathon Acorn practicing his unicycle skills at Sweetwaters, 1982

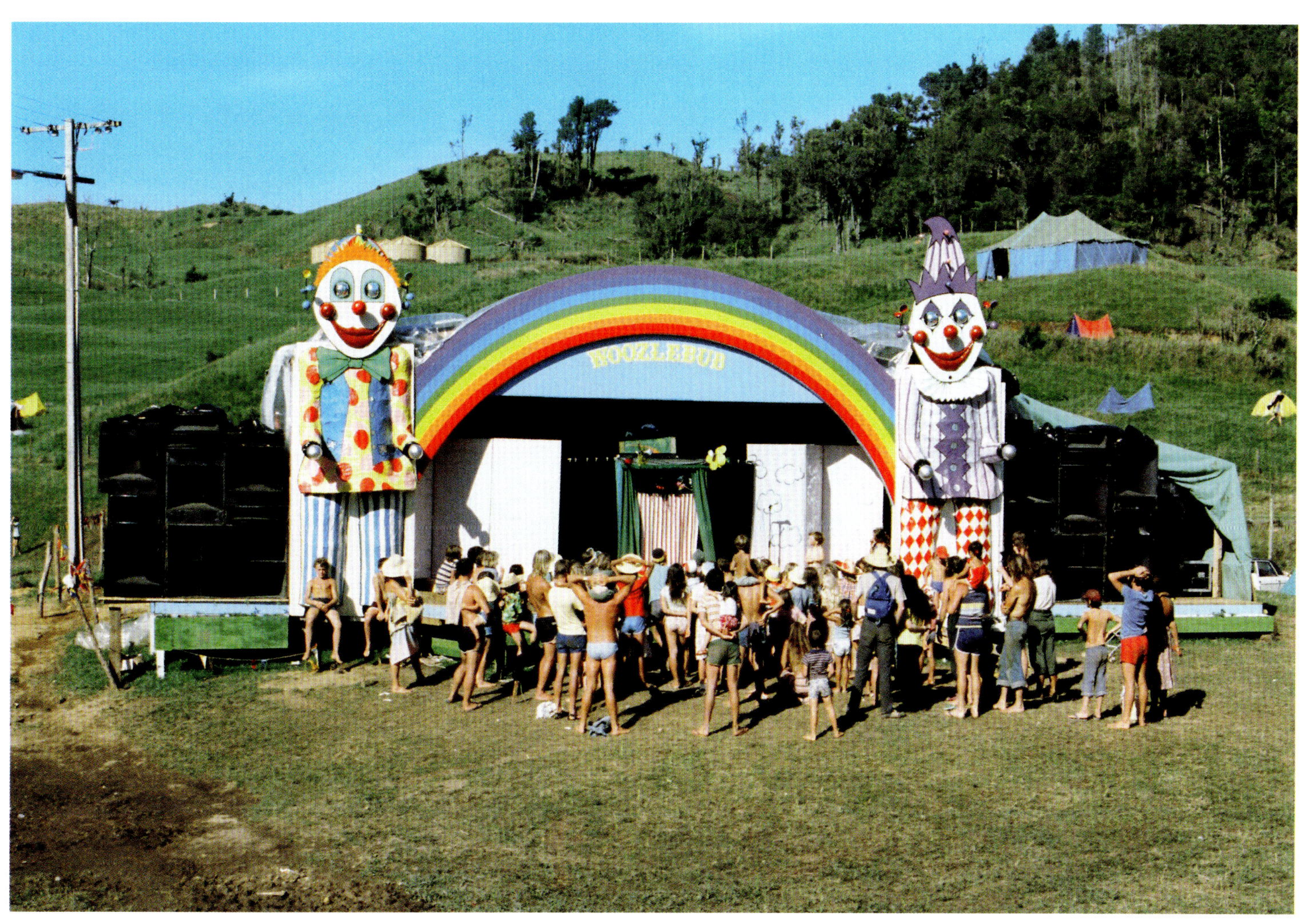

The Woozlebub Stage, Nambassa, 1981

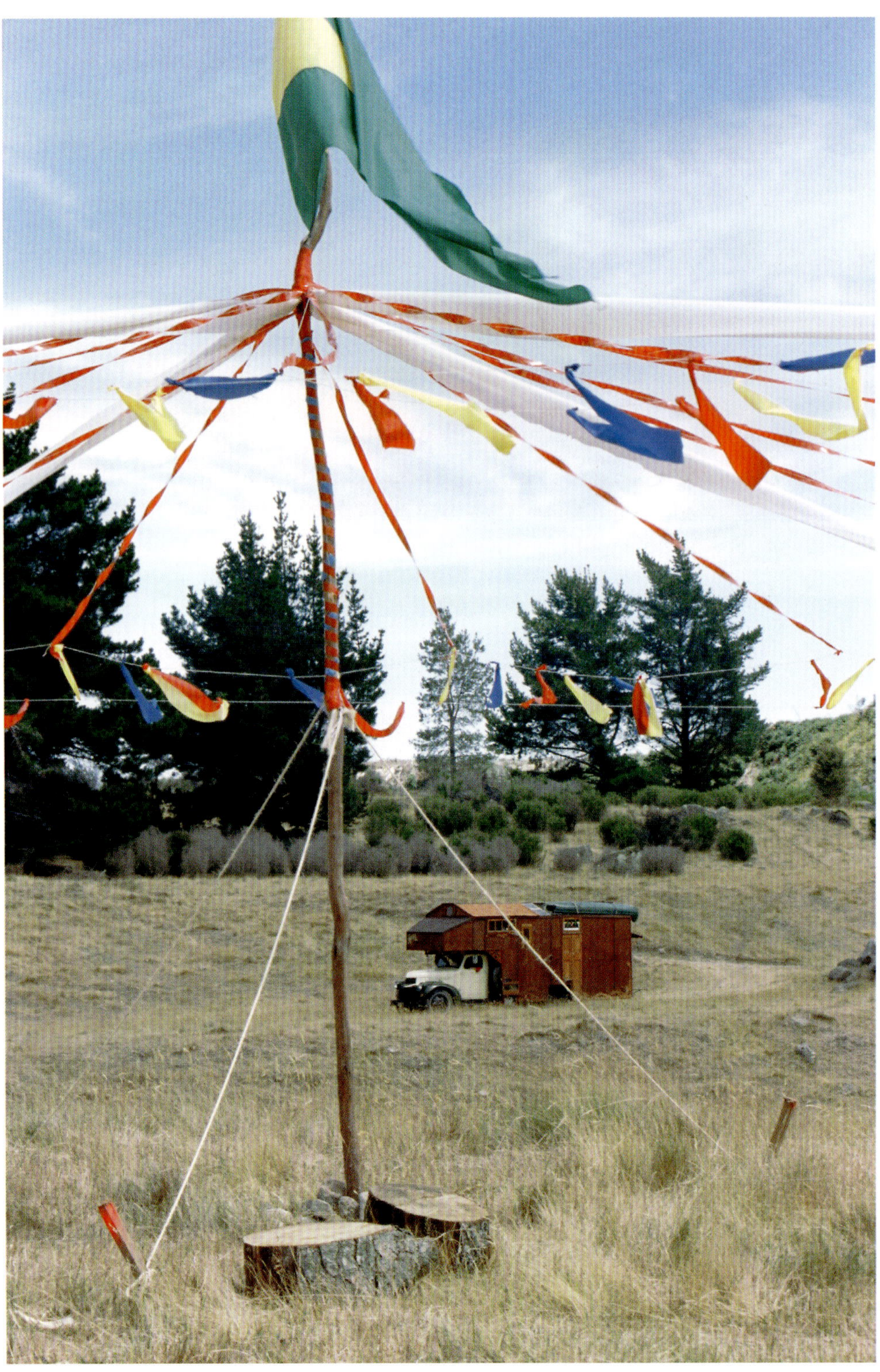

The ‘imaginary big-top’ tent, People’s Circus, Aratiatia, 1984

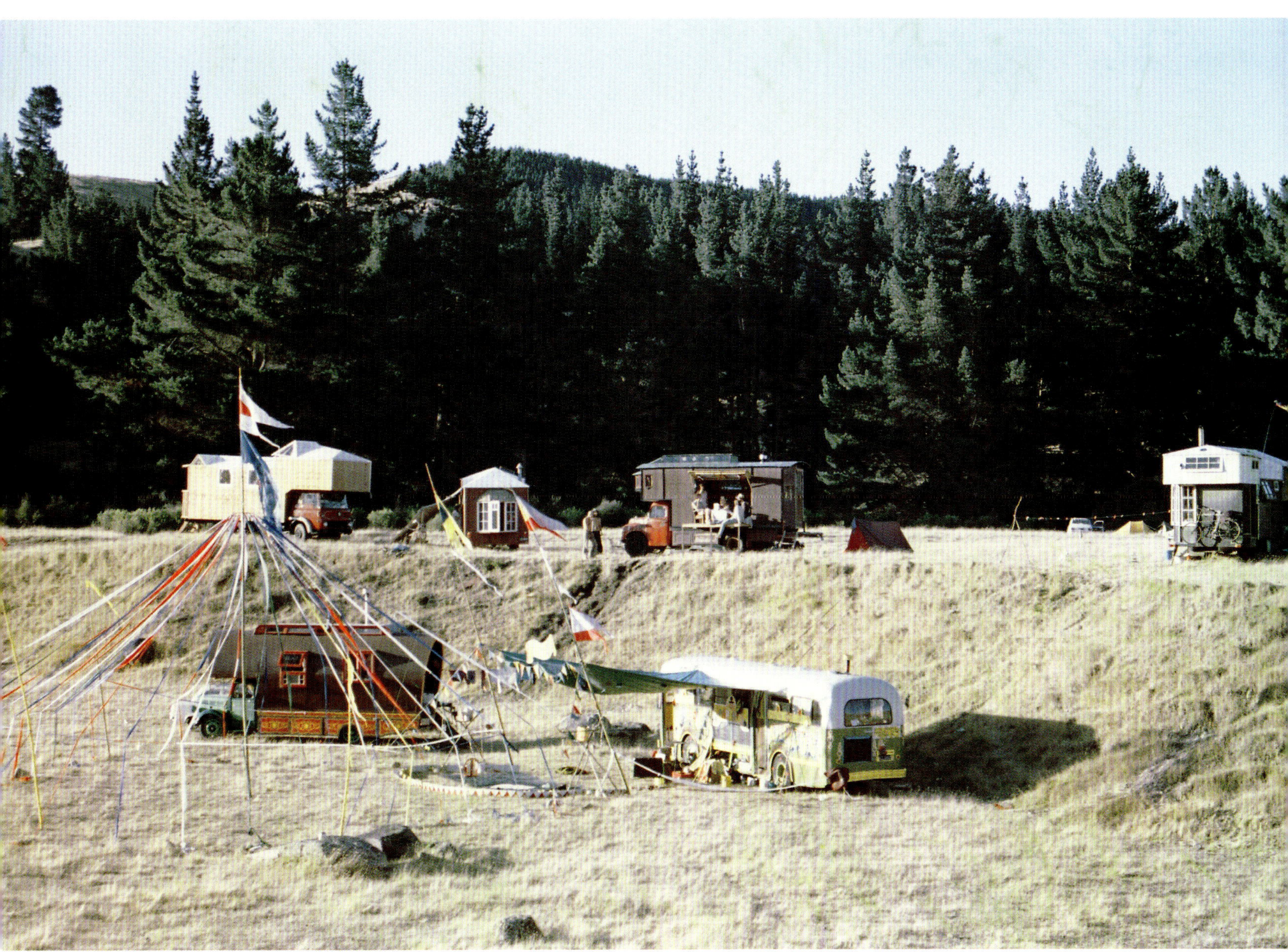

The 'imaginary big-top' tent, People's Circus, Aratiatia, 1984

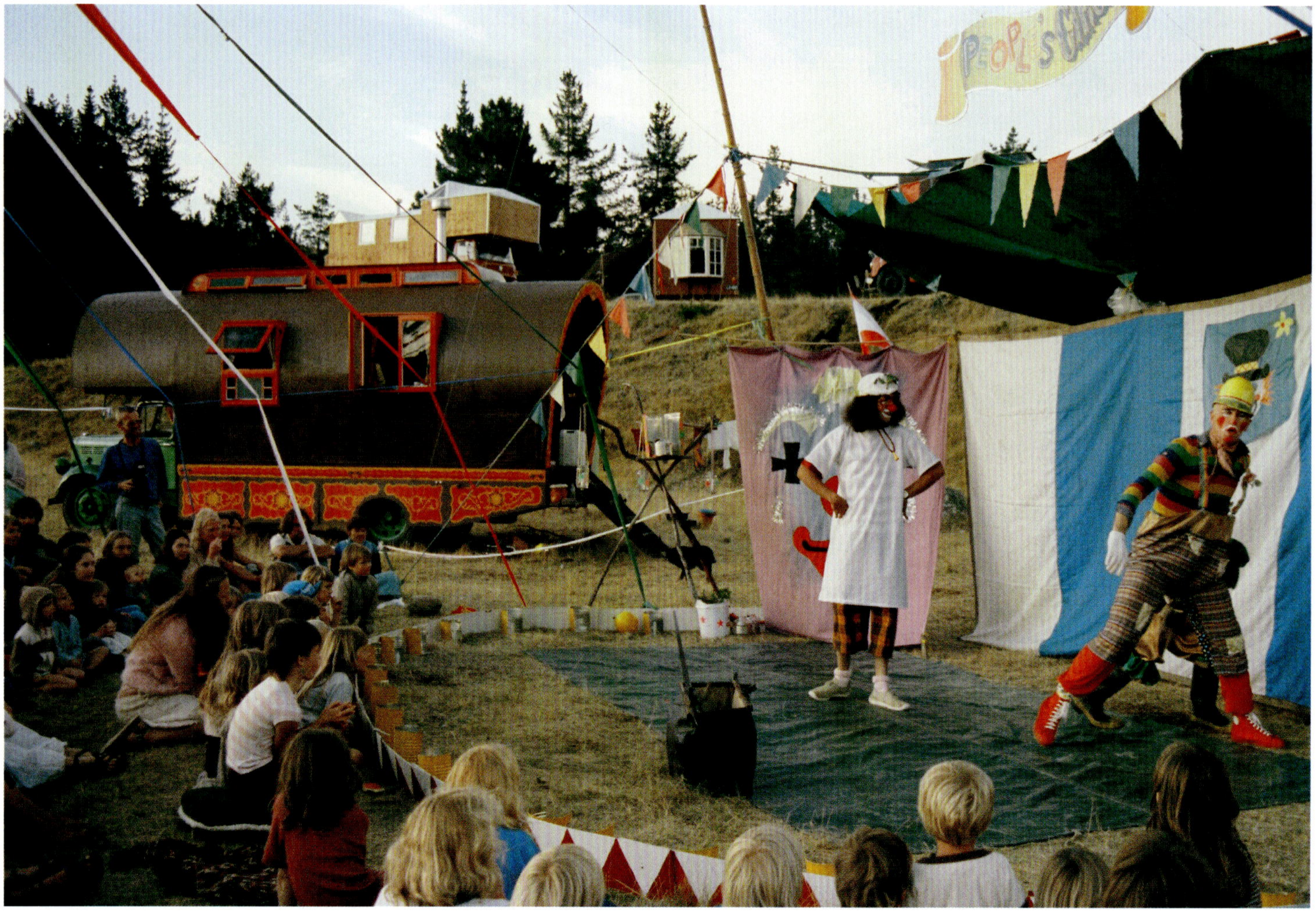

Noodle and PC (David Sheridan with his assistant Tui Sheridan) discuss the finer points of surgery at the People's Circus show, Aratiatia, 1984

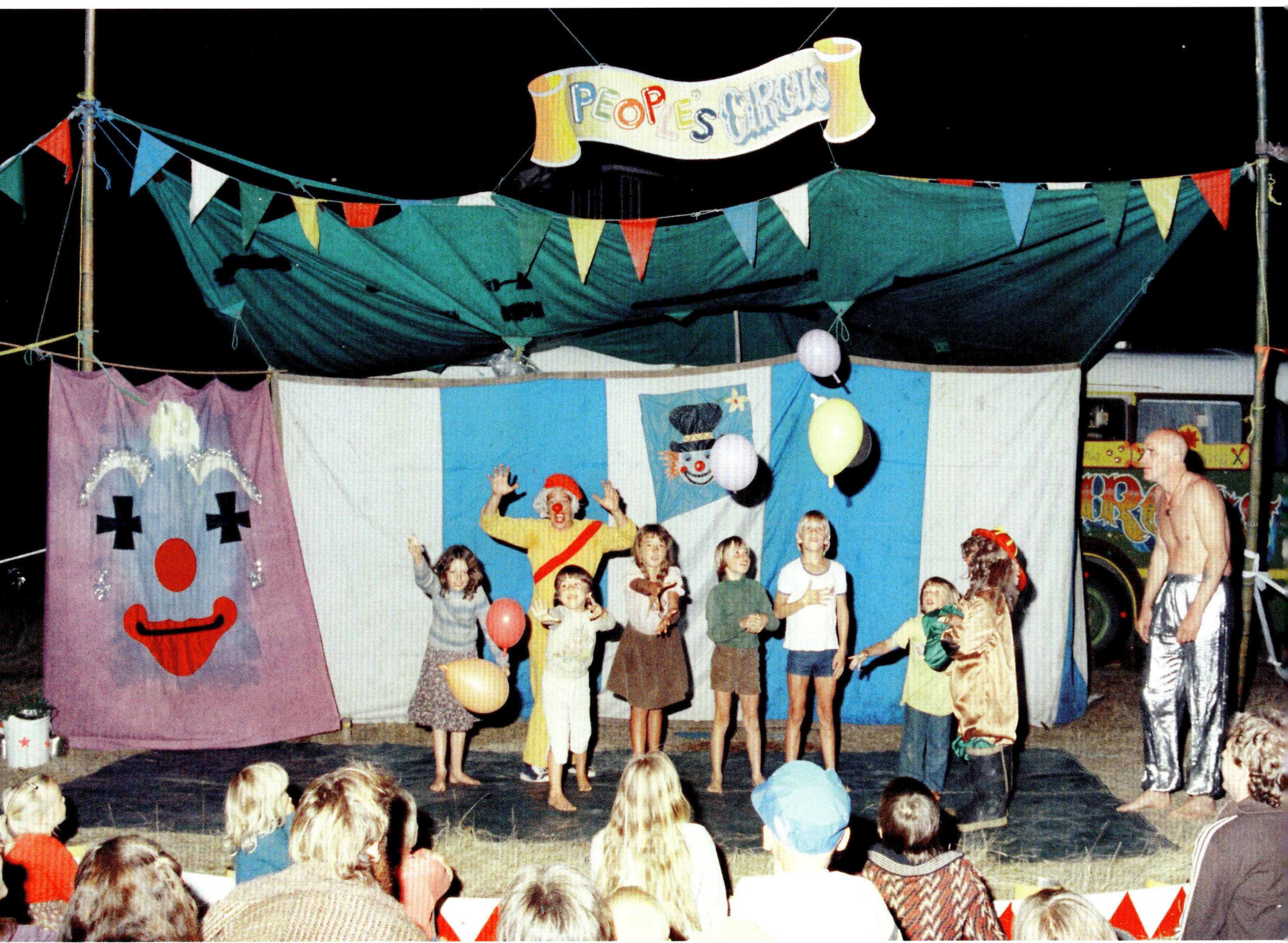

Stroodle (Jane Mitchell), Zoe Sheridan and PC (David Sheridan) with house truck kids as volunteers, People's Circus, Aratiatia, 1984

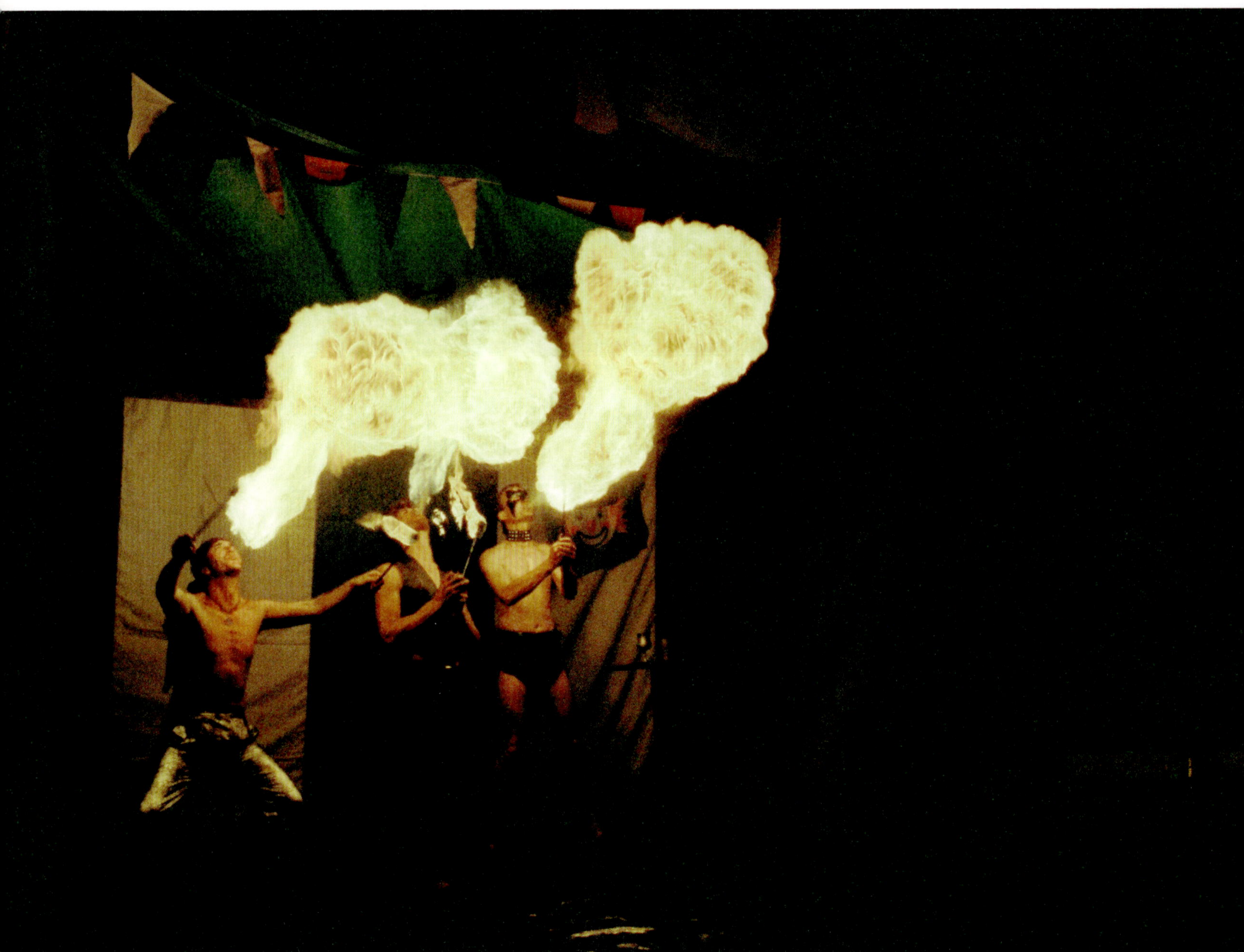

People's Circus show, Aratiatia, 1984

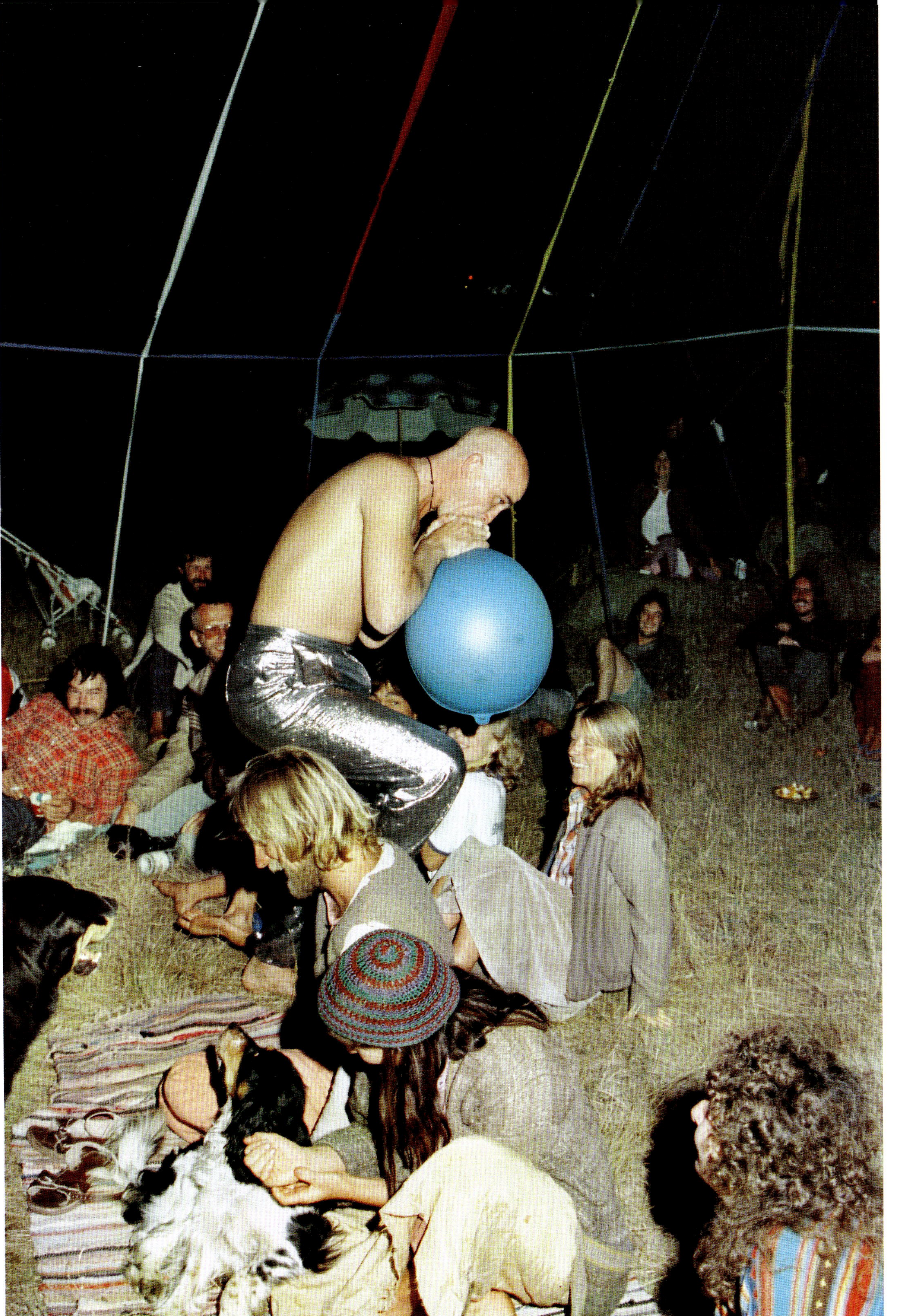

PC (David Sheridan), People's Circus show, Aratiatia, 1984

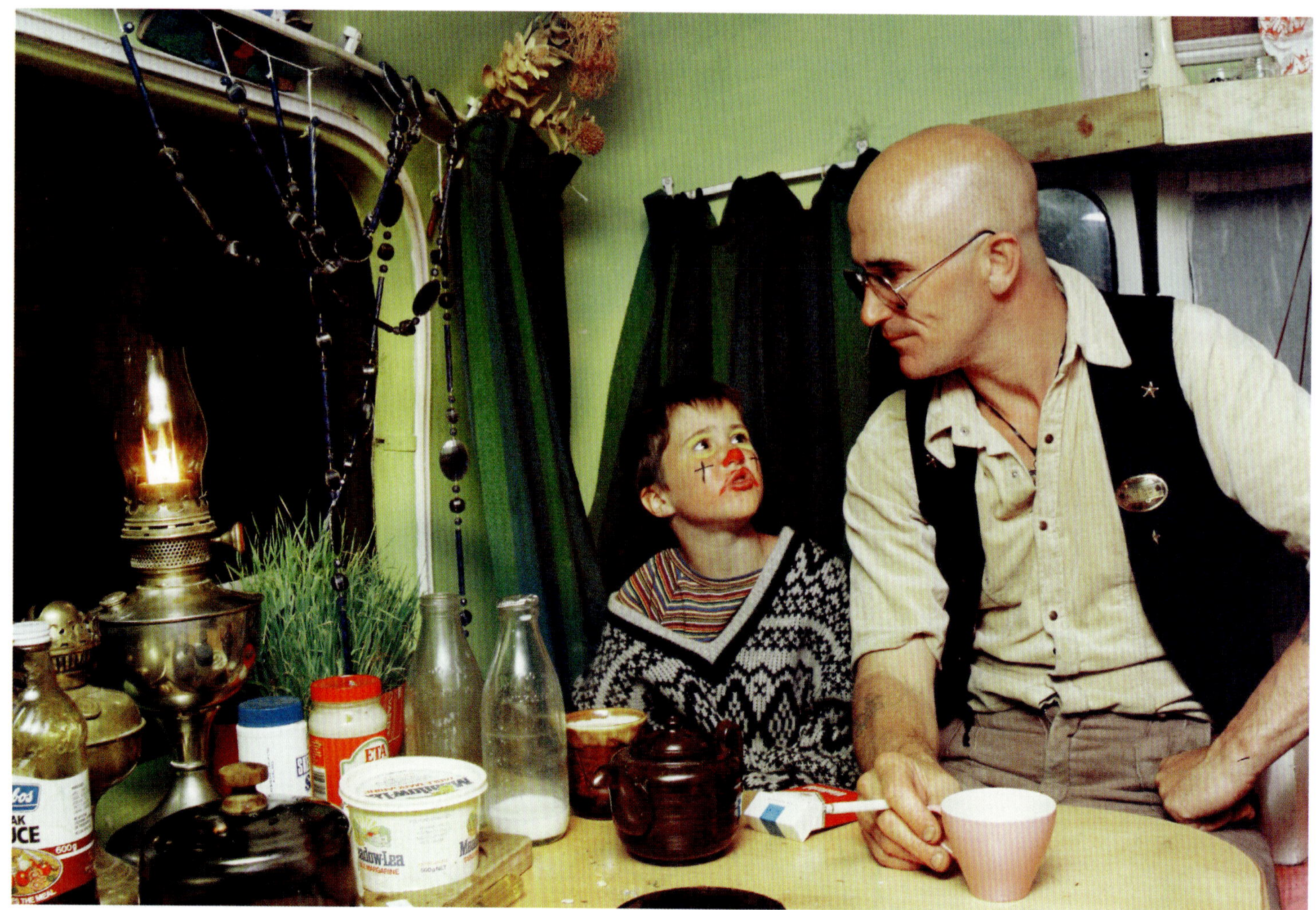

David Sheridan with daughter Zoe inside his bus, Aratiatia, 1984

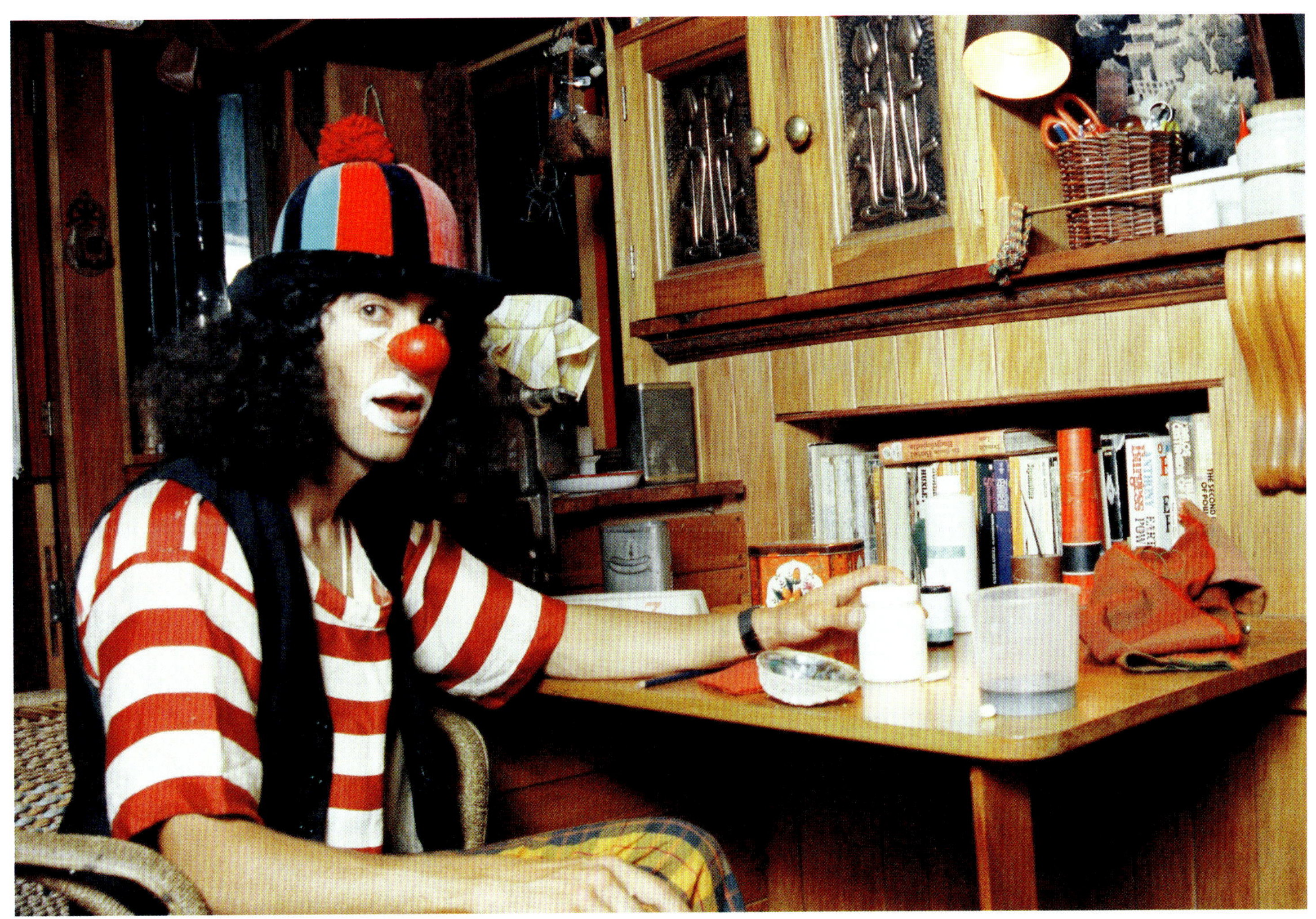

Dom Ferry as 'Noodle' in his truck, Aratiatia, 1984

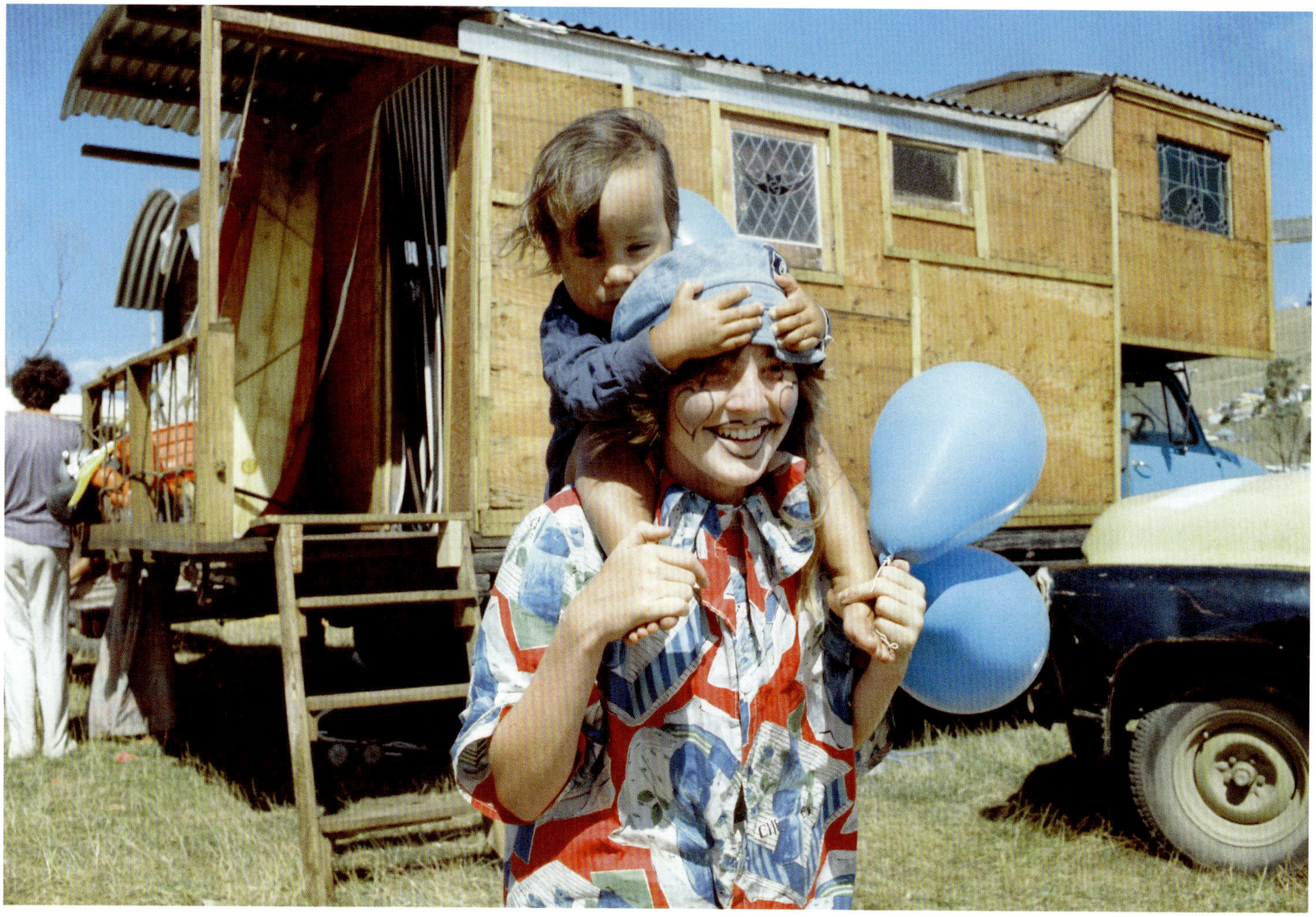

Wiremu Dixon riding unidentified clown, Nambassa, 1979

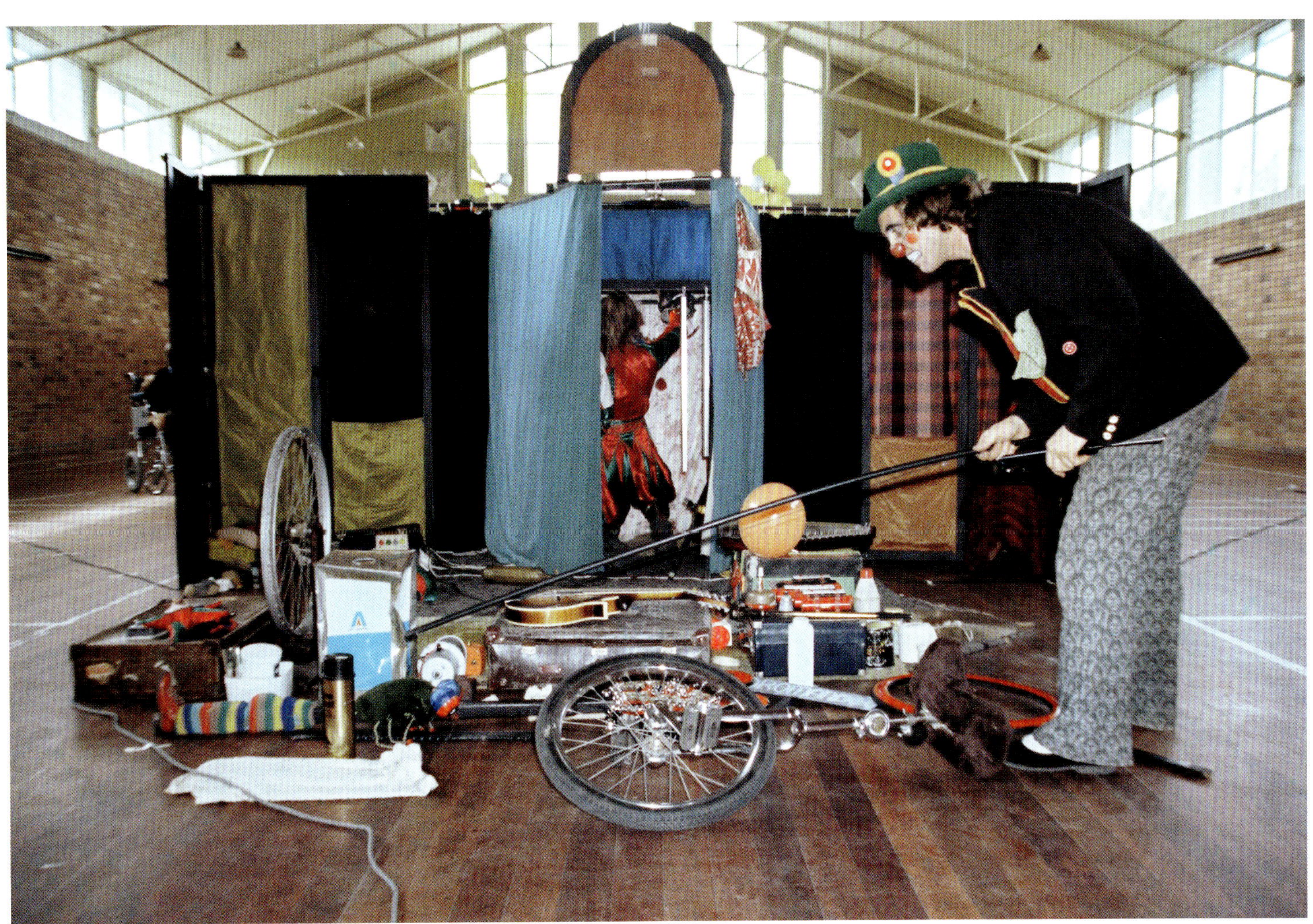

Walnut and Acorn set up the show, Waikato tour, 1981

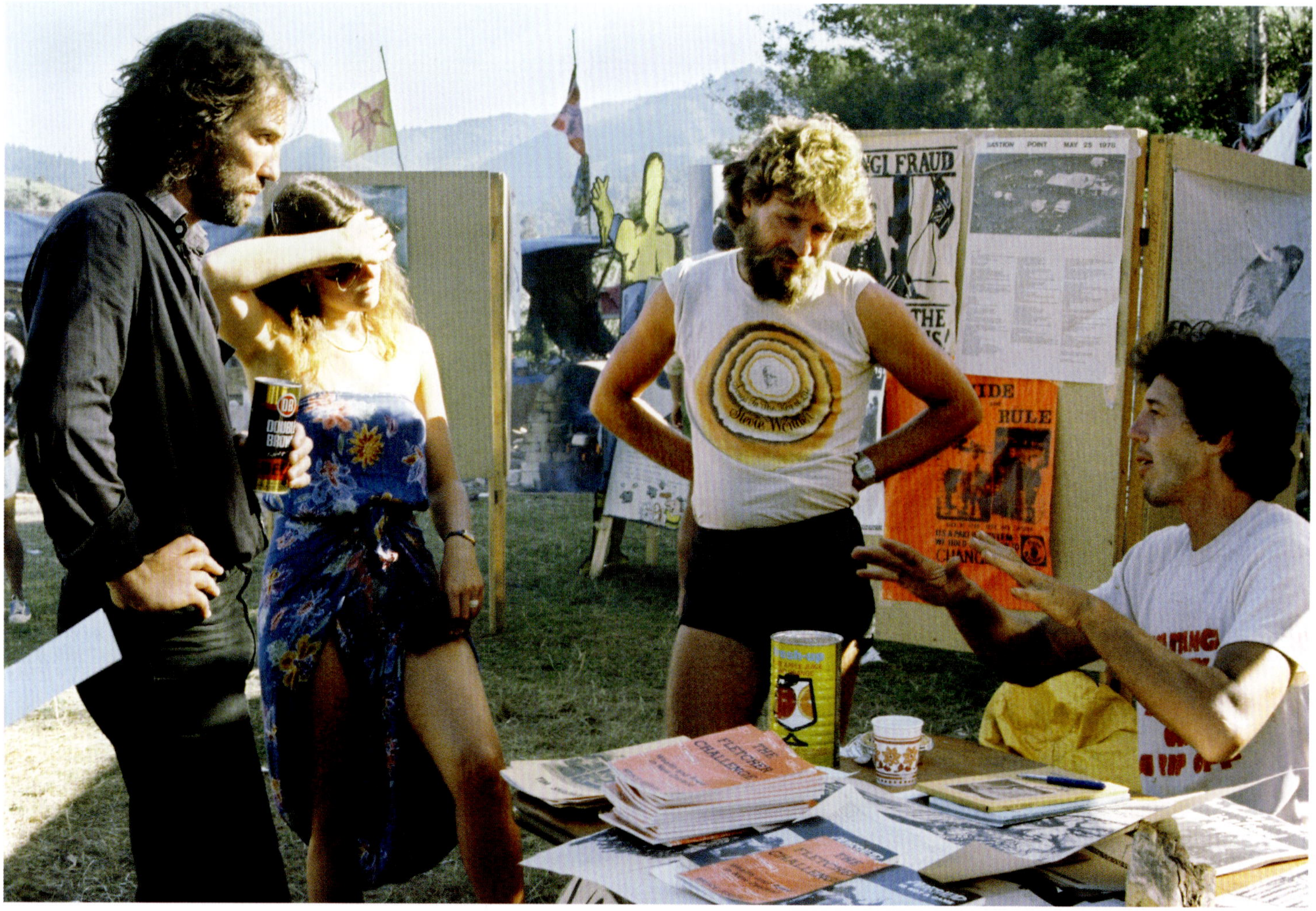

Tim Shadbolt's political stall, Nambassa, 1981

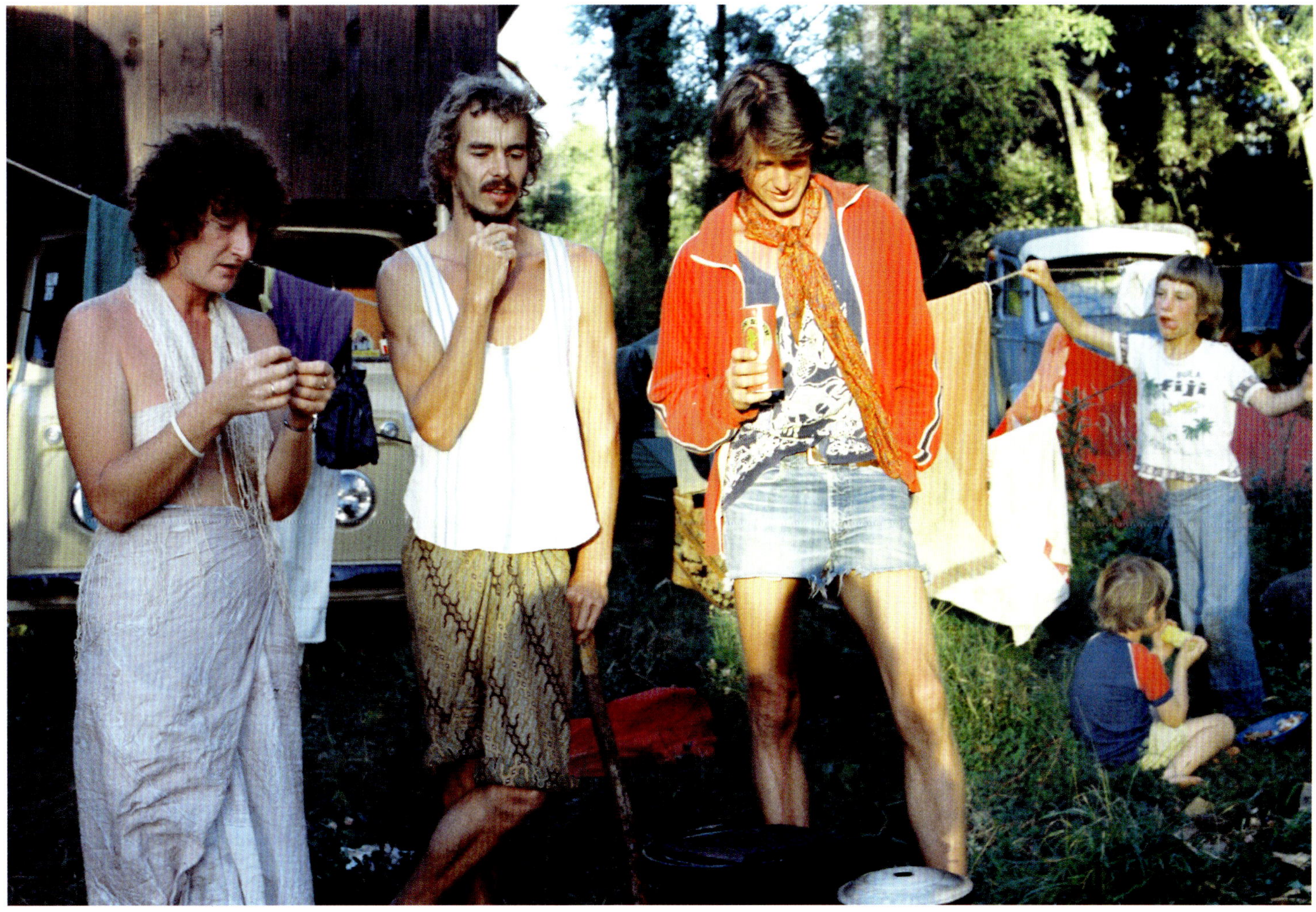

John Tucker nursing Lion Red with friends, Nambassa, 1981

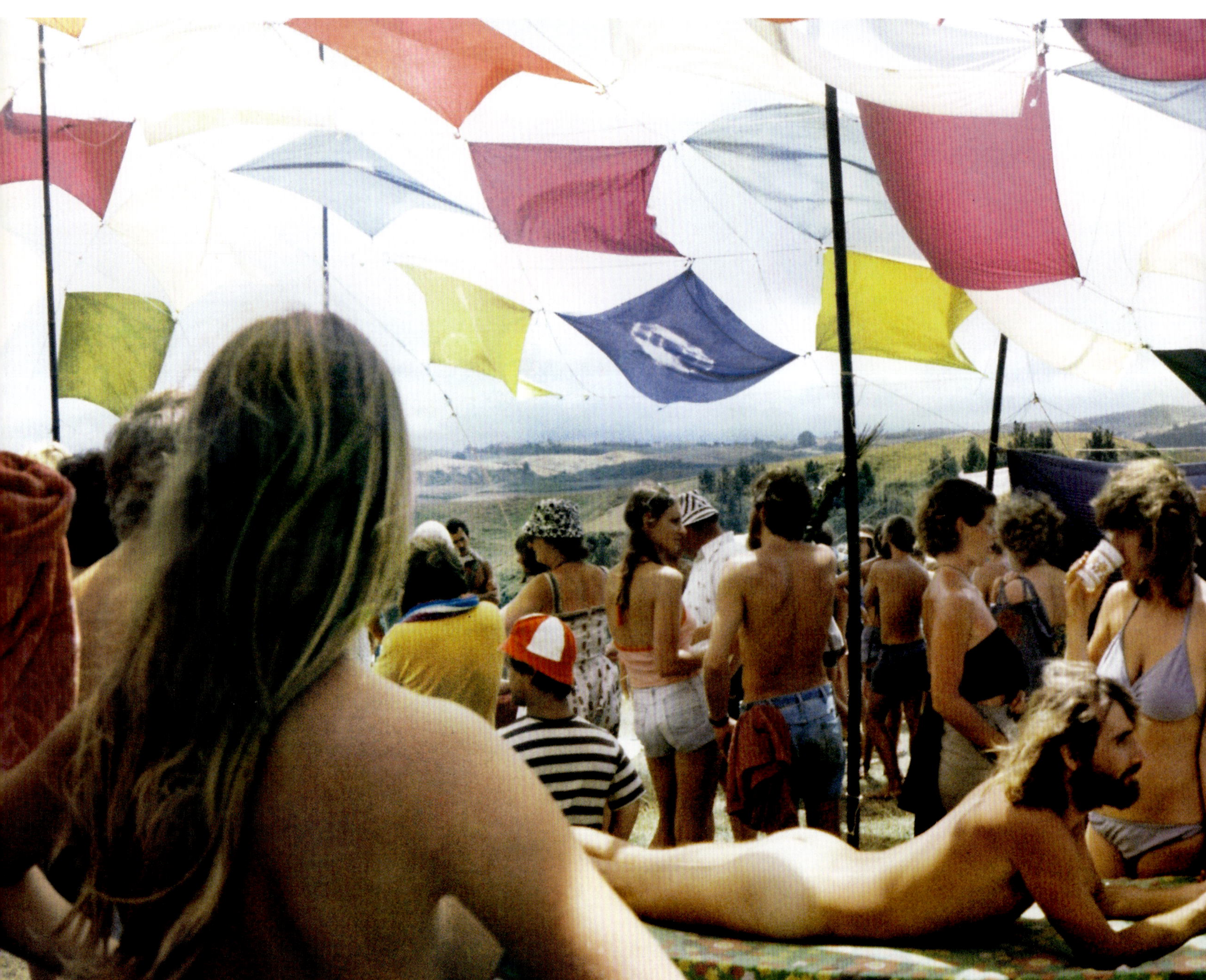

Nambassa, 1979

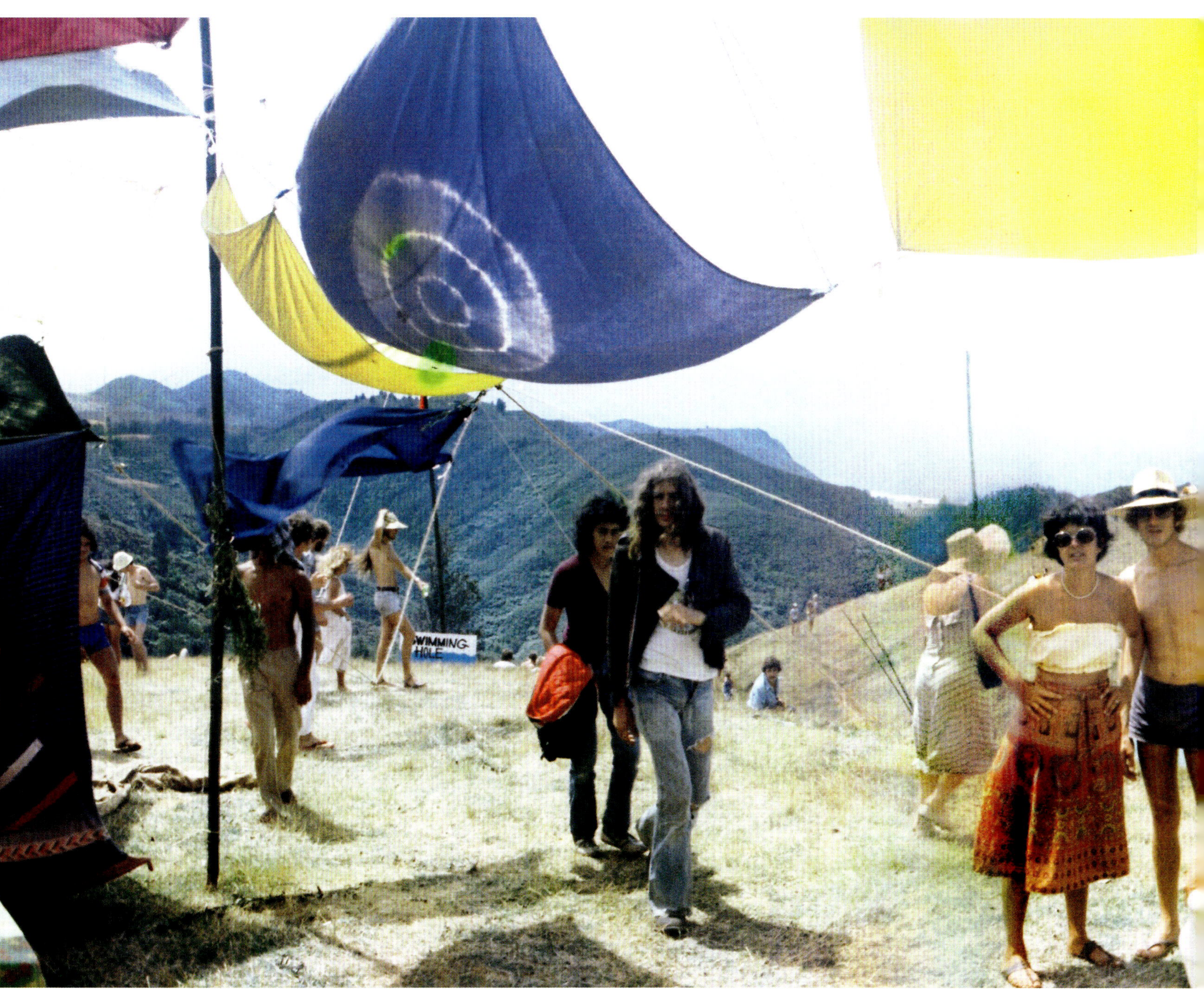
WIMMING
HOLE

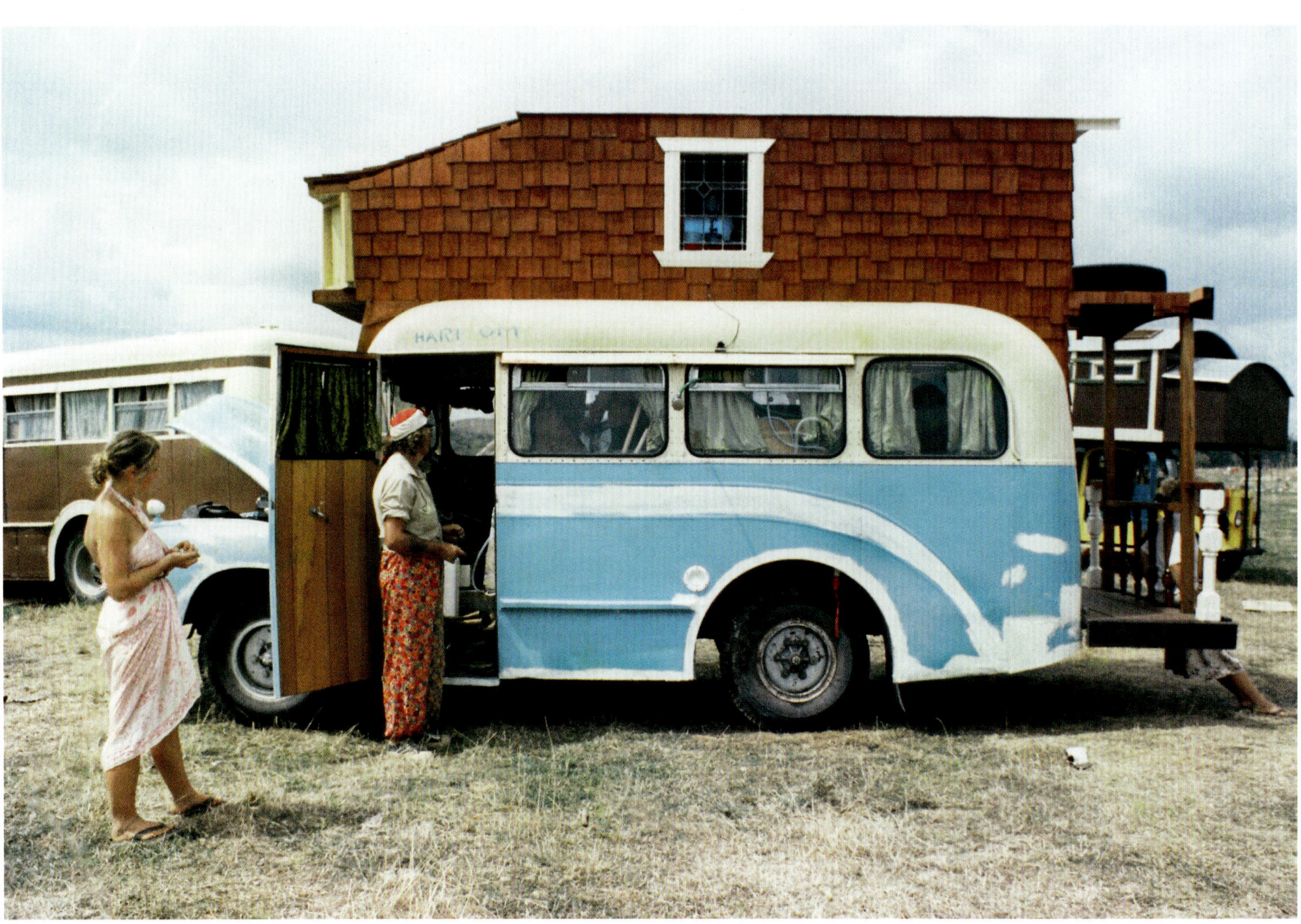

Sweetwaters, 1982

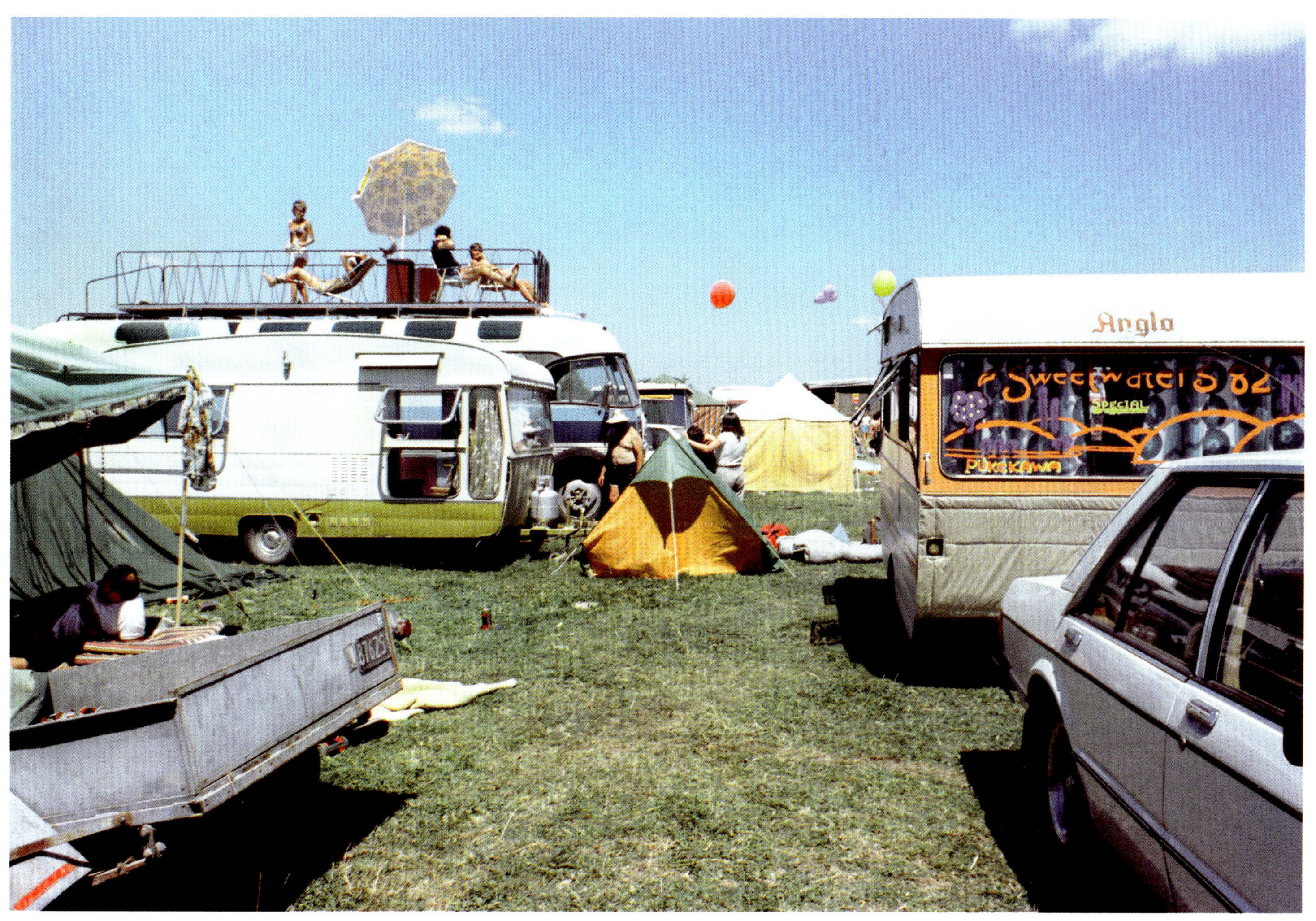

Sweetwaters, 1982

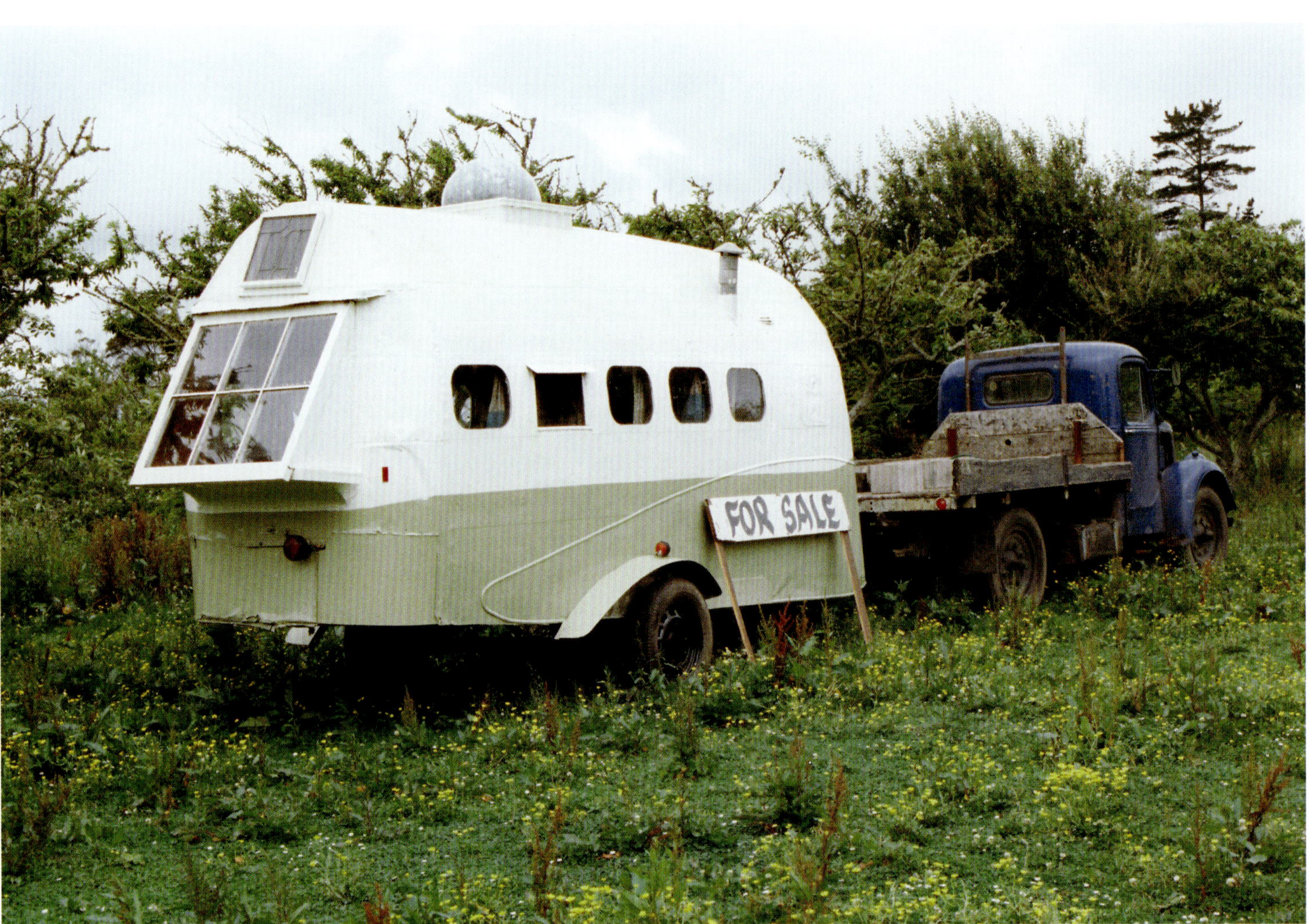

Aeroplane caravan at Taupaki, c.1981

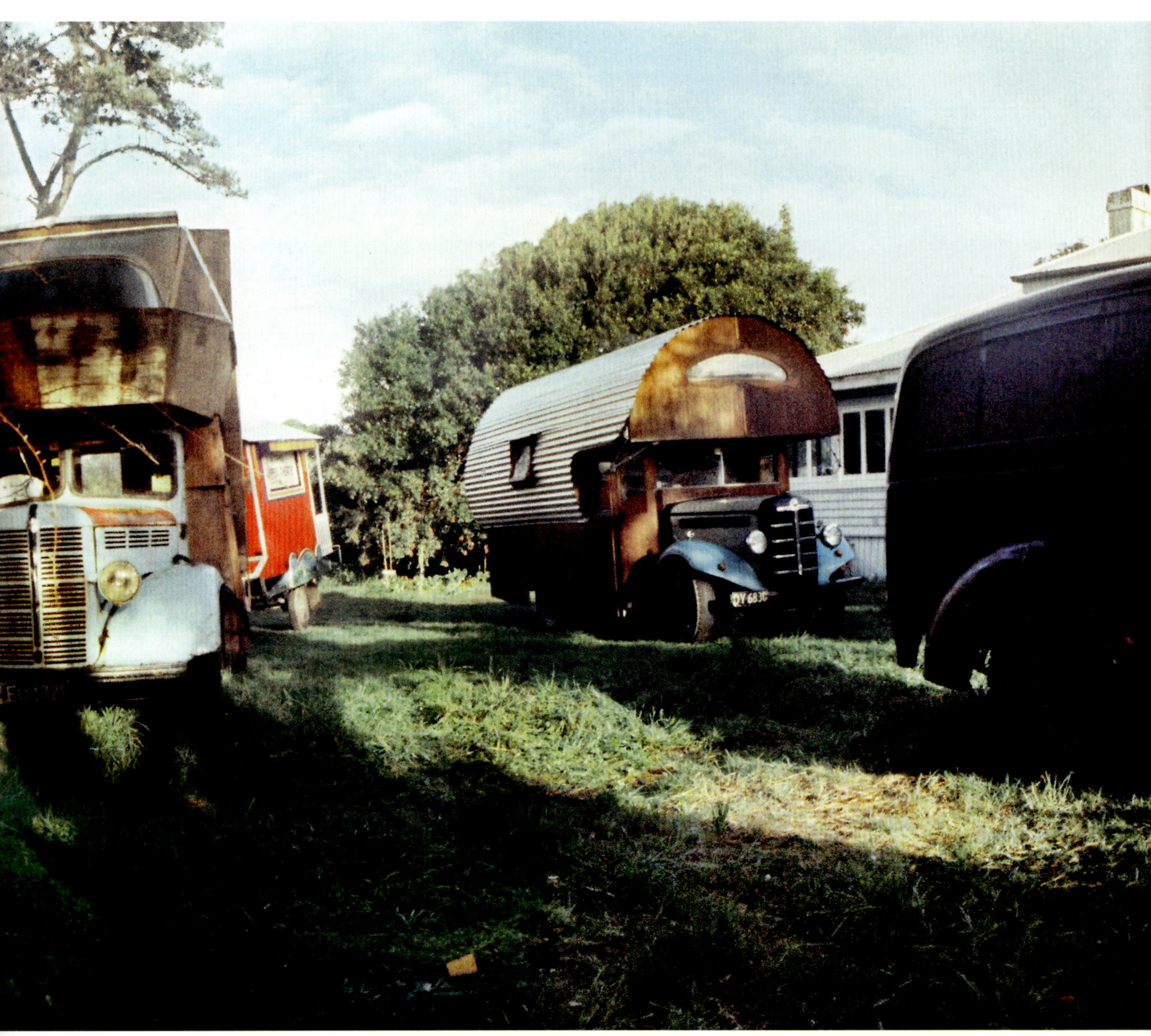

Some of The Original Travelling Road Show fleet at Clifton Road, Takapuna, c.1979

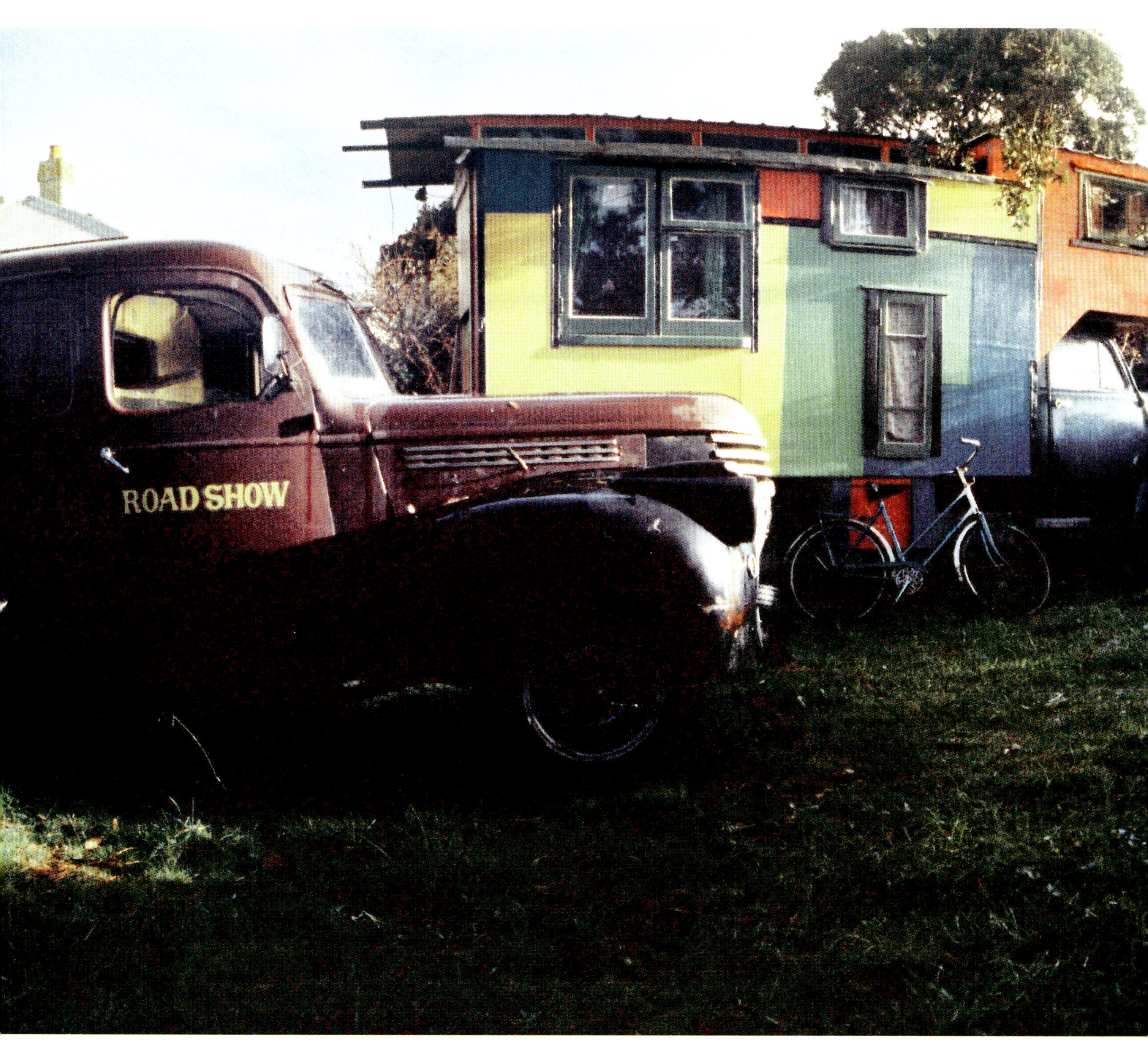
ROAD SHOW

Paul Gilbert's Road People

– Haru Sameshima

The photographic archive

Road People of Aotearoa is a photographic essay of 156 images by Paul Gilbert, selected posthumously from the unannotated collection of around 3500 photographs he made over seven years from 1978 to 1984, in the northern half of Te Ika-a-Māui, the North Island of Aotearoa New Zealand. Paul's colour negative films were neatly filed and preserved in three large ring binders by Sue Honeyman, who Paul met at the 1981 Nambassa Festival.

Paul arranged for this entire collection to be machine scanned in August 2019 and entrusted me to publish it as a book. Sadly, he succumbed to his illness before the scans were ready and never got to see how well the images from the forty-year-old negatives blossomed. During the short time we had to discuss this project, Paul expressed his wish that his photographs be accompanied by minimal text, with captions limited to names, places and events when known, and a short description if it helped to anchor an image. He wanted it to be a 'photobook', with the pictures providing the main narrative, rather than being used to illustrate someone else's story.

I asked Megan Jenkinson and John B Turner, Paul's old photography teaching colleagues from the Elam School of Fine Arts where he worked from 1995 to 2008, to help choose the images, based on the strength and quality of each photograph. As none of us knew the people depicted, or experienced the events in the photographs, we decided to select and arrange them loosely by subject, including making the house trucks, their wonderful interiors, the road people in everyday activities, children and families, the circus and musical performances in various festivals, and so on.

To help identify the photographs, Paul passed on a handful of names and contacts for the key people with whom he had shared this journey, so I could offer them the opportunity to view the photographs and perhaps add their voices to give context to this photo essay. Many responded quickly and with enthusiasm, and the list of names expanded through their contacts. They generously identified the names, places and events whenever they could, and shared their memories relating to Paul's images. Their stories are featured in this essay, to provide a semblance of chronology and context to the events, and names of the people Paul photographed – all the while celebrating his photography and the power it has to evoke a sense of time and place on its own terms, in this book.

Paul's archive is not that of a commercial photographer intent on accomplishing a job, whether academic or journalistic, to document the 'grand narrative' of a cultural phenomenon of its time. There is an abundance of such records already in documentary film footage, exhibitions and books. Wikipedia and YouTube entries are filled with official photographs commissioned by the festival organisers and media film clips depicting the increasingly large-scale music and fringe cultural festivals, culminating in Nambassa and Sweetwaters, and seek to define the alternative lifestyle movement of the period. Instead, Paul's photography cherishes personal encounters, localised events and the grassroots connections that he made on his journeys. His project is about these individuals – the musicians and performers and their friends – who came together through the Nambassa Festival of early 1978 with a belief that a sustainable alternative life on the road was possible, desirable, and preferable to a life tethered to a grounded house.

House trucks on the road, Down Under

The lead characters of Paul's photo essay are undeniably the beautiful handmade house trucks that were devised as an alternative to buying a house – with a mortgage and a steady job alongside – in which to raise a family. These house trucks reportedly appeared on the roads of both islands of New Zealand in the mid-1970s, as if by osmosis.

John Tucker, whose meeting and friendship with Paul in 1977 led toward this 'Road People' project, made one of the early house trucks of this period in the North Island.[1] Robyn Harris-Iles was John's partner at the time. She recalls that when she and John were living in the Pohangina Valley, a small rural community in the Manawatū region around 1973, John purchased a 1948 Bedford K-type from Te Horo in the Horowhenua. He put a plywood box on it for sleeping in and to store their gear during their frequent trips away (to concerts in Wellington and Auckland). The box had a bed in it but no door, and dust would blow in whenever they travelled on a dirt road, which was often in those days. They decided to put a proper house on the back of the truck that they could live in permanently: they wanted security but could not afford to buy a house and land. They began gathering materials: a small wood-burning stove, stained-glass windows, rimu beams and so on. They read up about Romany varda (horse-drawn wagons) in Britain, and discovered that there were six traditional designs, each one

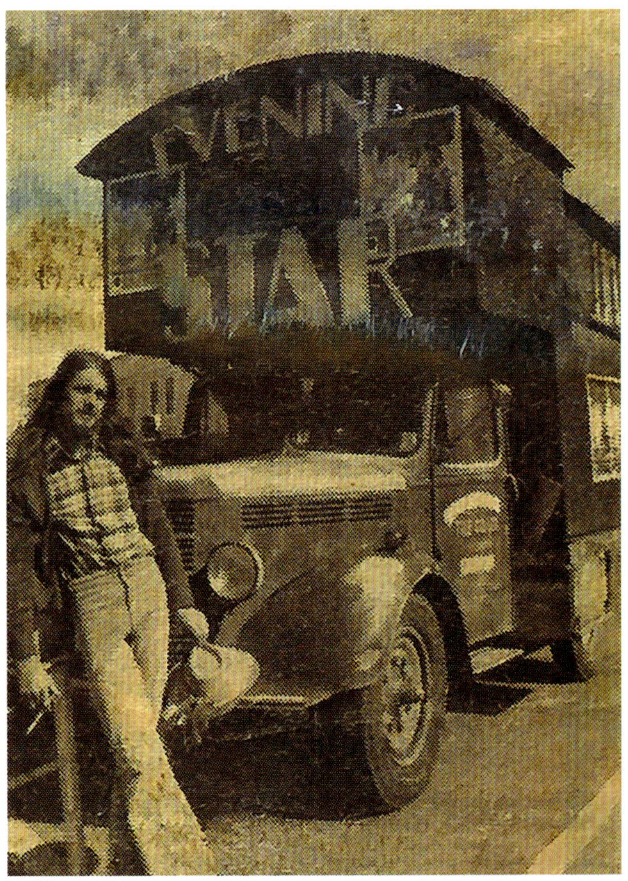

Unannotated newspaper clipping from the *Guardian* [Palmerston North weekly] 'Where Similarities End', Mr John Tucker. A portable home. Photo: *Guardian* [Palmerston North]. Private collection.

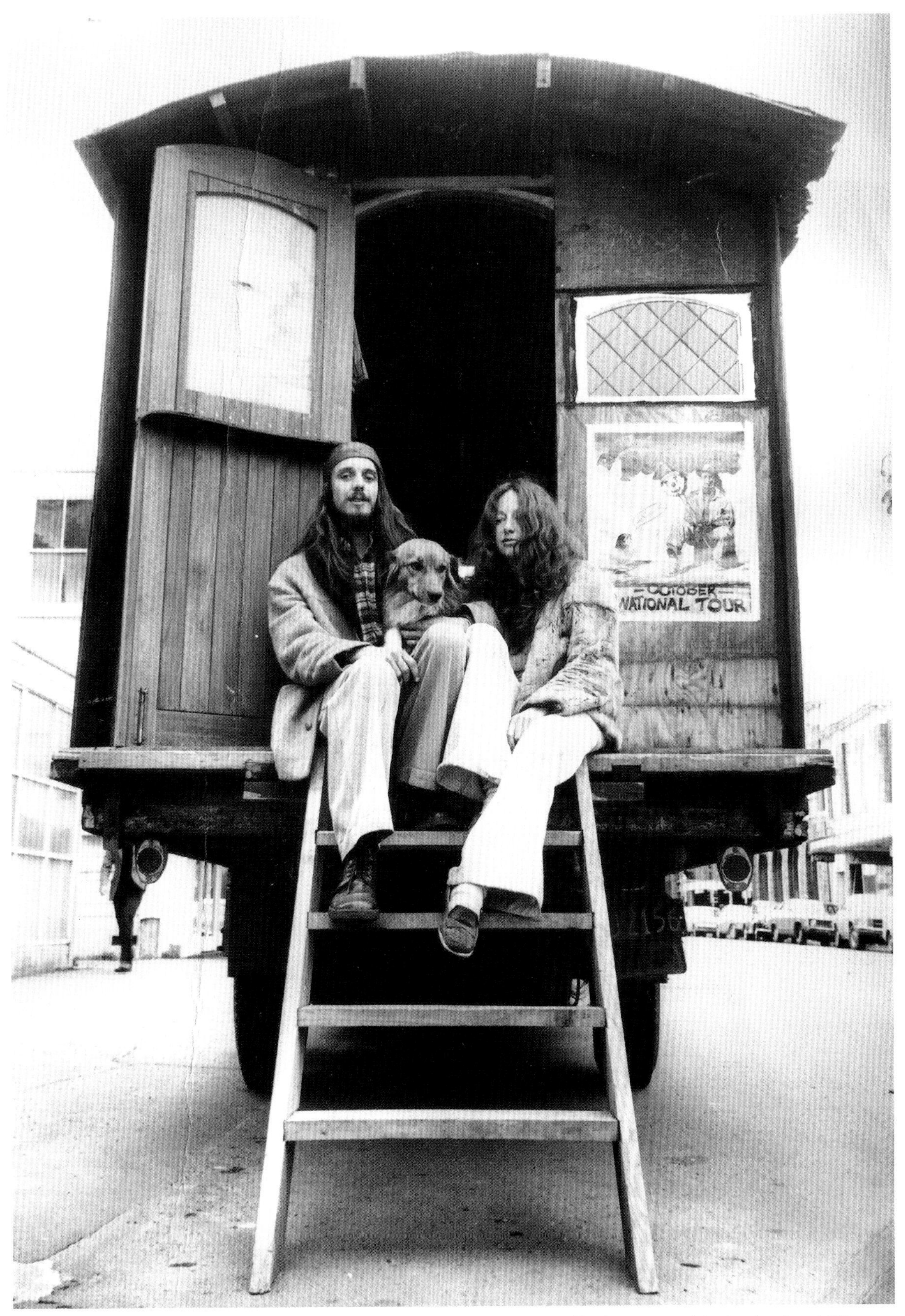

John Tucker, Roger (the dog) and a friend, with the newly converted 1948 Bedford K-type truck in Hawkes Bay, 1976. Photo: *Hawkes Bay Herald-Tribune*. Private collection.

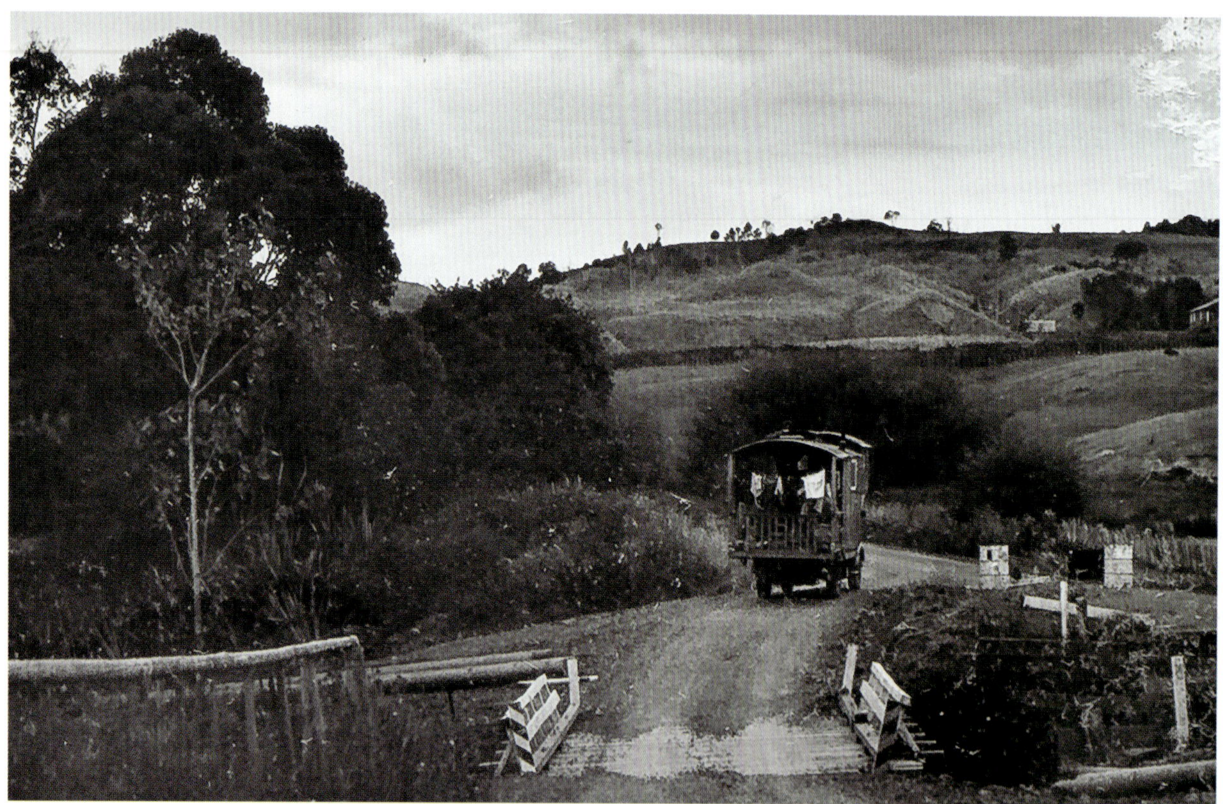

Near Tuakau on one of many travels north. On the way to Auckland for a Rod Stewart concert on March 3rd 1977 at Western Springs. Photo: anon.

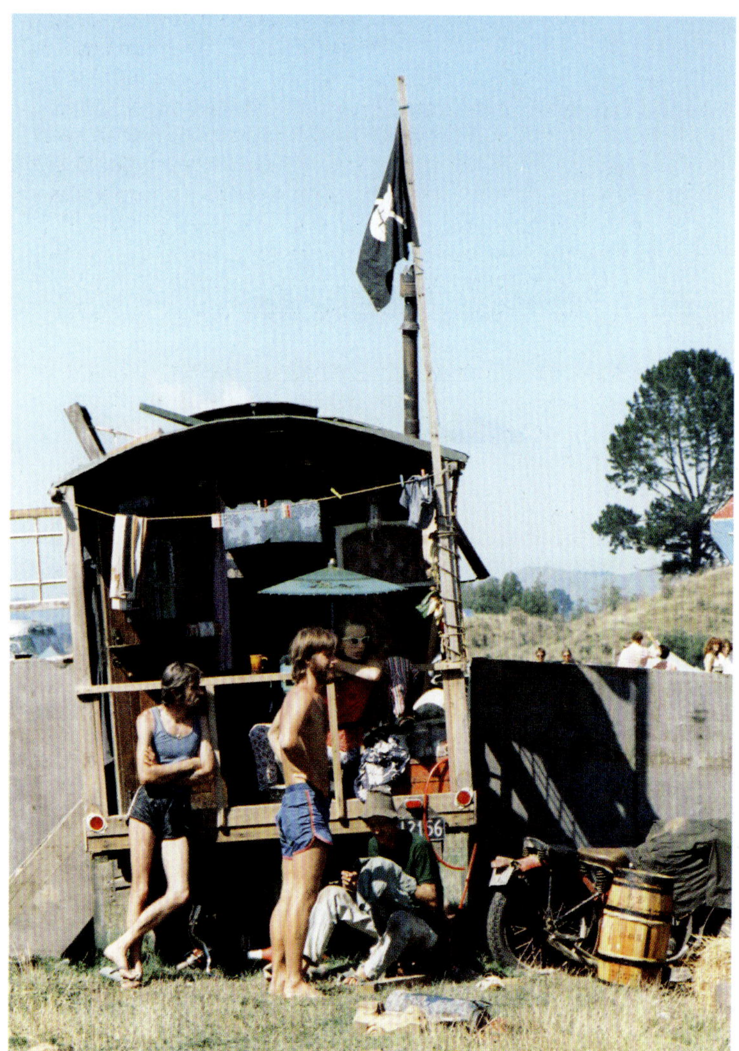

Number one Road Show house truck (on a slope) at Sweetwaters Festival, 1982. This was the first house truck built by Johnny Tucker. The house was a complete unit that could be removed from the truck. This made it a bit top heavy when cornering. Later house trucks built by Johnny had the deck removed and the house constructed directly onto the chassis. This house truck had a Shacklock 00 woodstove inside with the tall chimney extension removed and stowed on the verandah for travel, and a gas ring for cooking or boiling the kettle in summer. There were two skylights to assist airflow in summer and the house could be entered from the cab. The house had lights connected to the battery, stained glass windows and a beautiful barn door among its many original features. It could accommodate three people. The truck carried two chocks to level it when parked on a slope so I'm not sure why they weren't deployed here. We often flew the Jolly Roger at festivals. (See also page 107.)

— Robyn Harris-Iles

An artists' rendition of the second stage development of the original 1948 Bedford K-type, built in late 1975.

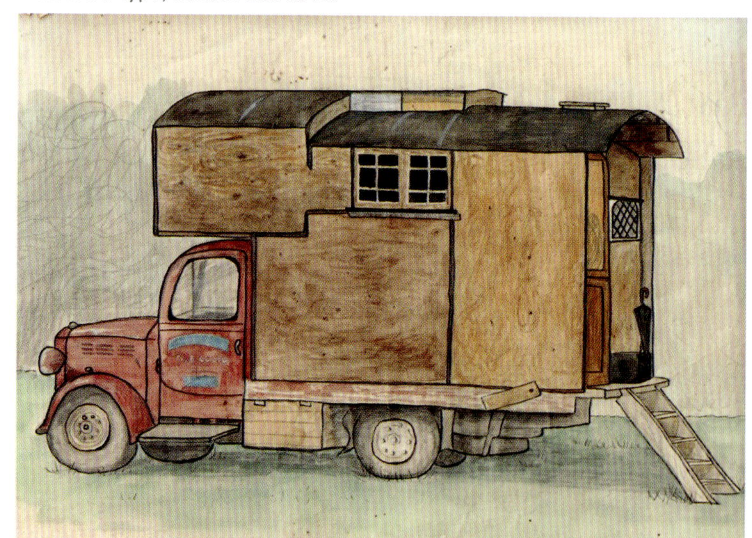

decorative and stylish. They were also inspired by the way Romany people lived, as craftspeople and musicians on the road.[2]

John finished building the truck to a more liveable state in late 1975. He named it 'Evening Star' and took to the road on his nomadic life, attracting attention from the media – and from friends who found his lifestyle appealing and asked him to convert a truck for them.

The renowned street performer and musician Dom Ferry, another veteran house trucker whose wonderfully distinctive home on wheels features on pages 36-37, remembers his first house truck encounter:

> In late 1976, two close friends from the 'Alice's' & later the 'Mt P' Wellington urban communities,[3] John Wenn and Chantal Roux, bought an old Bedford van and headed off to the South Island. Eight months later they drove off the Picton ferry with a house truck built on an old 1950s Bedford 'K'. It completely blew me away (I can still smell the engine–linseed oil mix on the timber walls in my mind) and right there and then both Jane and I decided that this was the way for us as well. We threw a rudimentary home together on the same kind of truck and left the big smoke in late 1977. John & Chantal's truck was the only one I knew of at that time and it wasn't until Nambassa '78 that I was suddenly in the midst of other house truckers.[4]

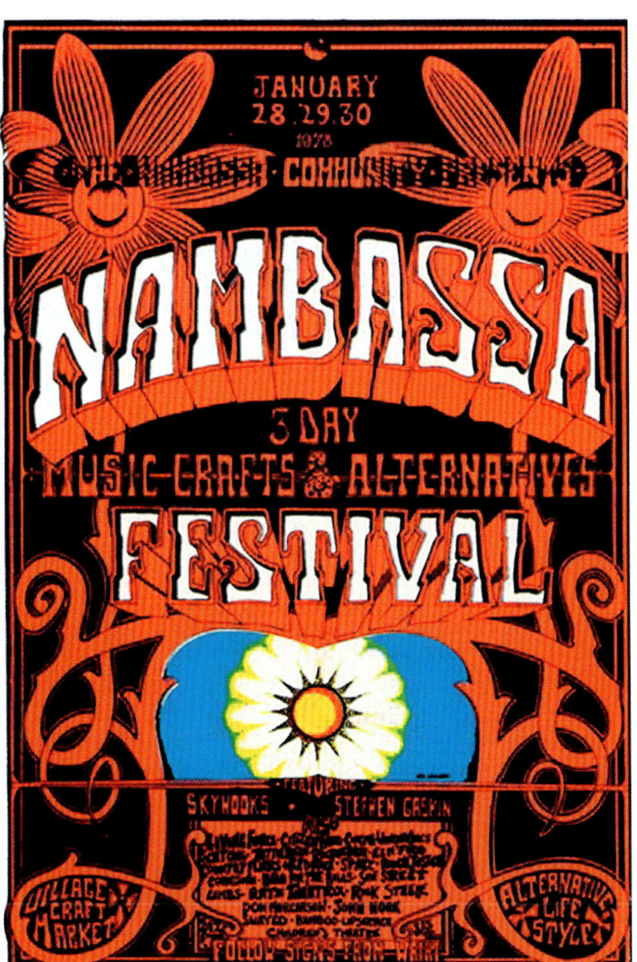

A poster for the first Nambassa festival of arts, crafts and alternative lifestyles in 1978, which drew between 15,000 and 25,000 revellers to Golden Valley, near Waihi. Image: Nambassa Trust and Peter Terry.

Alternative lifestyle, communes and Nambassa

> The first thing you need to know about the popular Nambassa festivals of the late 1970s and early 1980s is that they weren't Woodstock or Sunbury in Australia, or whatever convenient, but ultimately clumsy, cultural equivalent the media and authorities came up with.
>
> — Andrew Schmidt, 'The Nambassa Festivals and the counterculture movement', 2014[5]

As Paul Gilbert indicated in his introduction, the idea of an alternative lifestyle in a nomadic mobile home in New Zealand has its roots in the international 1960s counterculture movements, taking cues from American and British hippie communes. While communes in New Zealand – the 'intentional communities' – date back to the nineteenth century,[6] many more were established during the 1960s and early 1970s in rural and urban backdrops in response, including the government sponsored Ohu scheme.[7] Nambassa festivals in the late 1970s, popularised the cunterculture ideas by introducing the philosophies to the wider New Zealand society in the form of music festival. Peter Terry, (p119) the 'visionary and organiser' of all Nambassa-related festivals starting with Waikino Music Festival in 1977, conceived the series of events while visiting and staying at some of the communes in the Coromandel in 1975. In 1976, he published *Nambassa Newsletter 1* to rally support for his festival concept that embraced alternative lifestyle principles.[8]

> There can be no doubt that we are living in confusing times. The social pressures alone are enough to start one thinking about an alternative lifestyle . . . The present-day economic and social system offers us no true incentive – unemployment in New Zealand is extremely high, and life is becoming for the everyday 'man in the street' and his family an effort just to survive . . . We are using up our natural resources at an ever increasing rate – they are not going to last forever. Consider this and ask yourself. Is this a natural way of life, is this how we were meant to live? . . . Now, more than ever before, there is a need and growing desire for people to learn to live outside the collapsing economic and social system, with its greed and avarice, and its denial of individuality.[9]

Stephen and Ina May Gaskin, the co-founders of The Farm, a spiritual community in Tennessee, USA, were invited to open the three-day Nambassa Festival in 1978. The Gaskins had led 300 dissatisfied city dwellers from San Francisco on a four-month speaking tour in a convoy of 60 renovated school buses, trucks and vans around the United States in 1970. After the road journey, they purchased land in Tennessee, where they established a commune based on principles of nonviolence, ecological sustainability and kindness to the earth, while farming to sustain the community.

> We try to make it so our work is integral with our life: what we do for a living shouldn't contradict what we believe in or hamper our vision of the future.
>
> — Stephen Gaskin at Nambassa 1978[10]

The Farm was also renowned for Ina May Gaskin's revival of midwifery and the homebirth movement. The Gaskins were among the many international guests who gave seminars and workshops on alternative living at the Nambassa Festivals.

Michael Colonna (later known as Walnut the Clown, p141) arrived early as a volunteer to help set up the 1978 festival.

> I remember taking a young woman to the caravan of Stephen

and Ina May Gaskin a night before the festival began. The woman had walked into a barbecue plate at night and gashed her leg. The Gaskins had all the gear needed to sew the wound up because Ina May was a midwife. But the thing that sticks in my mind was the atmosphere. My father was a doctor and my mother a nurse and the front half of our house was a surgery. I grew up watching all the hospital dramas. But there was no urgency, no hurry. Ina and Stephen spoke quietly as they worked together; there was a strong feeling of peace and love there. It strongly affected me that such an operation could be done with such gentleness.[11]

Music, performing arts and the Aerial Railway Music Co-operative
New Zealand was (and still is) a small village: many of the people in this cluster of 'road people' had family or friends who were connected or travelled with Blerta[12] in the earlier part of the 1970s. They also knew the people involved in, and were inspired by, Red Mole Theatre (NZ), and read up on Merry Pranksters (USA), all performance groups who lived on the road from time to time. Many of the road people were avid readers of the *Whole Earth Catalogue*, both the US and New Zealand editions, and *Mushroom Magazine* (1974-1985) that featured what was going on in the counterculture movement here and offshore, along with instructions on how to succeed in leading an alternative lifestyle. The concept of a nomadic life based on music and street performance in the form of rock opera, clowns, mime and circus entertainment to earn a living while on the road, had its own attraction. After all, the travelling entertainment troupe has been a time-honoured occupation since the Middle Ages.

Synergy within the thriving communes and creative counterculture of the 1970s through music and performance art, was perhaps most evident at this time within the Moehau Community. This community began as a group of people from Auckland who were looking to purchase land in Moehau on the Coromandel Peninsula. Partially to raise funds for their campaign for nuclear disarmament,[13] they formed the Moehau Magic Mountain Band (MOMMBA) in 1974. This travelling road show (albeit sans house trucks) of '45 adults, 6 kids, 3 dogs, 14 vehicles' performed in Auckland and around the Coromandel, presenting a two-hour counterculture performance that was designed to hold the attention of an audience of all ages and types.[14]

After the MOMMBA tour, the Moehau Community organised a three-day hui, Moehau Celebration, with 2000 attendees. At the hui they offered workshops, demonstrations, talks, raves and music. The community's focus was on the production of original material and, importantly, casting their net wide to function as the communication hub for creative performers across the country, offering to manage concerts, and providing stage crafts and equipment. This expanded group, called the Aerial Railway Music Co-operative, connected and helped many original New Zealand music and performing arts groups such as Ratz Theatrix, Limbs Dance Company, Cohesion, Swayed and Head for the Hills. They also recorded music for films and documentaries, including the production and music for the celebrated film *Sleeping Dogs*.[15] It was their cooperation with Peter Terry, who attended Celebration, that led to the alternative 'Aerial Railway Stage' at Nambassa Music, Crafts and Alternatives Festival, in 1978, and presence at all other Nambassa and the new Sweetwaters festivals. Their call, through the Aerial Railway newsletter for grassroots musical and theatre groups to perform, attracted road people, among them Dom Ferry and John Tucker's Manawatū fleet, to converge in Golden Valley north of Waihi in the summer of 1978.[16]

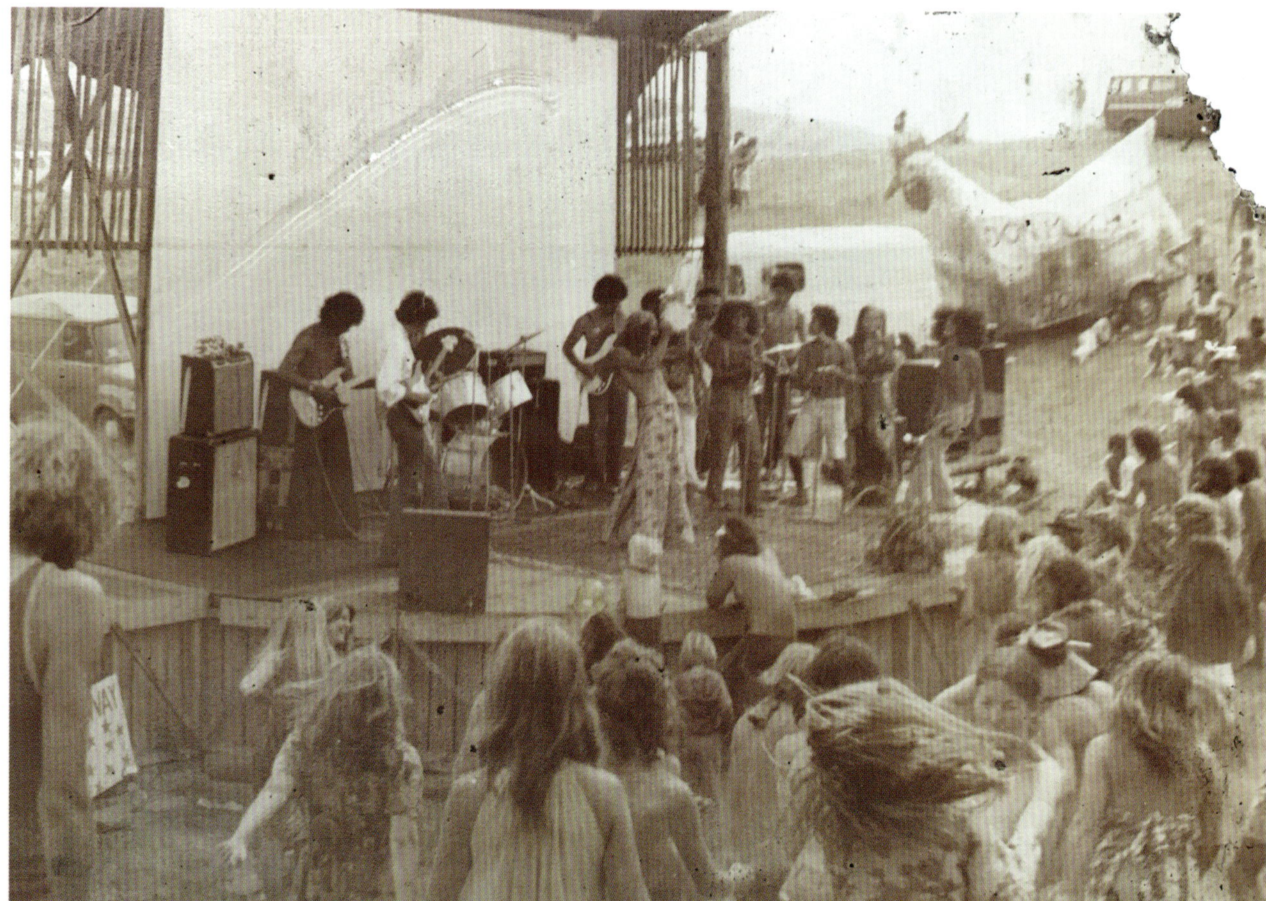

Mahana on Aerial Railway Stage, and the audience, Nambassa, Monday 30th January, 1978. Photos: anon

The origin of the Mahana Rock Opera in Manawatū

When Robyn Harris-Iles and John Tucker moved to the Pohangina Valley in 1974, they soon got together with a number of other talented local musicians and formed the Manawatu Musicians and Artists Club. They rented an old RSA building in Palmerston North, complete with a large hall and kitchen, as the club headquarters and music venue. They undertook renovations with the help of band members from the Dixon and Sweet whānau, and Bud Hooper. Bud's partner Gabrielle, and Chris Cole ran the bar, and John and Robyn ran the kitchen. The club hosted guest bands from out of town, such as Highway and the Country Flyers. Cadzow Cossar and Jimmy Dixon, who were mainstays of the house band, held a jam session every Sunday night in which musicians from near and far took part.

In 1977, when John was on the road alone in his 'Evening Star', he learned about the call out for 'acts' for Nambassa. He immediately returned to the Manawatū to connect with his musician friends. They gathered at Poupatatē marae on the Rangitikei River near Halcombe to form a band, later called Mahana.[17] They already had four or five house trucks, and they made other preparations for the 'act' at the marae. Members of the original line-up conceived a rock opera that would portray the impacts of colonisation in Aotearoa. Cadzow Cossar was the musical director, with support from Jimmy Dixon and John Tucker. The music for the rock opera was written mainly by Cadzow, including a song written by Corben Simpson, 'No Tresspassing', and rudiments of what became the climatic song, 'War Sequence', originally written by Billy TK and arranged by Cadzow. They completed the rock opera act over one month in December to January 1978. Lyrics for the performance were written collectively by Mahana members John Tucker, Dale Whitcombe, Jeff Whitcombe, Cadzow Cossar, Jimmy Dixon, Stephen Kereama, Bobby Simmonds, Waka Kimura and others.[18]

The first performance of the Mahana Rock Opera was scheduled as the last act on Sunday night on the main stage at the 1978 Nambassa Festival. But the programme was running late and it was four in the morning by the time they came on stage, after most of the audience had left for the night. The forty-odd sleepy members of the audience – and the rest of the festival attendees – soon realised they were experiencing something new and spectacular. The band was persuaded to perform again on the Monday afternoon on the Aerial Railway Stage, to an estimated audience of 3000.

> What happened that afternoon will always be remembered.
> — John Tucker, handwritten note in his scrapbook, n.d.

The programme for the Nambassa Winter Tour, later that year, promoted the Mahana offering as:

> The Mahana Band which is made up of both Maoris and Pakehas, came together musically in the early summer of last year, and made a sensational impact at the Nambassa Festival. Many will remember the Mahana rhythms of dance music, and movement on the Sunday night of the Festival; they made such an impression that they were asked to play again the next day . . . The actual opera was written by Mahana, and is a colourful and perceptive view of early New Zealand life. The story begins with the coming of the Maori in canoes, their meeting on the high seas, and subsequent voyaging to N.Z. It tells of the arrival of the White Man, politics, religion, settlers, and the colonialism that was to transform the country that has become dear to the Maori people. The opera moves to portray the wars that occurred between the Maori and Pakeha . . .[19]

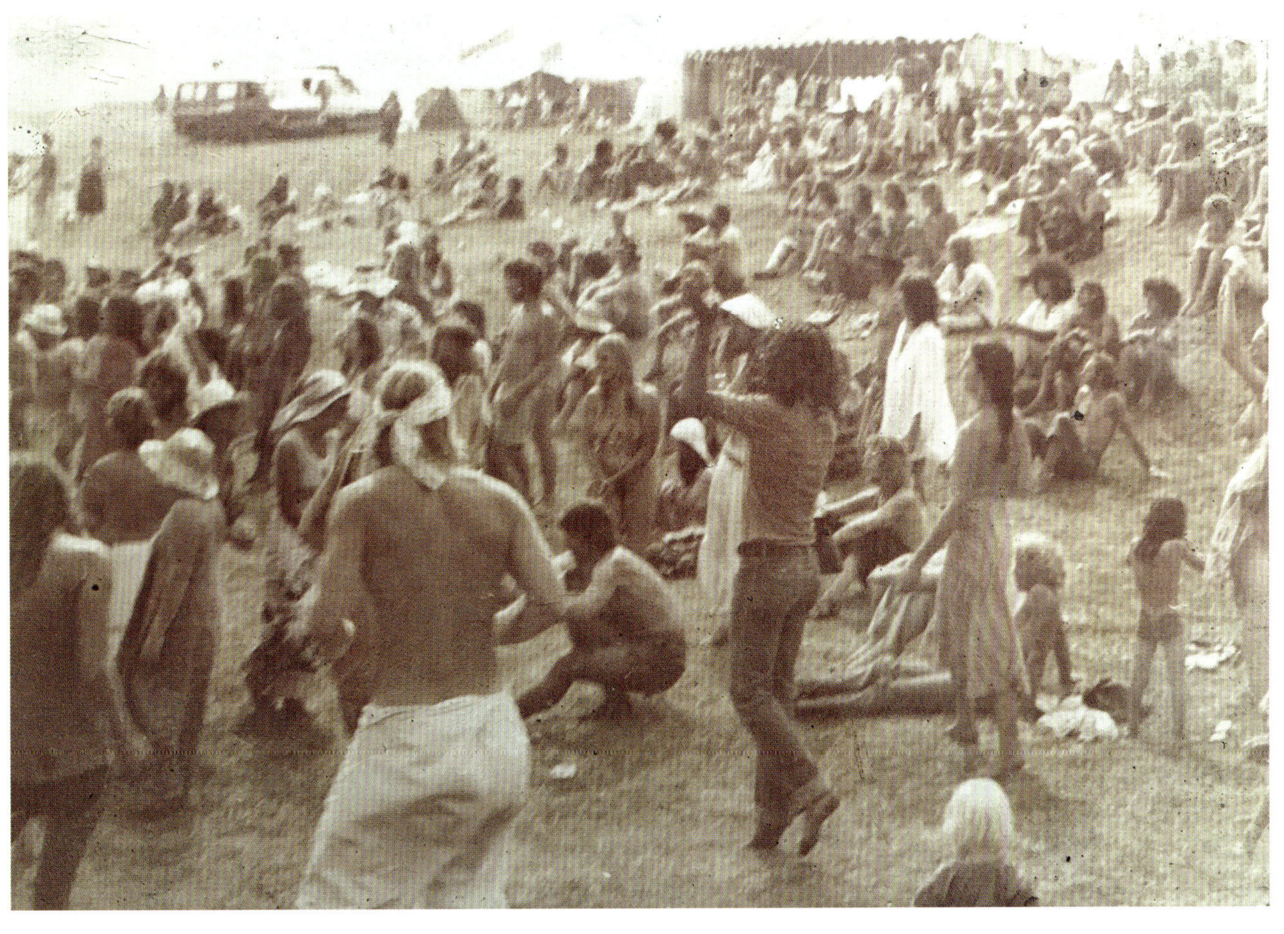

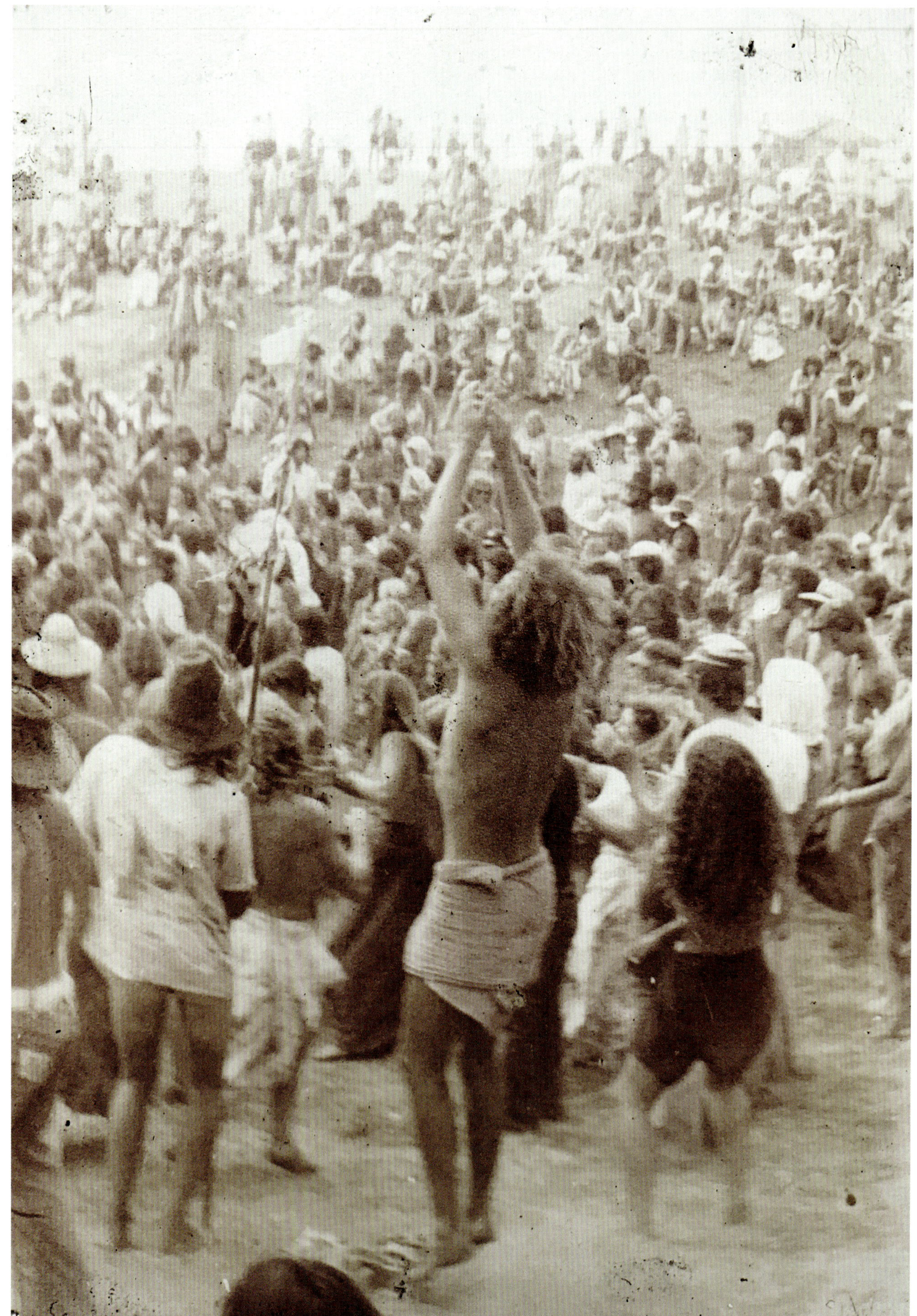

That performance on the Aerial Railway was absolute magic, with the band putting out energy and the crowd giving even more back. What an amasing experience it was. I had a humbling experience as I nearly didn't get on stage as I was having to share a mic with the Timbales and congas. Hey I just sang louder and it was one of the most amasing performance experiences.
— Dale Whitcombe

The Original Travelling Road Show – 'The Commune on Wheels with the Circus Spirit'[20]

Paul Gilbert's camera followed the artists and performers who converged at the 1978 Nambassa Festival, and had then taken up the option offered by Peter Terry to invest in a travelling road show that would take the festivals to communities. Michael Colonna, who met John Tucker and Mahana at the festival and joined The Original Travelling Road Show troupe, noted at the time:

> When the festival was over the feeling was still there. Peter Terry, the festival organiser announced that from a Youth Initiative Grant he had received for the festival; $1500 was to be given to set up a travelling road show. John Tucker jumped at the opportunity and immediately began approaching other artists who were interested in touring. From this was born The Original Travelling Road Show. When the Road Show left the festival site it was $1500 richer and 50 strong, consisting of musicians, actors, technicians, craftspeople, artists, clowns, buskers and dreamers. The Moehau Community, which had run the Aerial Railway at Nambassa, offered a campsite at Sandy Bay, Coromandel so The Road Show in convoy drove up there to organise itself. The idea was to put together a travelling Rock Circus encompassing all of the skills of the tribe.
>
> After a few weeks of dreaming and rehearsing the concept became a reality. The Road Show travelled to, and played at Mini Nambassa in the Maratoto Valley then did a show at Waihi.[21]

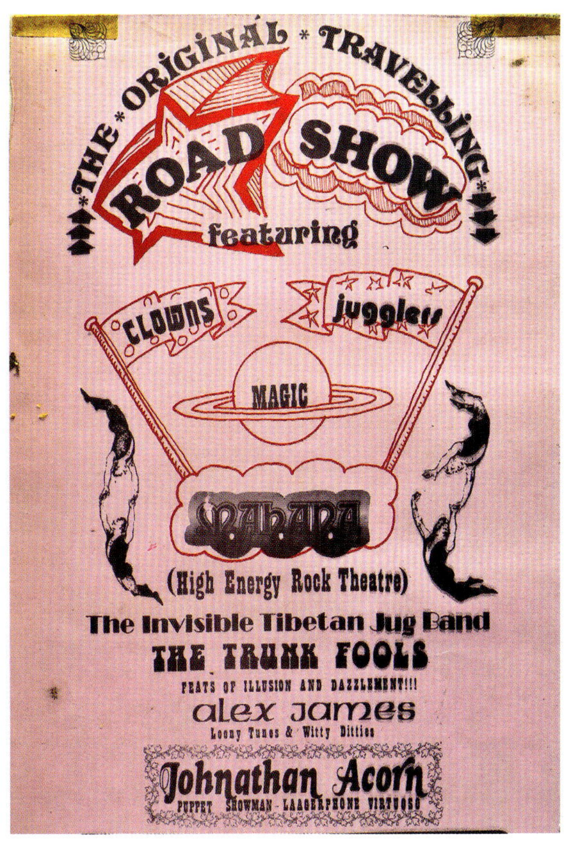

Poster used for the Waihi concert, 1978. Private collection.

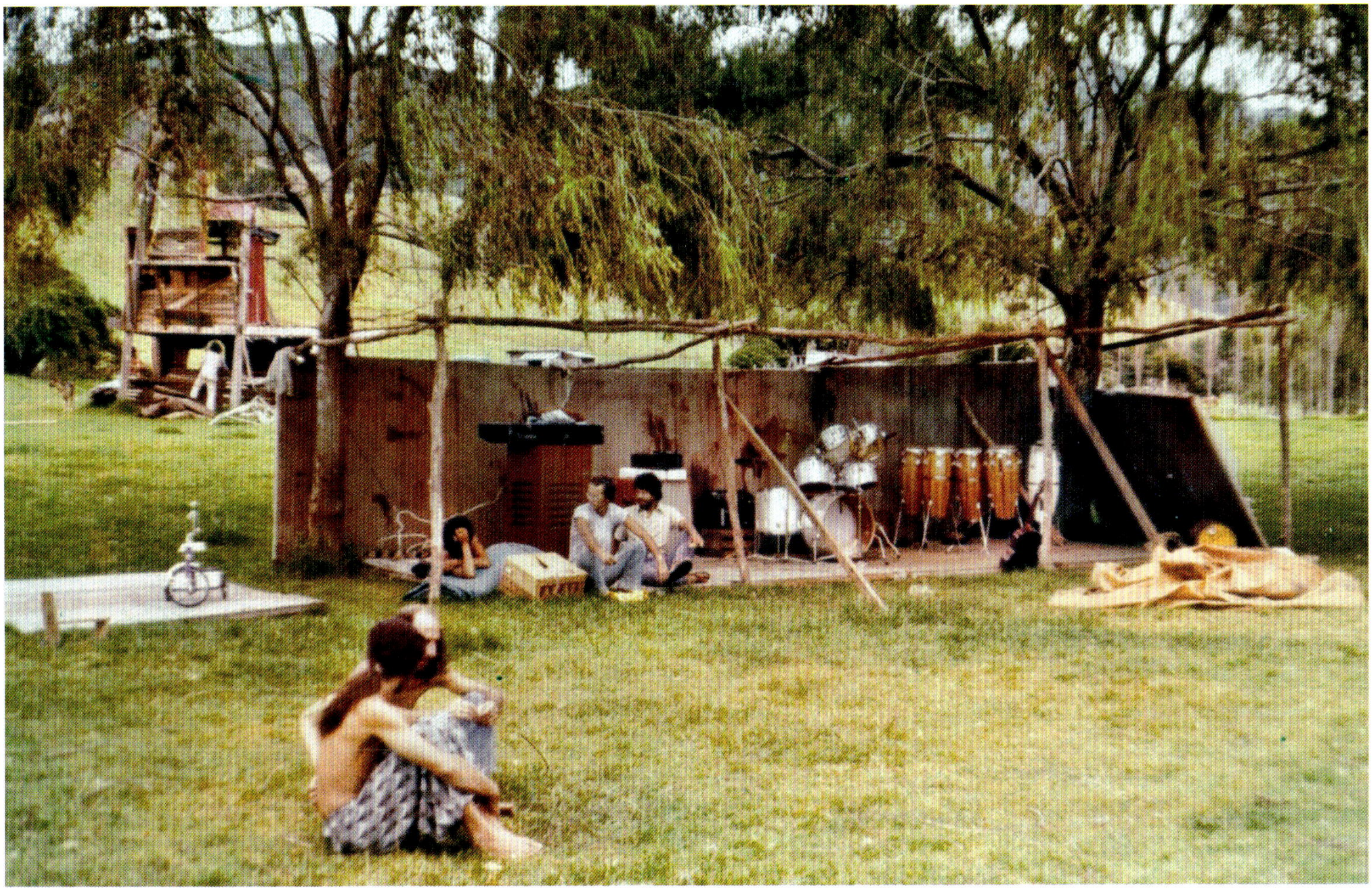

Mahana stage at Sandy Bay, Moehau community, 1978. The remnant of the 3-story structure from the Moehau Celebration in the back. Photo: M Colonna

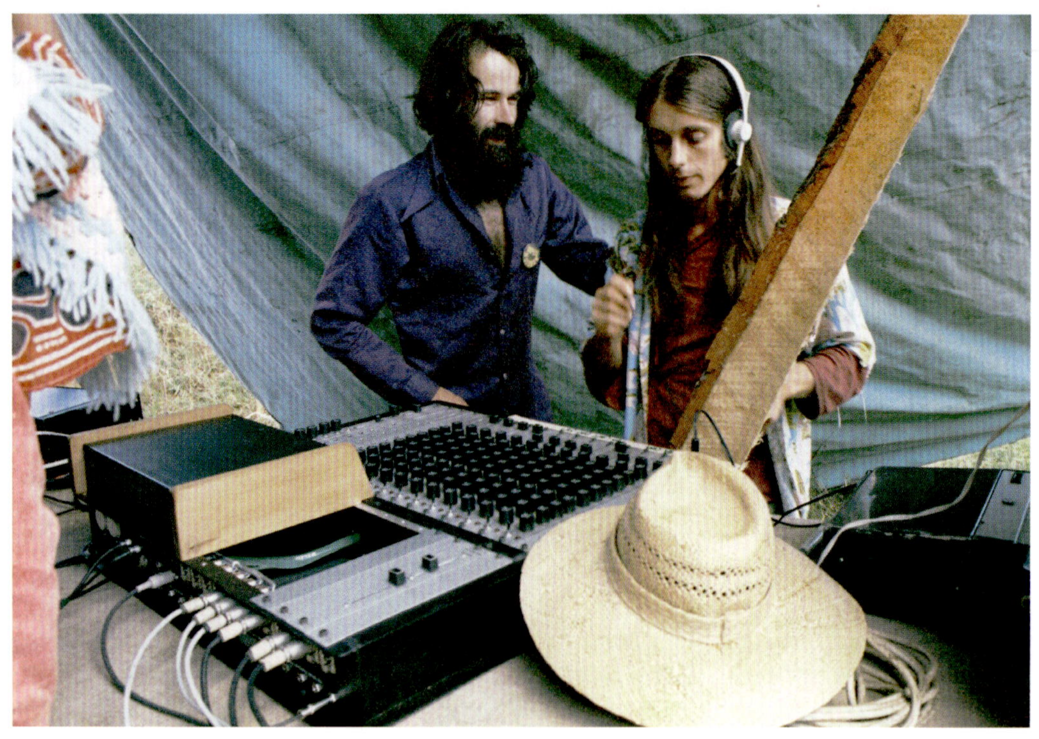

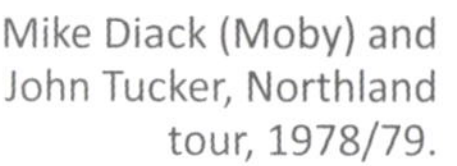

Mike Diack (Moby) and John Tucker, Northland tour, 1978/79.

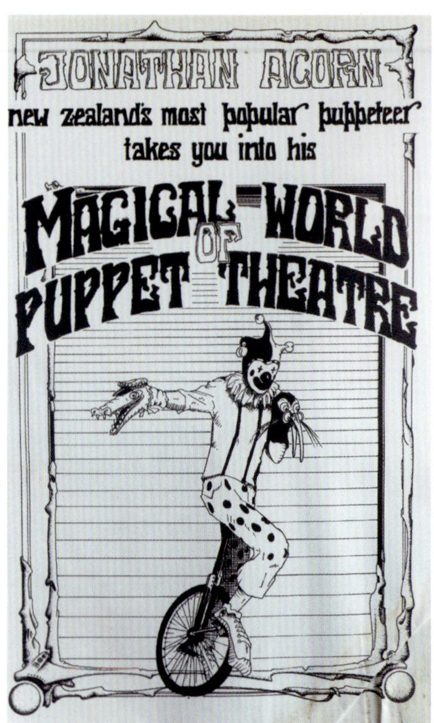

Jonathon Acorn Puppet Theatre poster c.1978. Private collection.

Jonathon Acorn's puppets and the Imperial Trunk Fools

The people who joined the The Original Travelling Road Show with Mahana, ranged from puppeteers, poets, mime artists and street theatre actors to fire-eaters, as well as people interested in yoga, pottery, alternative food, a mechanic, a school teacher and a carpenter – and an almost qualified med student.[22] Notable additions included veterans from the Moehau Community: a brilliant electronic and sound engineer Mike Diack (Moby), Jonathon Acorn (p123) and Hendrika Vaneveld (p80) and their child Tahi (p98). Jonathon joined the Moehau Community house in Ponsonby, Auckland in 1974 while he was developing Acorn, his new clown character. He was an experienced performance artist and had been performing fringe theatre with MOMMBA: he had developed his walking puppet theatre and put together a comedy act with Terry St George, called *The Butler's Table*, which they performed on the television talent show *Opportunity Knocks*. He had studied performance under the international mime–clowns Nola Rae and Francis Batten, and travelled around Auckland in a gypsy caravan converted into a puppet theatre, performing as Clip-Clop and Clown in parks and festivals. Later he wrote and performed *The Magic Hammer Show, The Sausage King* and *Captain Kiwi and the Coromandel Gold* – an anti-mining show.[23]

> I joined up with a group of travelling fools called the Imperial Trunk Fools led by Alan Clay. We wintered over in Auckland, rented a house so that Alan's wife (Kerstin) could give birth to their eldest daughter.[24]

The Imperial Trunk Fools Company Unlimited was originally formed in Sweden by expat New Zealander Alan Clay and his Swedish partner Kerstin Gronlund, who developed theatre workshops as clowns under their stage names Stewy and Franciss. They took part in the annual Festival of Fools in Amsterdam, and gathered more expertise at a six-month-long 'fool school' in Stockholm.[25] They took on new members when they arrived in New Zealand, including Jonathon and Hendrika, Peppe Olson, Bogdan Szyber, Brian Tracey, Monica Haar, Terry St George, Michael Colonna and more. The Trunk Fools ran many workshops, teaching local performers street skills, such as the art of fire-eating and 'the fine art of performing as a Fool'.

'Trying Hard to become a Fool' by Terry Bell. Unannotated clipping from *New Zealand Woman's Weekly*.

Michael Colonna and Shaquelle Maybury fire eating, Stewy and Franciss in the background, 1978.

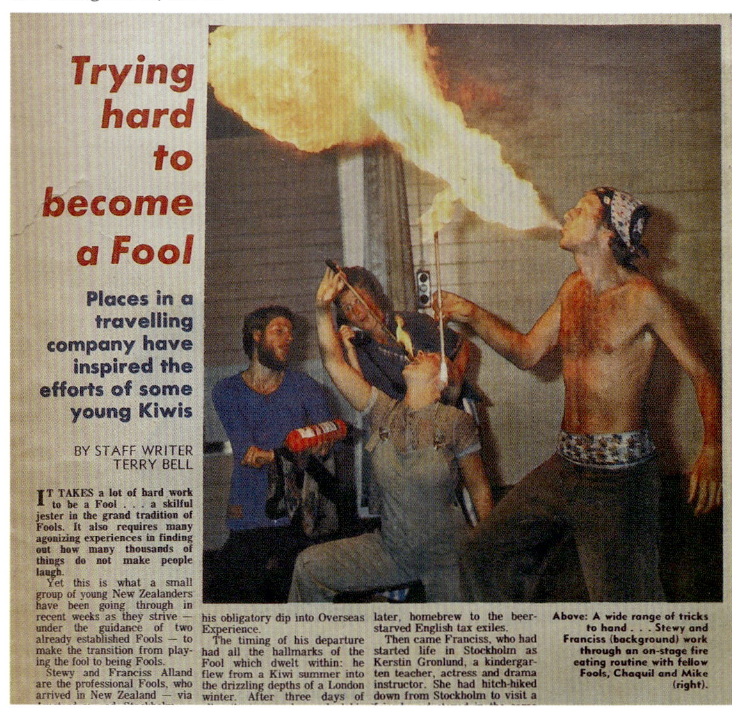

Trying hard to become a Fool

Places in a travelling company have inspired the efforts of some young Kiwis

BY STAFF WRITER TERRY BELL

IT TAKES a lot of hard work to be a Fool . . . a skilful jester in the grand tradition of Fools. It also requires many agonizing experiences in finding out how many thousands of things do not make people laugh.

Yet this is what a small group of young New Zealanders have been going through in recent weeks as they strive — under the guidance of two already established Fools — to make the transition from playing the fool to being Fools.

Stewy and Franciss Alland are the professional Fools, who arrived in New Zealand — via

his obligatory dip into Overseas Experience.

The timing of his departure had all the hallmarks of the Fool which dwelt within: he flew from a Kiwi summer into the drizzling depths of a London winter. After three days of

later, homebrew to the beer-starved English tax exiles.

Then came Franciss, who had started life in Stockholm as Kerstin Gronlund, a kindergarten teacher, actress and drama instructor. She had hitch-hiked down from Stockholm to visit a

Above: A wide range of tricks to hand . . . Stewy and Franciss (background) work through an on-stage fire eating routine with fellow Fools, Chaquil and Mike (right).

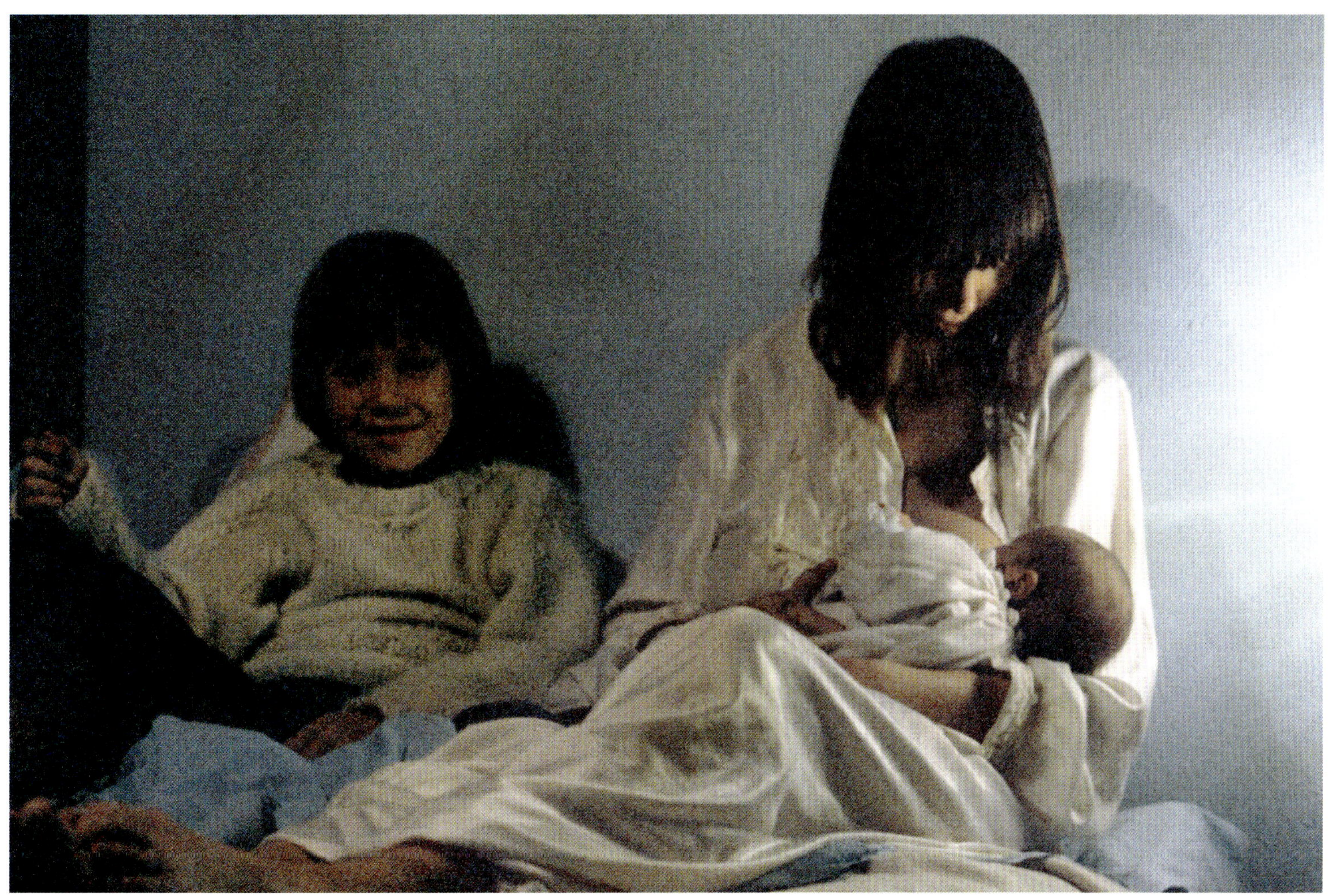

Marie and Briar with Matt, Epsom, Auckland, 1981.

My beautiful leadlight rolling home (p97) was built in 1976 entirely from recycled materials by my two good friends John Tucker and Neville Pope and a bunch of enthusiastic hammerhands, myself included. I was a sole parent at the time, raising my son and studying. I'd had enough of being grounded; a mortgage was the last thing on my mind. John and Neville were both proud owners of newly built rigs, and when these freewheeling folk showed up and offered to get me on the road, I had no hesitation. Within a week I'd bought a 1953 Morris Commercial flatbed, and I set about accumulating windows and materials. John went on to inspire several other families, mainly musos from the Rangitikei region, to embrace the life of the road with kids in tow.

The early house truckers I travelled with included the Dixons; Mike and Barb Sweet and their tribe of kids, who went to live on the Mahanaland in their commodious mobile home; and Dale and Geoff Whitcombe, who promptly sold their general store at Koputaroa near Levin to set off in an old army truck converted into a house truck with their two kids, to join the Manawatū fleet to Nambassa. The original convoy to Nambassa there was a busload of talented Māori lads from Halcombe district, 'the Boys', wonderful vocalists and guitarists who would play around the campfire until dawn. Music, always, the campfires, shared kai, hordes of kids – a seemingly endless adventure. Which is not to say we were always mobile. The Mahana bus was forever breaking down, for one: all the old rigs needed ongoing maintenance. Dale went onto build her own house truck (red A2 Bedford, backcover), to be part of the Mahana roadshow. We all went from Maratoto Festival to Moller's Farm in early autumn, planning the Road Show north. Then came the rain, and several of the rigs became impossibly stuck in the mud. The Boys and Dale went off and did the Winter Show during those months. More rigs were built. Apple picking kept us solvent.

I was never particularly focused on settling down – I have always been a bit of a nomad. Trucking provided the 'stable roof' while I was raising my family, as well as the mobility and freedom and community I craved. For over 20 years my house truck was our home. For several years it was annexed to a bach at Waikawau Bay, and later to an off-grid cottage on Wharf Road, Colville. Always it was Home.

Some of us were inspired by The Farm, the spiritual community in Summertown, Tennessee, started by Californians Stephen and Ina May Gaskin, who were guest speakers at the early Nambassa Festivals. Ina May was the doyenne of the homebirth movement, a midwife herself in a community doing all the alternative back-to-nature stuff that we Down-Unders were getting into. I am part of that generation, of the subculture seeking an alternative way of life in Aotearoa.

Paul took a beautiful photo of Matt and me with my newborn daughter Briar in Epsom. I'd taken the truck up to Auckland so that Joan Donnelly, a well-known midwife, could help with the homebirth. A dozen old bungalows in Dunkerron Avenue housed a curious assortment of tenants, a neighbourhood of young like-minded people, including house truckers. That was where Briar was born in the spring of '81.

— Marie Ward, email correspondence with HS, 26 September 2020 (edited)

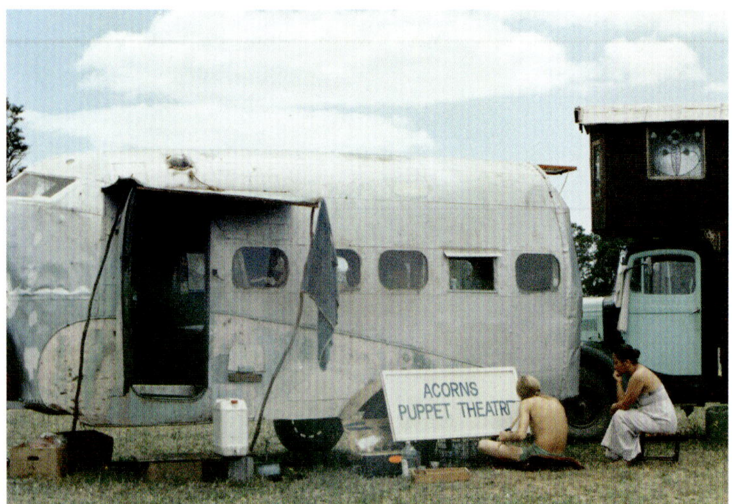

Ewan Morris signwriting Acorn's Puppet Theatre signage, with Lockheed Hudson bomber fuselage in the back, unknown location, 1978.

1978 road shows – Auckland Festival, the Nambassa Winter Show
After the short Coromandel tours, post Nambassa, the Road Show moved north and set up in West Auckland where they did casual work such as picking apples on a local orchard. Michael Colonna continues his 1979 account:

> Some brave truckers wintered over on Bill Moller's farm at Oratia, on the outskirts of Auckland, while others set to work to build themselves their own house trucks. With a large number of children involved, a schoolroom was set up in a caravan made from a Lockheed Hudson fuselage discovered at the Tapu camping grounds. A stage-truck was built, shows and concerts were organised, babies born, and by spring 1978, a tour of Northland had been prepared followed by another Nambassa festival for the summer of 1979 at Waihi.[26]

Wilma Pucci who lived at Moehau Community while teaching at Colville Primary School, taught the Road Show children open-air at Maratoto Valley, and in the Lockheed Hudson caravan at Moller's Farm. Moehau Community were involved in organising the 1978 Auckland Festival, and Aerial Railway Music Co-op facilitated events and concerts across the city, including one at the Maidment Theatre with the Limbs Dance Company. Mahana, Trunk Fools and Acorn performed in numerous music venues, streets and concerts, including a benefit concert at Bastion Point to support the protesters who were occupying contested ground there.

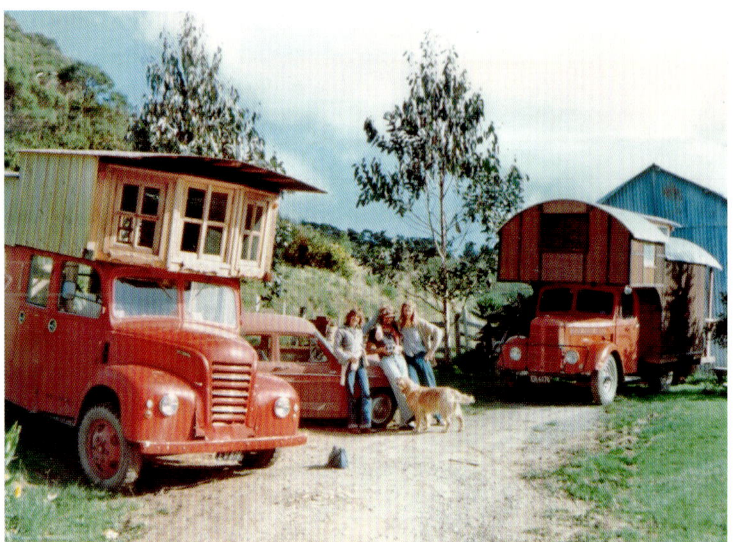
John Cameron's fire-engine truck. The other red Bedford is Dr. Stewart Wells' truck. Photographed during a visit to the farm on Waiwhiu Rd where Michael was building 'Snail', Dome Valley, near Warkworth, 1978. Photo: M. Colonna.

The winter months were hard on the morale of The Original Travelling Road Show as a group. Not everyone was equipped with a ready and functioning house truck, and many had to go their separate ways to prepare for hibernation. The paddocks at Moller's Farm – a longtime venue run by Bill Moller for folk and jazz festivals since the 1960s; it became something of a house truckers' sanctuary for many years – became a mud pool after hosting so many trucks. Some found a place in Wellsford, 70 kilometres north of Auckland, to build their trucks; Stewy and Franciss of Trunk Fools moved into a house in Northcote to have their child; and others stayed at Clifton Road, the clifftop premises that Paul Gilbert was renting in Takapuna, on the North Shore.

> Energy levels dropped slowly over the winter months until no one was sure if The Road Show was ever going to hit the road again.[27]

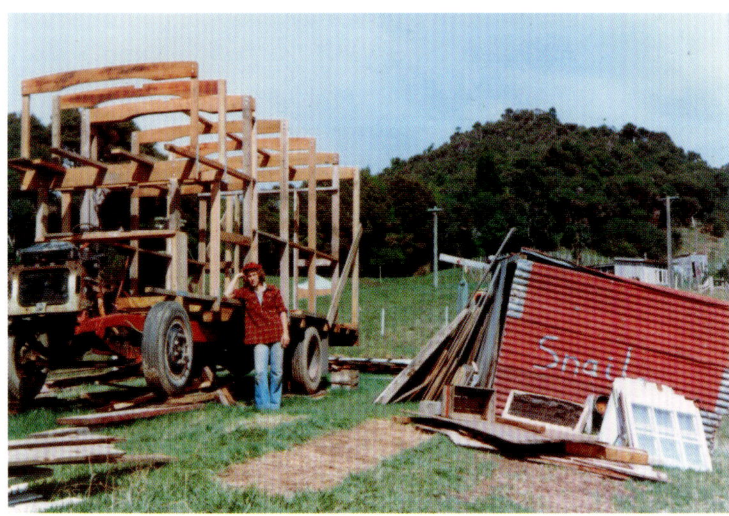

Michael Colonna's truck (Snail) being constructed, Warkworth, 1978. Photo: M. Colonna.

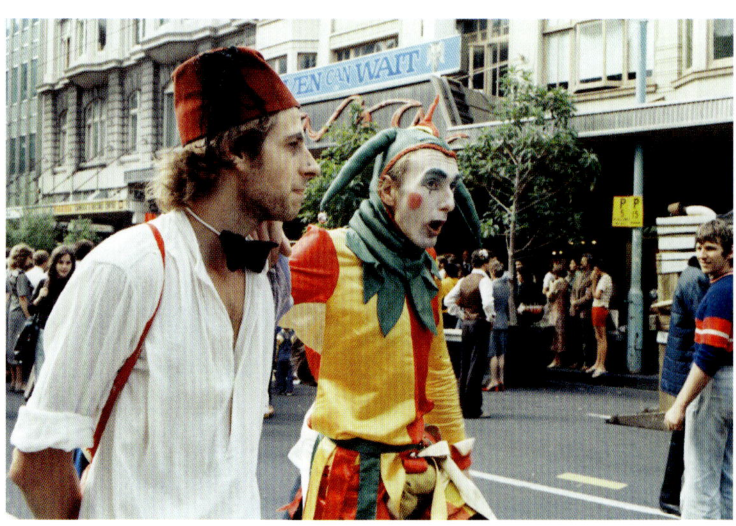

Alan Clay and Jonathon Acorn, Queen Street, Auckland Arts Festival, 1978.

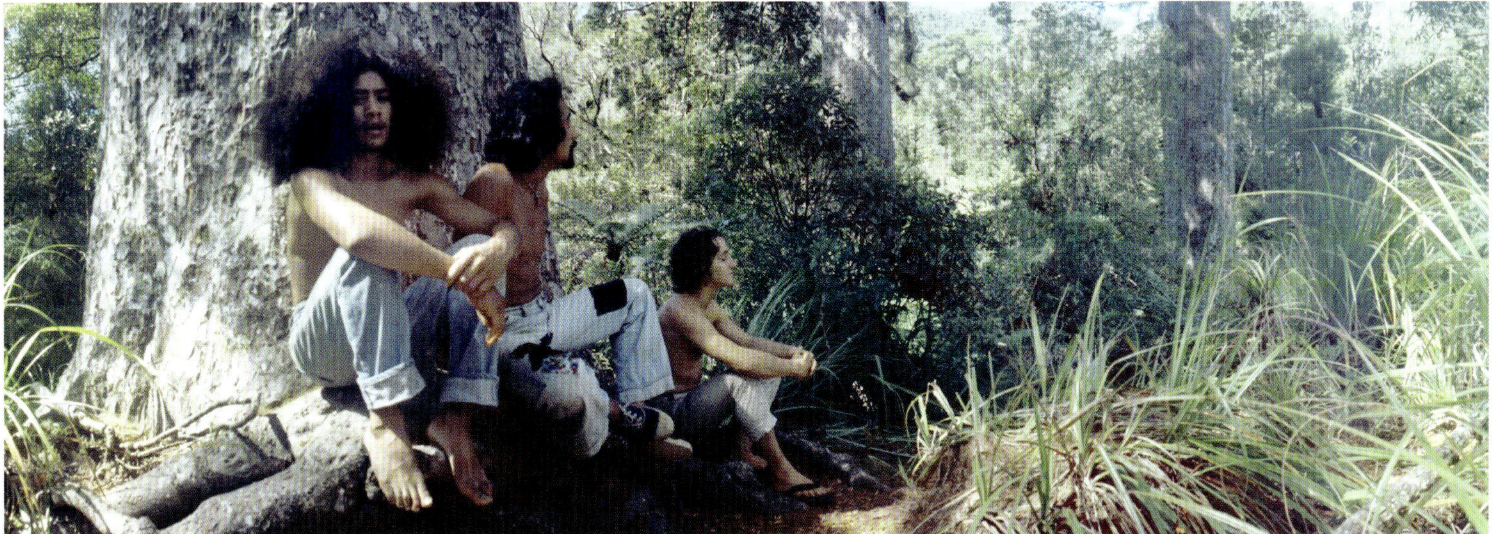

Waka Kimura, Jimmy Dixon and Stuart Crooks, Moller's Farm, Waitakere Ranges, 1978.

Posters featuring the performances by The Original Travelling Road Show and friends, Auckland Festival 1978. John Tucker collection.

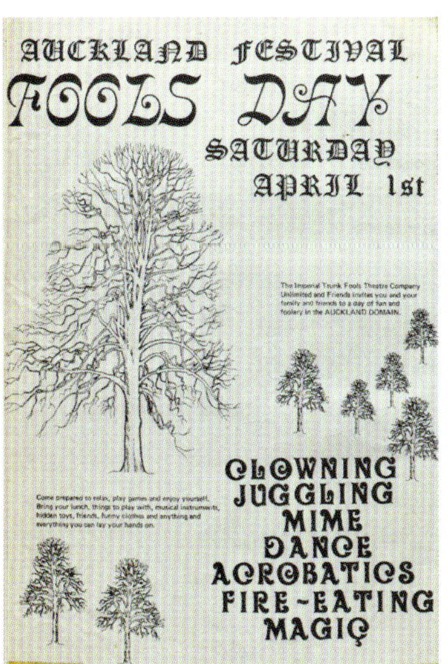

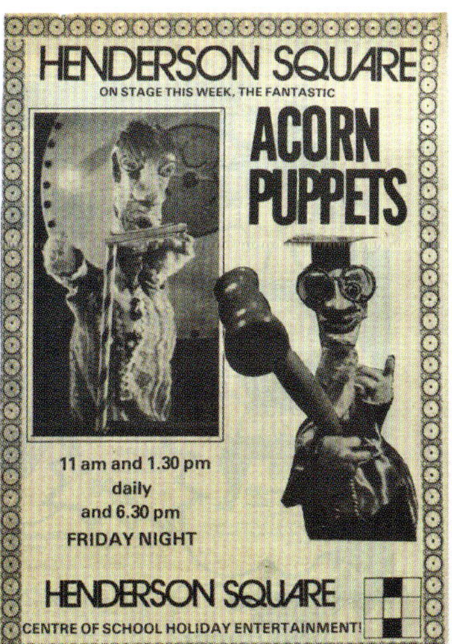

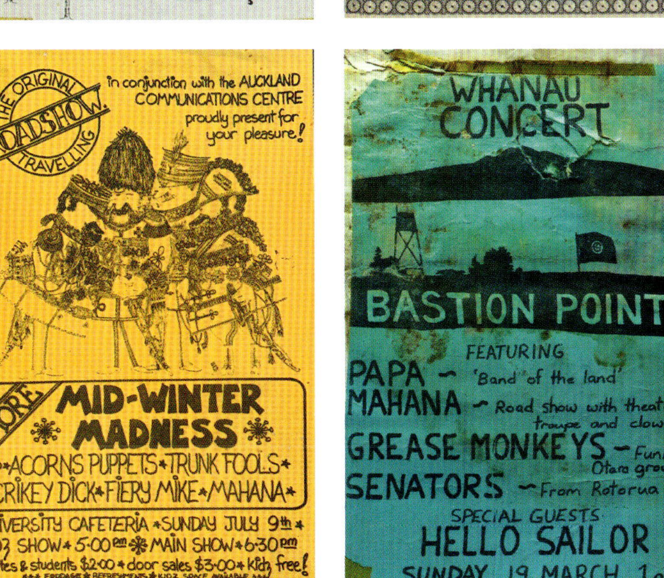

Born from the energy manifested at the Nambassa Festival, a Road Show has been created that embraces all the romantic and mysterious delights normally associated with such an enterprise. Imagine if you can, a tribe of 50 – 80 gypsies travelling slowly from town to town around Northland, setting up camp and turning on the most unusual style of variety show ever to be attempted in this country. . .

The first of the 'new age' road shows it is aptly titled 'The Original Travelling Road Show', or Ko Tahi Tatau (We're one together).

Nambassa Sun Vol. 1 March 1978

A Night at the Theatre – a diverse selection of entertainments for your perusal, featuring music, mime, dance and theatre. Presented to you by the Aerial Railway Music Co-operative.

When I saw this in the Orientation Programme I thought I might as well go along as I'd paid my $2.50 for Registration and wanted to make the fullest use of it. So I settled into a comfortable chair and soaked up one of the best shows I've seen for years.

From the clowns who kept the crowd amused with their antics & dramatics to the emotional intensity of Mahana. From the movements of Limbs to the beautiful originality of the Aerial Railway Music Co-op, it was a night to remember.

The clowns (one of whom played in both groups as well) warmed the crowd up with clever tom-foolery, dramatical effort and the aid of a water pistol. By the time Limbs entered everyone was in a relaxed mood. Limbs then proceeded on and off the stage in a series of dances & movements which dazzled the crowd. The energy coming forth from this professional group ...) as they worked together was fantastic . . .

The Aerial Railway came next after a short break and they were really something, Their music is mostly original – simple, complex and beautiful. Mathew Brown on piano with tunes like 'Song for Paris' and 'Anytime' must be one of the most talented pianists in N.Z. He doesn't sing too bad either. It's easy to see why this versatile group was allowed to control the small stage at Nambassa and did the music for 'Sleeping Dogs' and the Ed Hillary documentaries.

Next and last in the line was Mahana – a group who put a musical interpretation into a Maori land issue. The music was tight and charged with emotion enough to sweep anyone away just with its intensity apart from its great musical qualities. For those who missed Mahana, Limbs & Aerial Railway – tough, you missed a great show.

A NIGHT AT THE THEATRE: AERIAL RAILWAY & OTHERS
OLD MAID 8pm to 10:30. Ian Bach, *Craccum* 13/3/1978

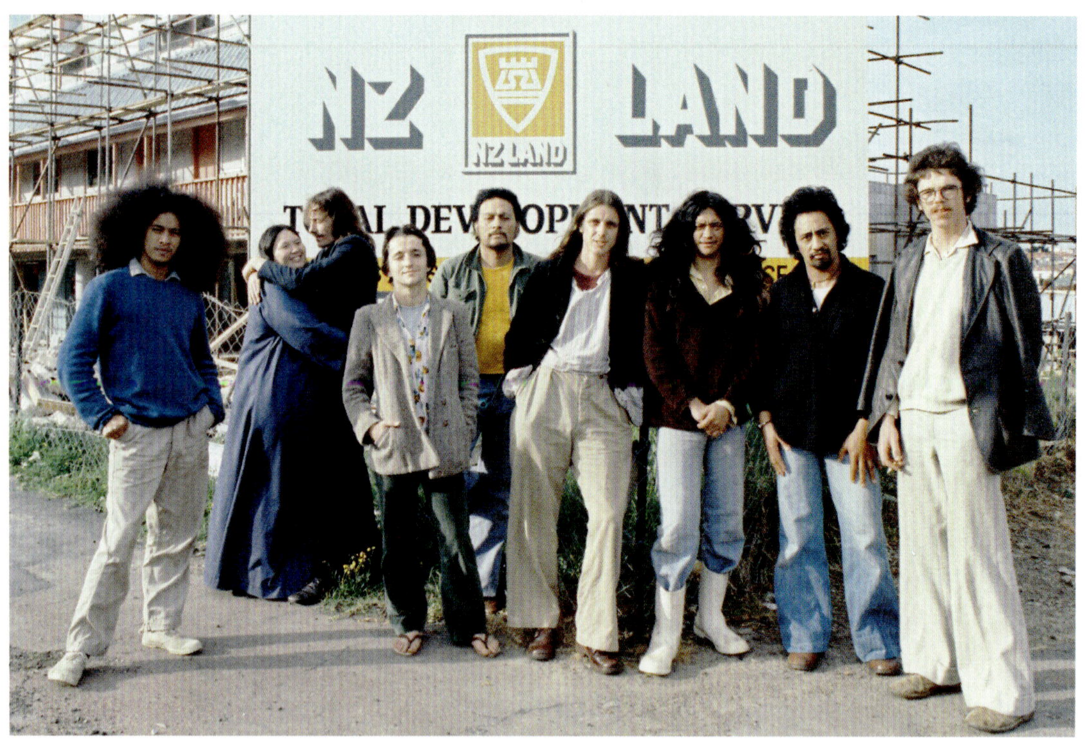

Mahana, Ponsonby, 1978.
Wakatau Kimura, Mahia Blackmore, Corben Simpson, Stuart Crooks, Ara Mete, John Tucker, Bobby Simmonds, Jimmy Dixon, and Cadzow Cossar.

Rehearsing for the Nambassa Winter Show. Matthew Robertson (the Mask Man) preparing to tour, at the St John's Camp, Parau, West Auckland. Podge's and Little Greg's trucks in the background.

Mahana at Nambassa Winter Show in Auckland, 1978. The first theatre production of the rock opera at His Majesty's Theatre. John Tucker, Bobby Simmonds, Wakatau Kimura, Dale Whitcombe, Norma Leaf, Mahia Blackmore.

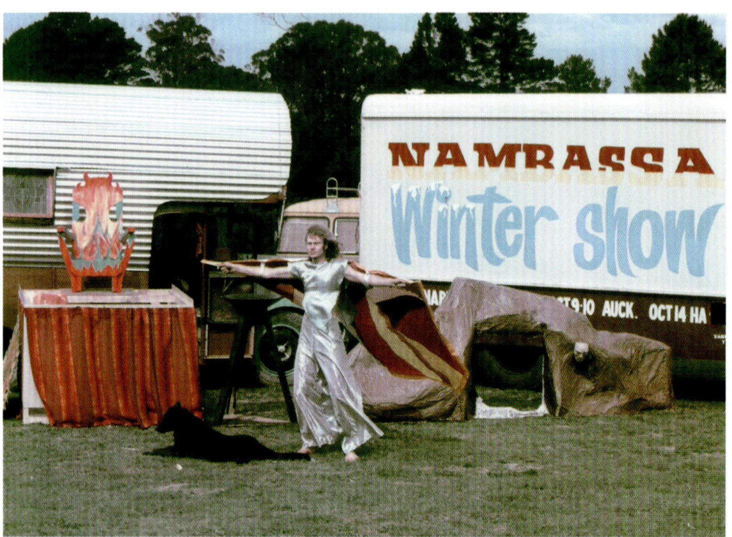

The original Nambassa fund was gone before the winter. The Original Travelling Road Show applied to the government's Youth Initiatives Fund for support in August, and made a tentative booking to take The Road Show to the Northland Arts Festival in November.

Meanwhile, the Nambassa Trust, led by Peter Terry, was rehearsing for the Nambassa Winter Show at the St John's Camp at Parau, West Auckland. Mahana was invited to be part of this North Island promotional tour for the January 1979 Nambassa festival, billed as a tour of 'theatrical extravaganza'. Paul did not accompany the tour, but he photographed the rehearsals at Parau and the Winter Show performance in His Majesty's Theatre in the centre of Auckland City.

There was some disagreement on the North Island tour, en route back from Wellington, between the Nambassa Trust and some members of The Original Travelling Road Show, as Robyn Harris-Iles explains:

> Part of an agreement we made with the organisers of the [Nambassa] Winter Show was that we would put on some free shows at prisons on the way back north. They reneged, so a van full of volunteers left the tour and performed variety shows at the two prison farms (Rangipo and Hautu) near Turangi. When we turned up to invite ourselves at Rangipo a senior prison officer asked where we were staying and we said 'in the van' (there were about 12 or 14 of us) so he took us to his local marae and gave us the use of it for a couple of nights. Beautiful hospitality. Our prison shows were called 'Escape to Anywhere' – a mixture of comedy, song, juggling and clowning around that was loosely put together with a lot of improvisation on the night. When we arrived at Hautu I discovered a music promoter friend was serving time there. He told me after our show that it was fantastic and we should tour it in other prisons as well as in towns.[28]

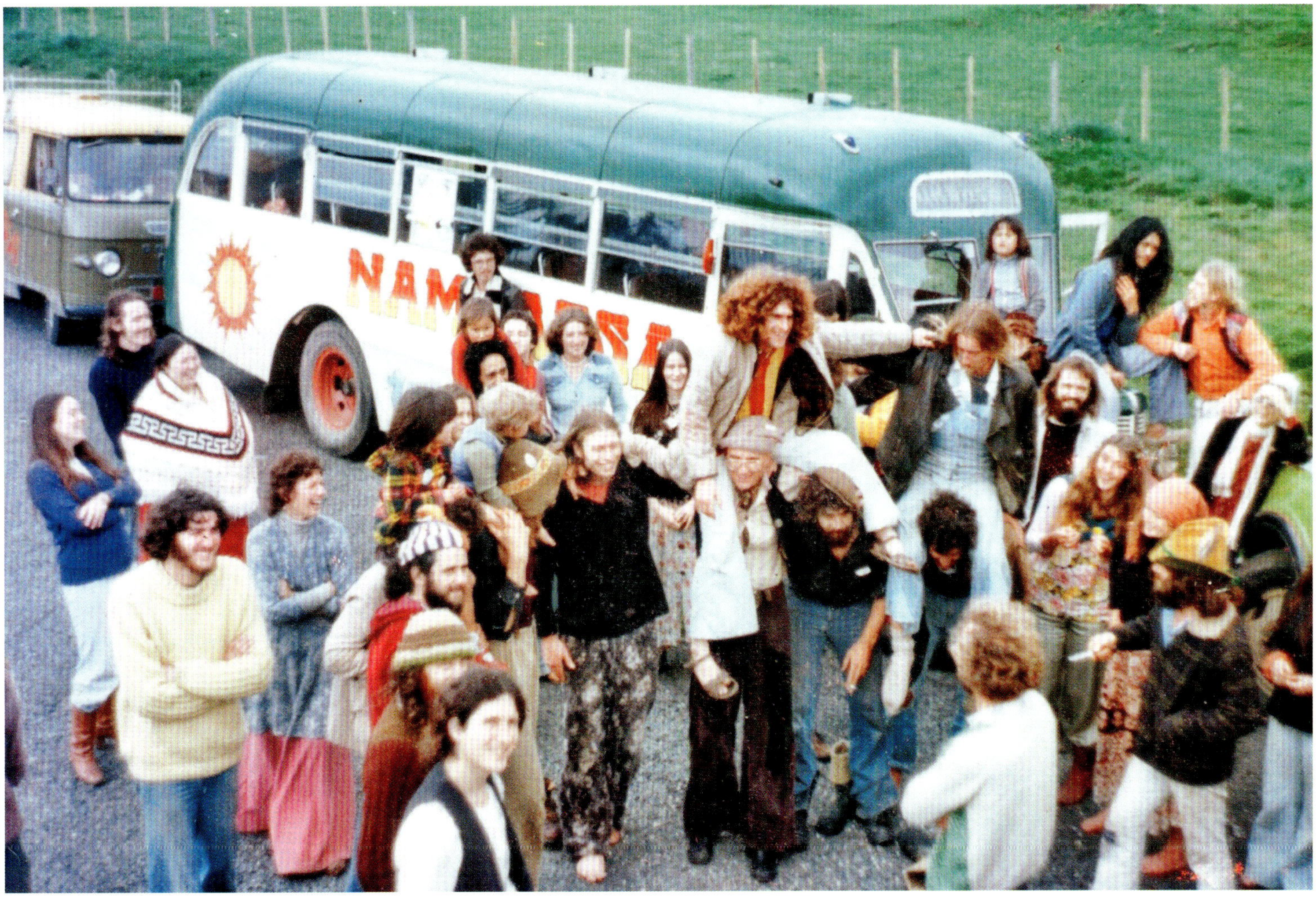

The start of the Winter Show. The crew and performers, 1978. Photo: M.Colonna.

Escape to Anywhere break-away group at Hirangi Marae, Turangi. Matthew Robertson in the middle, Ted Chapman, Little Greg (owner of the Nambassa Winter Show Truck), Robyn Harris-Iles, Brian Tracey (owner of Eager's Washing Van), Pixie Fox and others. Photo: M. Colonna.

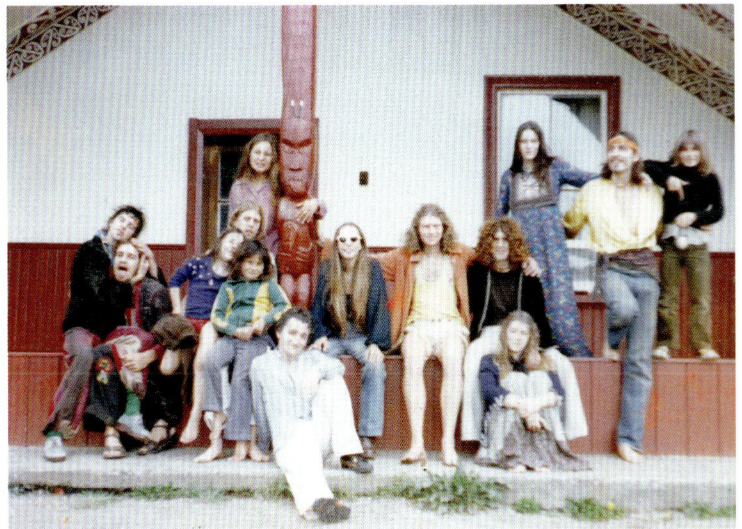

Podge repairing the Nambassa bus with Rob Fryer and Trevor McGlinchie fixing the Bedford bus brakes. The photo was taken at the camp in Parau preparing to go on the road with the Nambassa Winter Show.
Photo: M. Colonna.

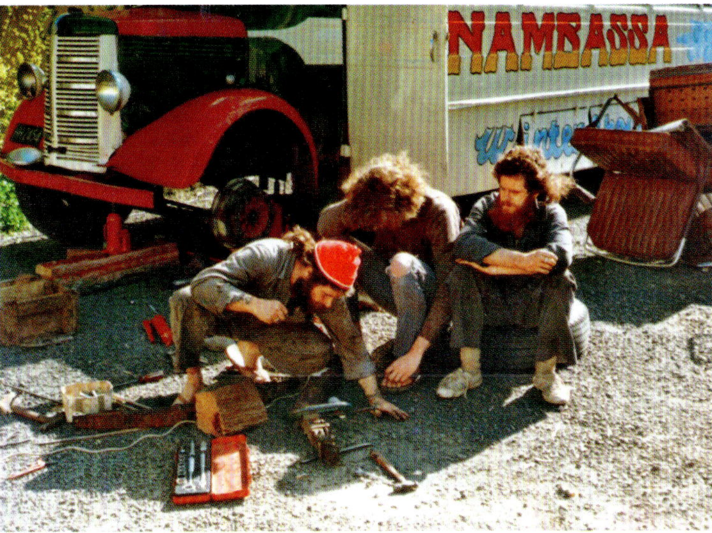

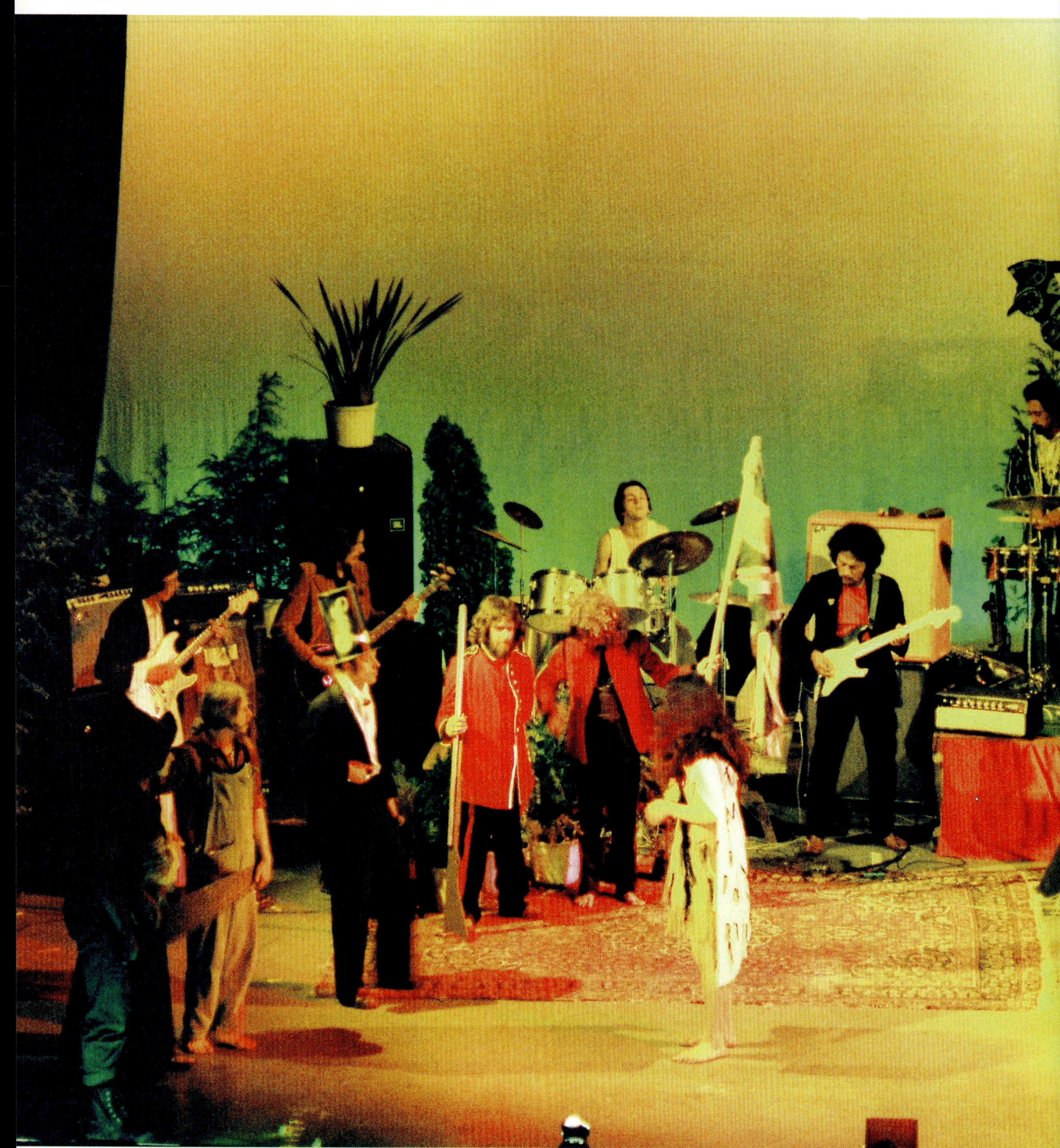

Mahana at Nambassa Winter Show at His Majesty's Theatre, Auckland, 1978. Back row left to right: Cadzow Cossar lead guitar, Corben Simpson bass, Stuart Crooks drums, Ara Mete guitar, Jimmy Dixon percussion & saxophone, Ray Roberts percussion, Ben Grubb keyboards. Front row left to right: Unidentified & Greg Norman, John Tucker, Podge, Ted Chapman, Bobby Simmonds, unidentified, Waka Kimura, Dale Whitcombe, Norma Leaf, Mahia Blackmore.

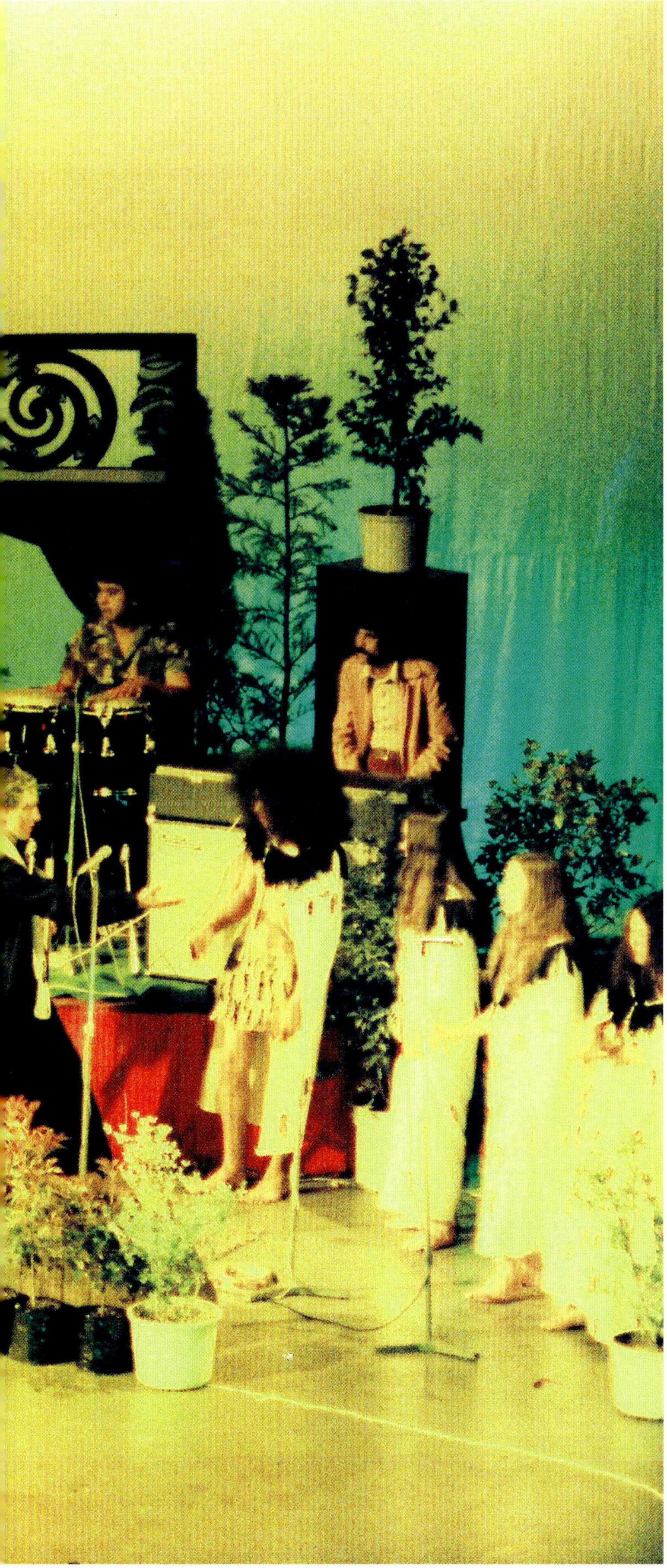

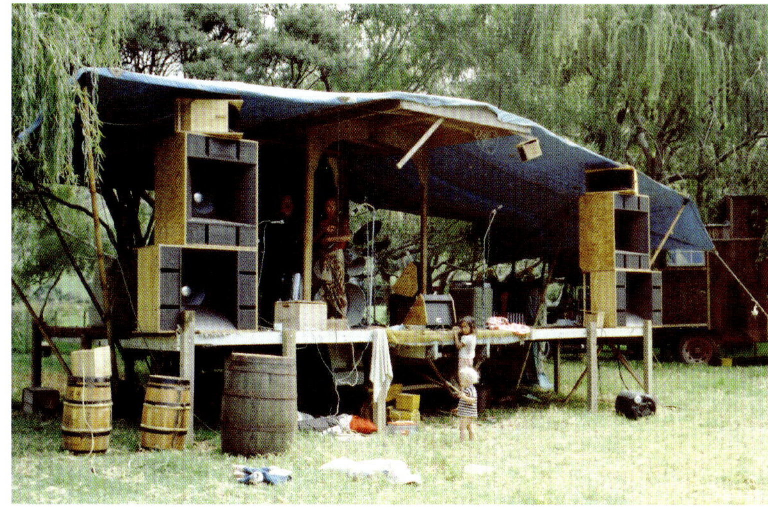

Stage-truck being utilised on tour with The Original Traveling Roadshow, Moehau Community, Sandy Bay, 1979.

Resetting the stage-truck from the travelling mode to the stage mode, Peria, Northland, 1978/79. Photo: anon.

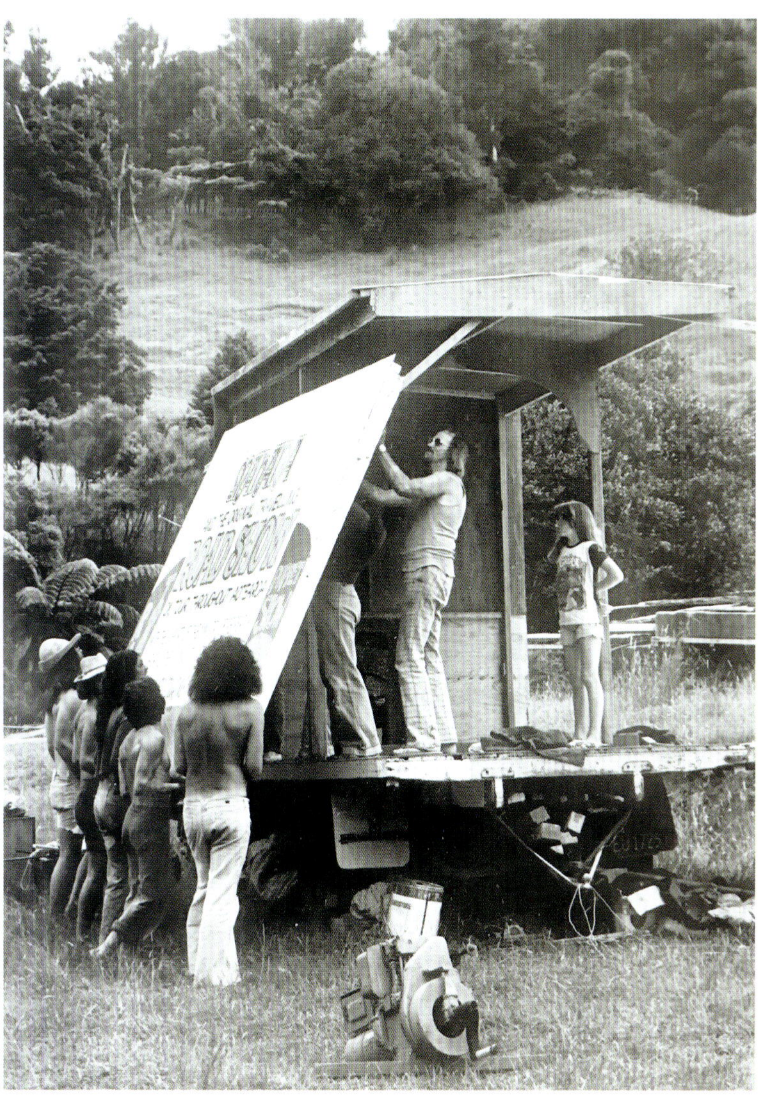

The Northland / Coromandel Tour and 1979 Nambassa

While travelling with the Nambassa Winter Show, John Tucker received welcome news from the Ministry of Recreation and Sport that their Youth Initiative Fund application had been successful. The Ministry would fund $7,000 worth of sound and lighting equipment, according to The Road Show specifications: the equipment would be managed by the Auckland Festival Arts Society, and would be loaned to The Road Show at no charge for the duration of their tour in Northland and Coromandel.[29] With letters of endorsement from the Ministry, Auckland Arts Festival and Nambassa Trust, The Original Travelling Road Show secured its place in the Northland Arts Festival programme. They performed in various venues in Whangarei, including a concert in the Town Hall on 14 November and a six-hour outdoor concert in Okara Park on Saturday 18 November with Whangarei's local musicians and bands LA Hooker, Labyrinth and Cockroach. Security for the concert was provided by members of the Black Power gang. The programme included a workshop held at the Henry the Eighth Cabaret, for the intermediate school kids doing a project on social life and pop music, where Mahana members gave music demonstrations.[30]

The extended Northland tour over the summer of 1978–79 epitomised the lifestyle and ethos of the Road People, with performances in small-town Aotearoa New Zealand sustaining life on the road. The Trunk Fools, with Acorn's Puppet Theatre, travelled separately in their own house trucks, towing the gypsy caravan puppet theatre. They performed at local schools and fairs, occasionally joining forces with The Road Show. They all made a camp base in Fern Flat near Peria, north of the Hokianga Harbour, from which they toured the nearby Northland townships, including performing at Dargaville's Empire Theatre in December, and at Paihia in January 1979. They performed at a mini-festival in Peria and at Kaitaia's Great Far North Music Festival in mid-January, then moved down to Golden Valley in Waihi, Coromandel to join the three-day 1979 Nambassa Festival.

At this largest of the Nambassa festivals (45,000 paid the entry fee, plus unaccounted multitudes of an estimated total of 60,000), The Original Travelling Road Show formed their own self-contained encampment incorporating dozens of house trucks and buses, and named it the 'Road Show Fayre'(p118). Within this large circular enclosure, they managed continuous entertainments, including fire-eating and theatrical events spontaneously provided for children and adults.[31]

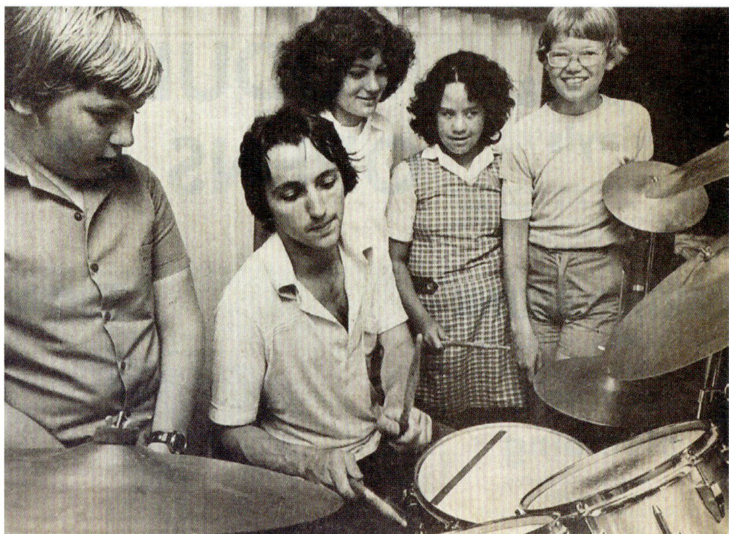

Stuart Crooks gives some aspiring drummers from the Whangarei Intermediate School, a quick demonstration.
Photo: Northern Advocate, 24 November 1978.

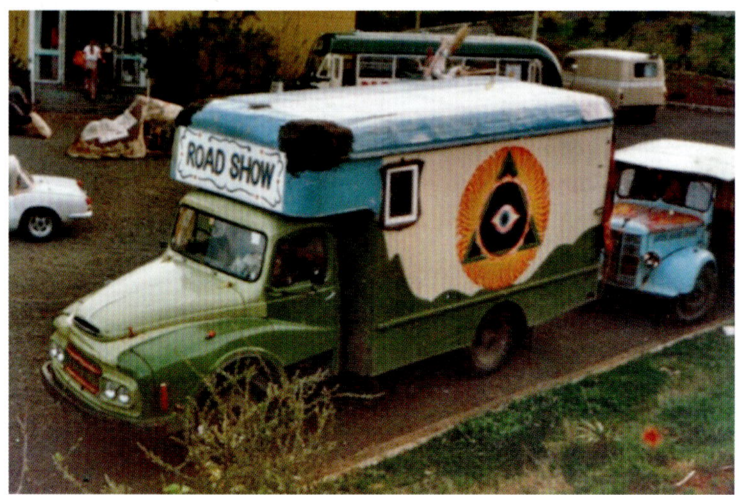

Little Greg's truck, 1978. Photo: M. Colonna.

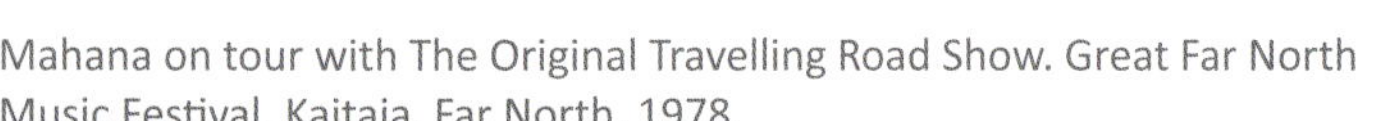

Mahana on tour with The Original Travelling Road Show. Great Far North Music Festival, Kaitaia, Far North, 1978.

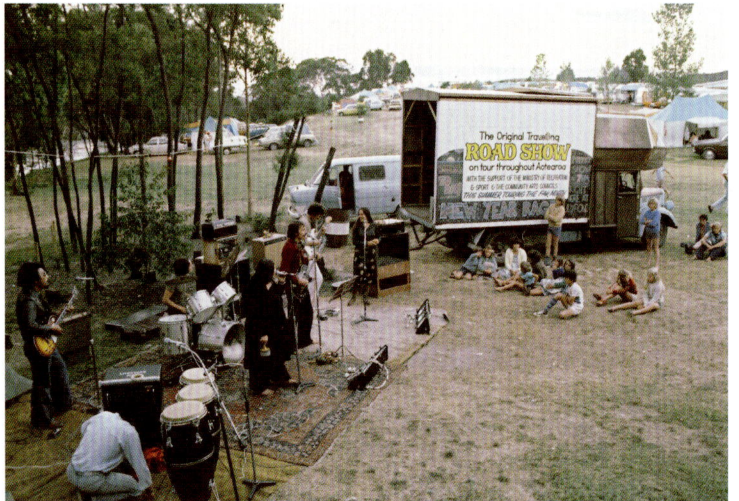

Jonathon Acorn announcing the arrival of The Original Travelling Road Show in Kerikeri, 1978.

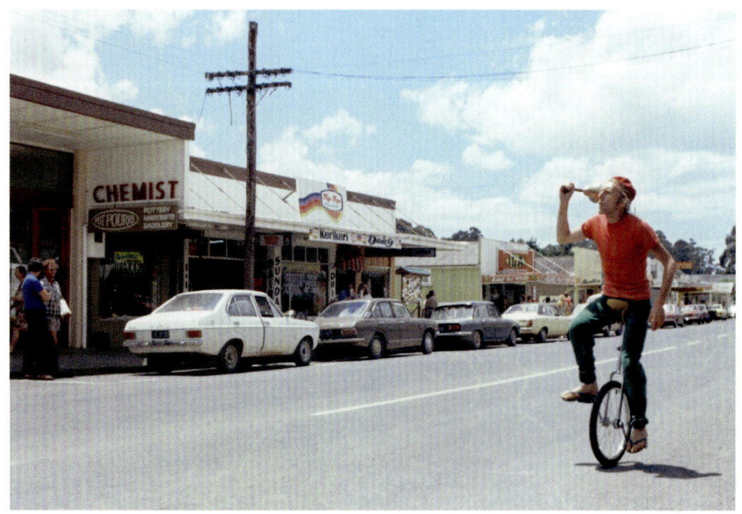

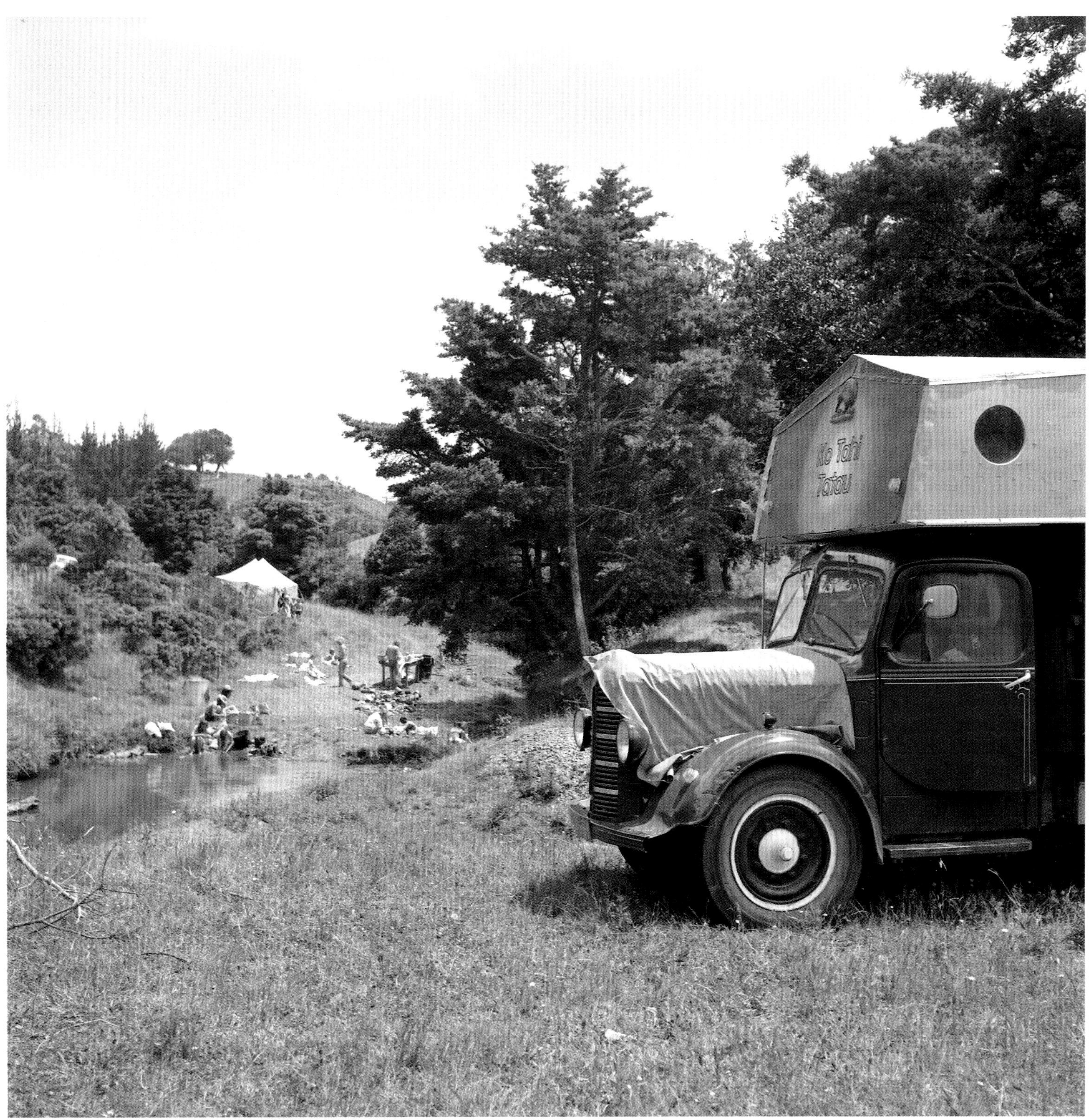

Fern Flat campsite at Peria, Far North (near Kaitaia) with Goober's house truck in foreground. They camped here for four weeks over the Christmas/New Year holiday period, 1978/79.

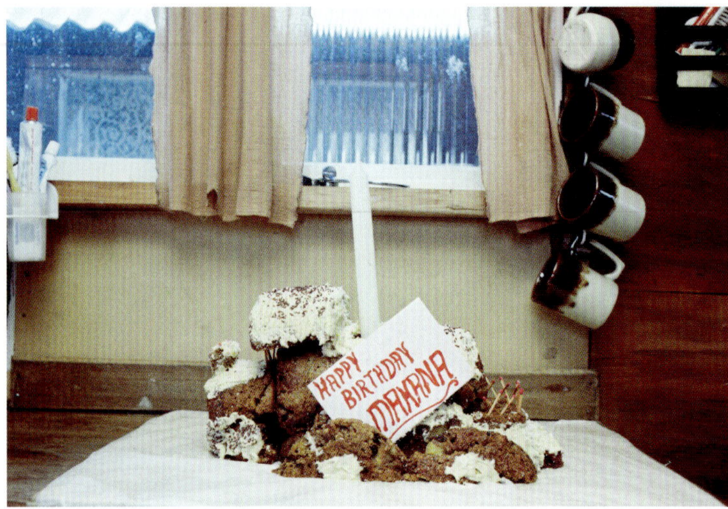

Mahana's first birthday cake, Nambassa, 1979.

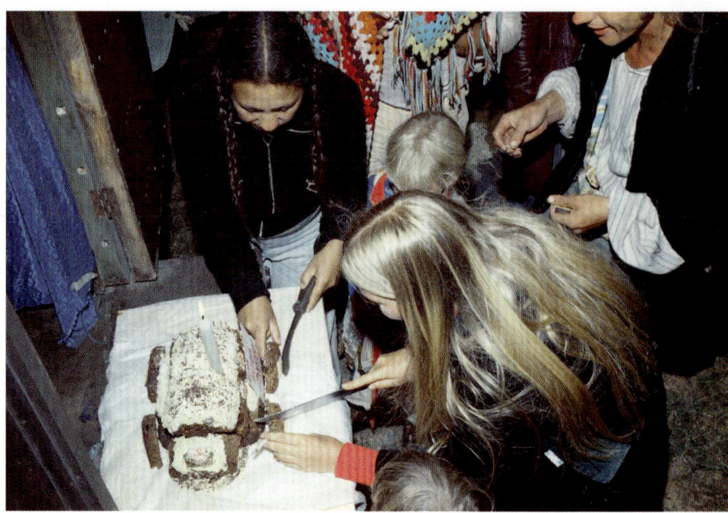

Cutting the birthday cake. Nambassa, 1979.

During the evening of 28th January, the Mahana Rock Opera played on the main stage, in front of an audience in excess of 40,000. The first birthday of Mahana was celebrated with more music and a cake.

The 1979 festival returned a profit of $200,000 to the Nambassa Trust, which announced plans for establishing cooperative communities in Waikato's Waitekauri Valley at the base of the Coromandel Peninsula, with a new school and businesses for up to 50 families. To achieve this, a year was skipped until 1981 for the following festival.[32]

Early 1980s shift of culture, for festivals and road people

After the excitement and success of their work at the 1979 Nambassa, there was a lull in activity. The Road Show did a weekend gig at Moehau Community in Sandy Bay in March, which was billed as The First Annual Coromandel Communities Celebration, but otherwise there is no record of substantial Road Show or Mahana activities until the 1981 Nambassa festival. The prospect of going through another winter downtime may have weighed heavily on the troupe. Their original plan was to establish a Road Show community on land where they could establish a base for planning new acts and preparing for touring shows over summer.

After the 1978 Nambassa, a group of people procured land in northern Coromandel, with the idea anyone was welcome to come and live on the land. Geoff, Brett and Nathan Whitcombe with Alan Wright, were the first to move onto the land, and others followed to establish a commune, Mahanaland, named after the band.[33] When the members of the now disbanded Road Show and Mahana troupe came together for what turned out to be the final Nambassa festival in 1981, they visited with a view to join Mahanaland, which by then, was well established. However, only a small number of the house truck families moved onto the land, due to the steep inaccessibility for their trucks.

The ill-fated 1981 Nambassa Festival after the hiatus in 1980, went head-on with the second Sweetwaters music festival by running on the same dates. Sweetwaters drew 65,000 to Ngāruawāhia, and Nambassa drew less than 10,000 to Waitawheta Valley. It was the beginning of the 1980s shift of the festivals from culturally rich, alternative lifestyle workshops with a spiritual focus, to more blockbuster big-act gigs that depended on drawing large crowds as an entertainment business.

Meanwhile, Dom and Jane, on their house truck journey since 1977, were continuing their nomadic life on the road, attending and helping out where they could at many of the large and small festivals.

> Mid-winter that year (1979) Daniel Keighley found us passing through Waihi and hired us as site workers for the first Sweetwaters Festival. We subsequently spent three months a year for the next 4 years on the two Sweetwaters Festival

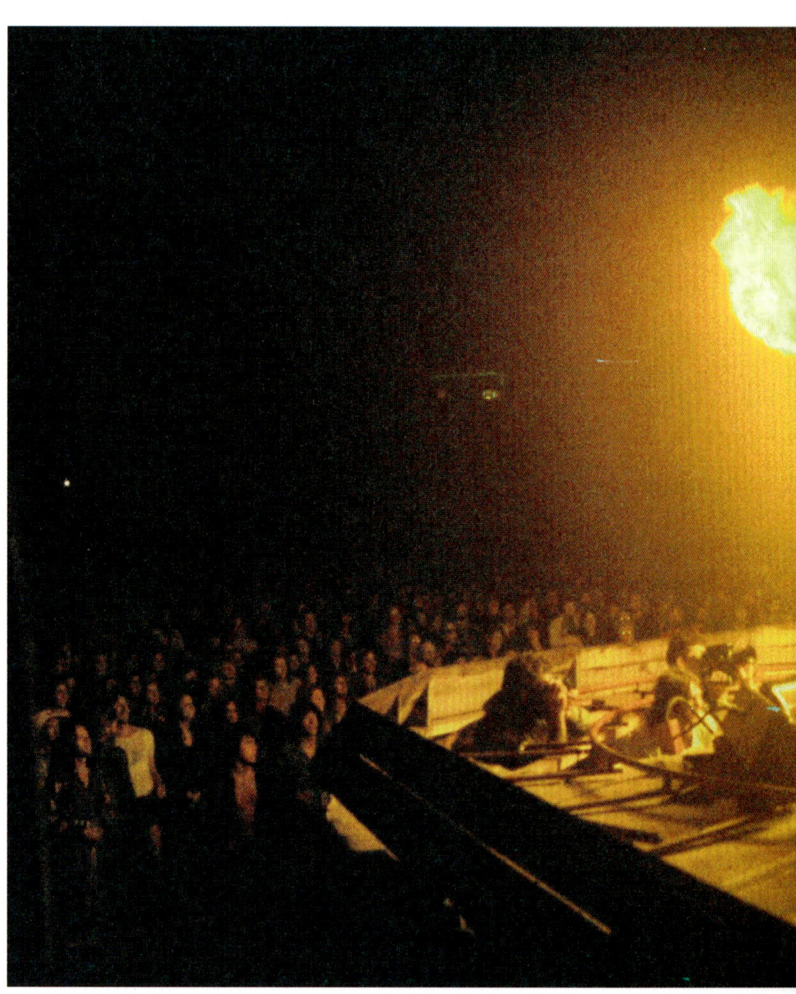

Michael Colonna (as 'Zarak') breathing fire during Mahana gig on Nambassa main stage, 1979.

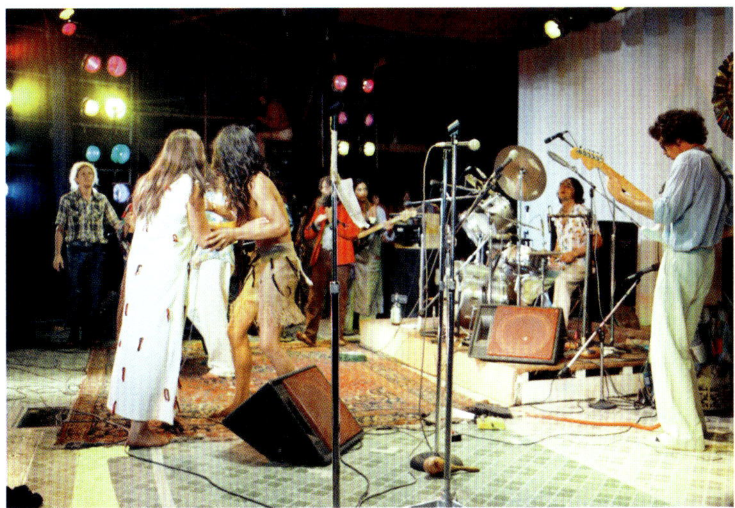

Mahana in full force on the main stage, Nambassa, 28 January 1979.

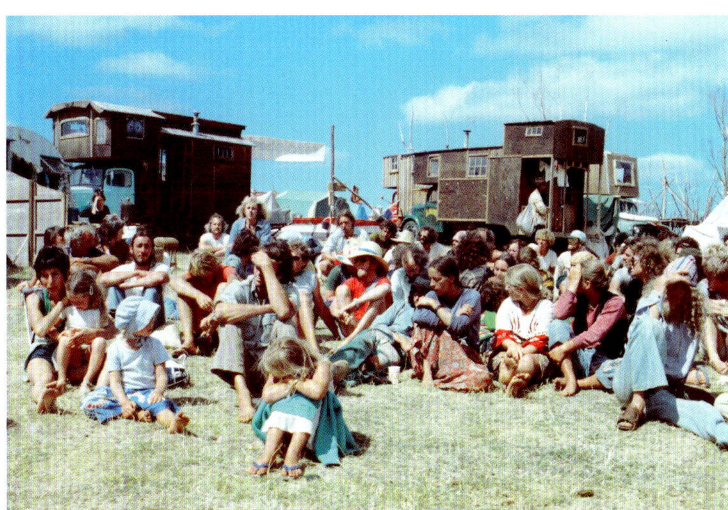

Nambassa Road Show Fayre, 1979.

sites becoming more and more involved in its operation whilst still spending the rest of the time free camping on the side of Kiwi back roads . . . The changing of the guard for the last Sweetwaters in the '80s was dramatic. Previously, although business people, Paul McLuckie and Helen McConnachie were also human and approachable. And Daniel Keighley was one of us. They all lived with us, ate with us, laughed, sat around the fire with us. The days were long, hot and hard

pp168–169 overleaf: Road people arrive to view the Mahanaland community, after the final Nambassa Festival 1981. Only a small number of them, however, moved onto the land in northern Coromandel, due to inaccessibility for house trucks during the Winter. Also, the community's complete open door policy and lack of rules at the time, did not suit everyone's needs.

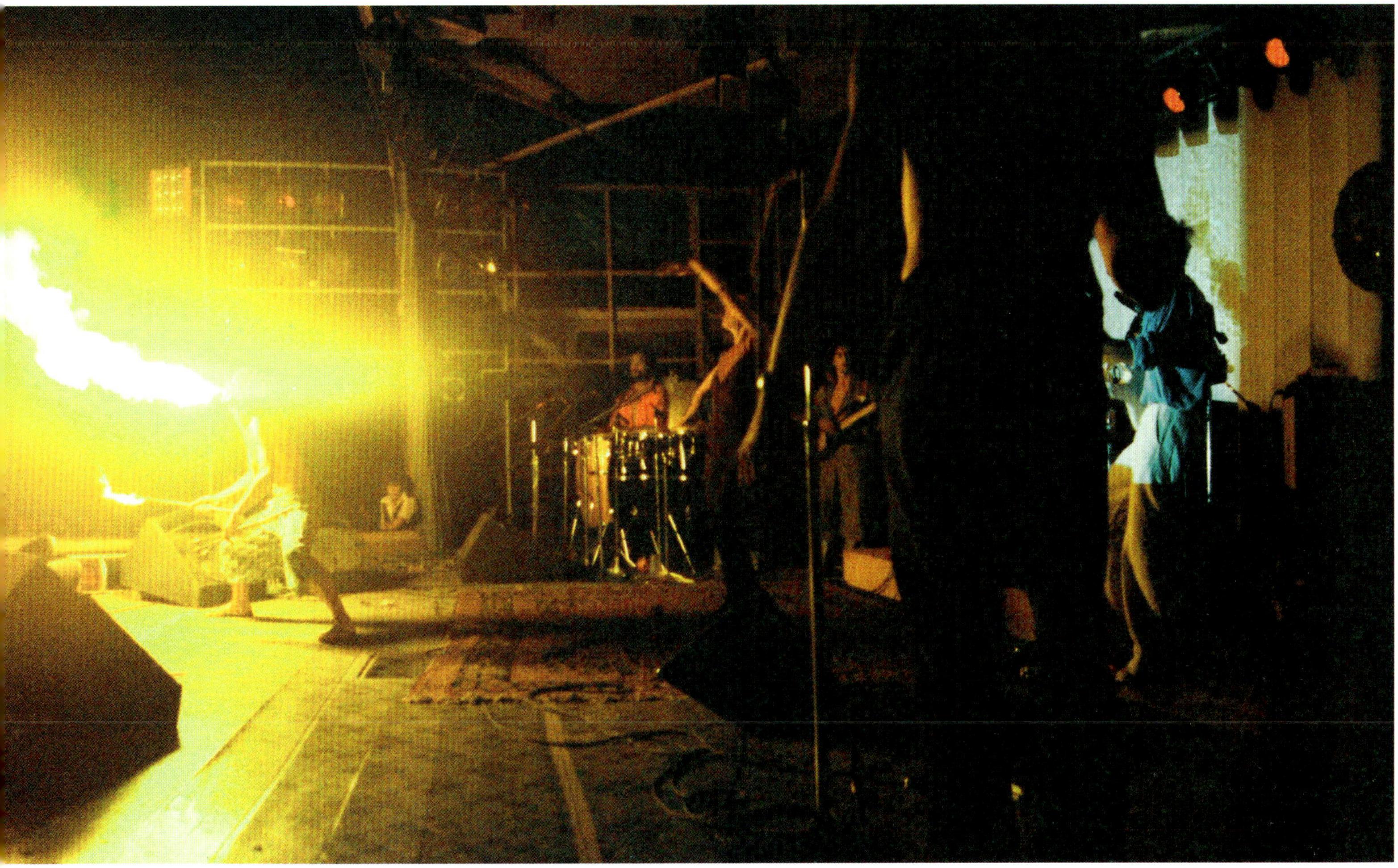

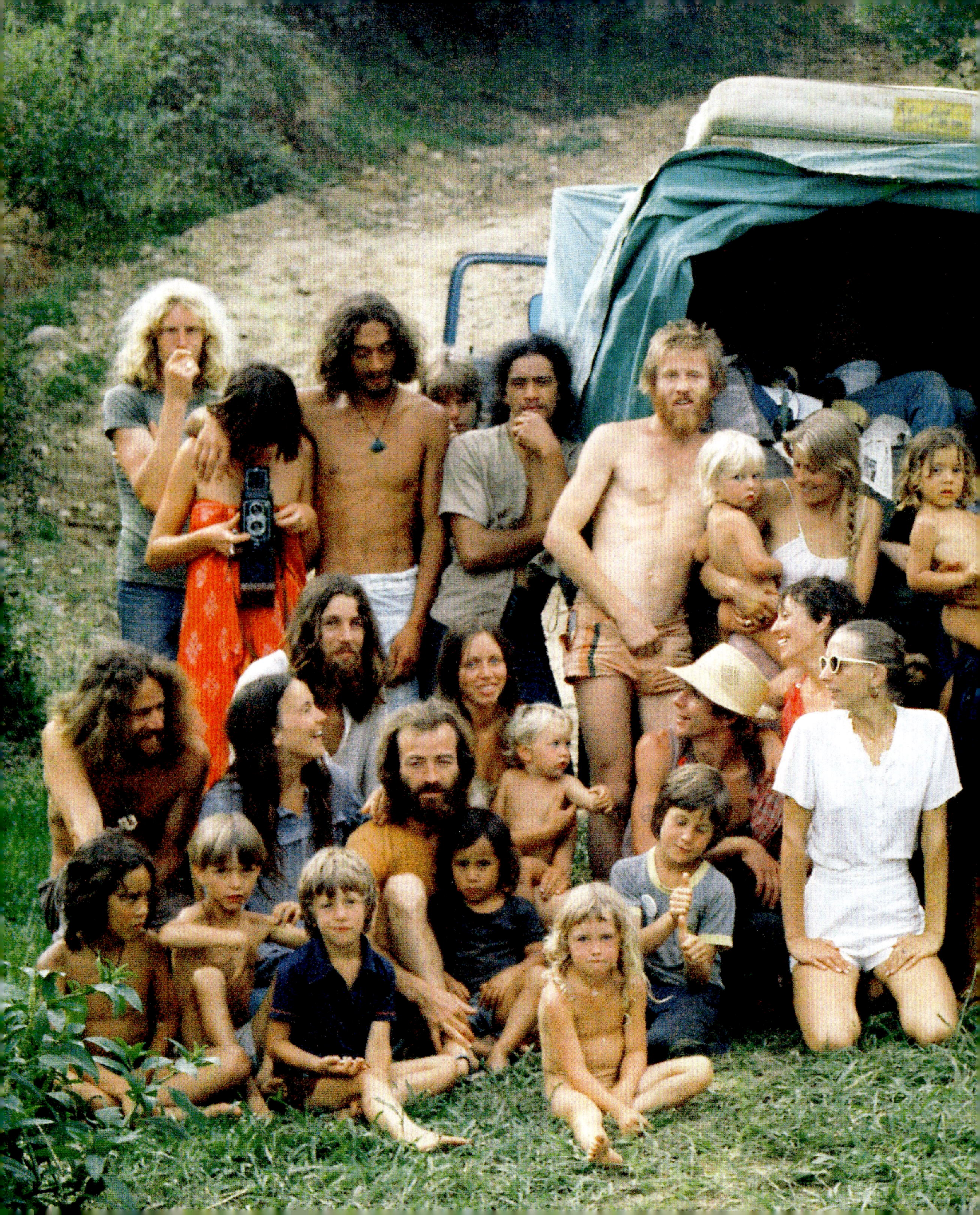

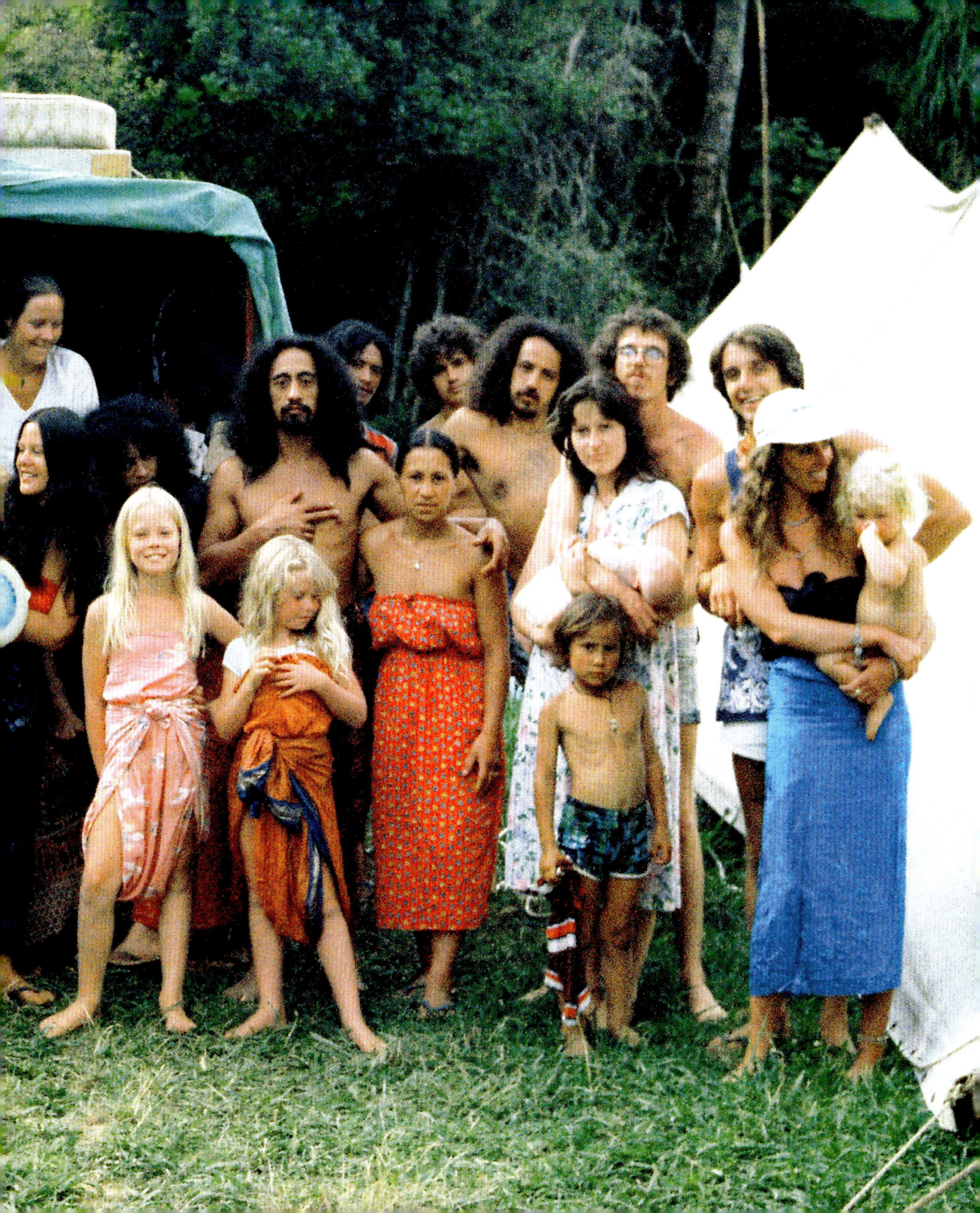

> but it was one big happy family. They always made sure we all got paid even when the first one went broke. When they sold it to a corporate stocks company we understood but all that vanished. The new mob were just accountants bent on recuperating the money they'd invested in their purchase. Which was their demise as after 1984 Sweetwaters became history.[34]

Entertainment troupes: Acorn & Walnut, Midnight Circus, and the People's Circus

As the festivals gave way to the big entertainment business, many of the road people found jobs backstage, and continued their own circus and street theatre activities. Jonathon Acorn built and managed the children's area for the 1981 Nambassa, and the large Woozlebub stage (p131) ran well for the wide range of musicians, clowns and children's acts that entertained audiences throughout the five days. After pull down Acorn set up a two term tour of Waikato schools and hit the road again; travelling first in convoy with a charcoal fuelled Model A Ford and then Michael Colonna's mobile home. Together they developed a weekend stage show called *Acorn & Walnut*, and they produced a series of repertory shows, *Just Walking the Goldfish* and *Lets Rep it Up*. They built a large portable puppet theatre and staged *Sleeping Beauty*, which was officially opened by their patron, the Mayor of Tauranga, and they later toured the show through the Coromandel.[35]

Midnight Circus offered spectacular shows in the Big Tent at the 1982 Sweetwaters Festival. It was organised by the collective core of the road people, with many chipping in as performers. They erected a large circus tent in which The Robinson Family Circus performed their shows during the day, while Midnight Circus performed in the evenings for the adults. (p129)

Brian Tracey, who built the children's playground with Pixie Fox (Alan Rogers) at the 1978 Nambassa and joined the Road Show immediately after as a member of Trunk Fools, and also played the 'Prince of Darkness' in the Nambassa Winter Show, recounts his version of the Midnight Circus:

> The ringmaster was John Tucker. The opening act was myself with a woman I met at the festival dancing to a Tim Buckley song as we changed clothes. Also on the programme was Jonathon Acorn, the Topp Twins . . . The crowd stunner however, was when the magician staggered onto the stage to announce that he had been robbed and beaten. Although I thought it was part of the show, Johnny was more aware and called on the men in the audience to defend the situation. Accordingly the big tent was vacated and outside there were fire blowers and a woman with an axe. But like a tide we retreated into the tent and the show went on, with Acorn being the true professional meeting the needs of the occasion.[36]

Dom Ferry joined up with David Sheridan to form the People's Circus ('without the animals'). As he recounted:

> In the winter/spring of 1982 by sheer coincidence, Dave Sheridan parked his van next to us in Thames. Next to the river over some local fish-and-chips, the People's Circus was born. He went off to Auckland to purchase/fit-out his PC bus and we met up again at Sandy Bay in the hills of the Coromandel to put it all together. The People's Circus did its first run of summer shows on the Whitianga footie oval in Jan 1983. We all went broke and went our separate ways before reuniting in Auckland in spring 1983 where Jane and I became participants in Clown School run by Alan Clay.[37]

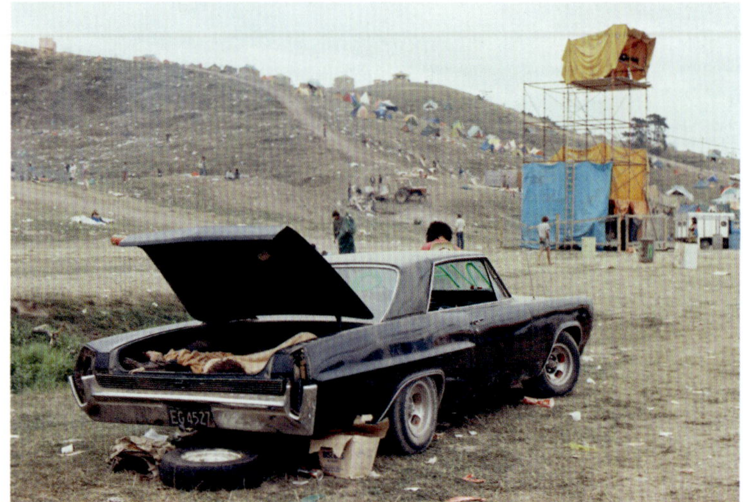

This page: The aftermath of Sweetwaters Festival, 1982.

Paul's 'Iris' bus, Walnut's 'Snail' Bedford K-type, and Acorn's Puppet Theatre trucks, outside Waikato Museum, 1981.

Showtime! Walnut and Acorn, Waikato Museum, 1981.

Opposite page:'Noodle' with candle in mouth intently lighting the footlights, Aratiatia, 1984.

David Sheridan and Stephanie Burns pose at Civic Cinema, Auckland, n.d.

David and Stephanie performing in the Midnight Circus, Big Tent, Sweetwaters, 1982.

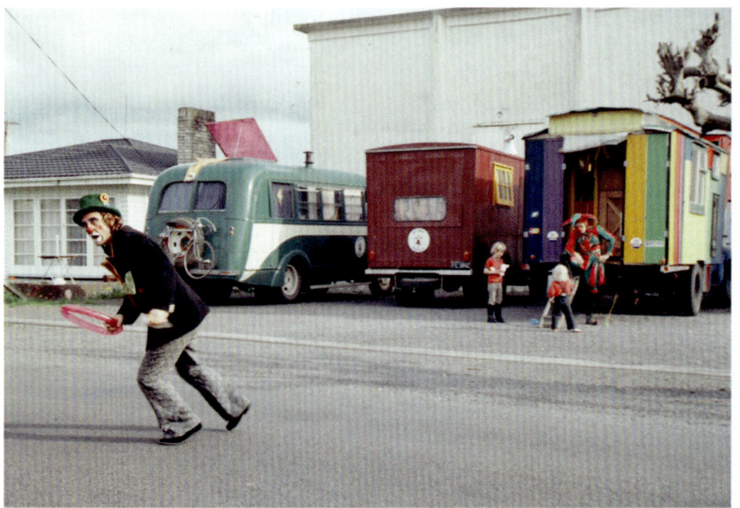

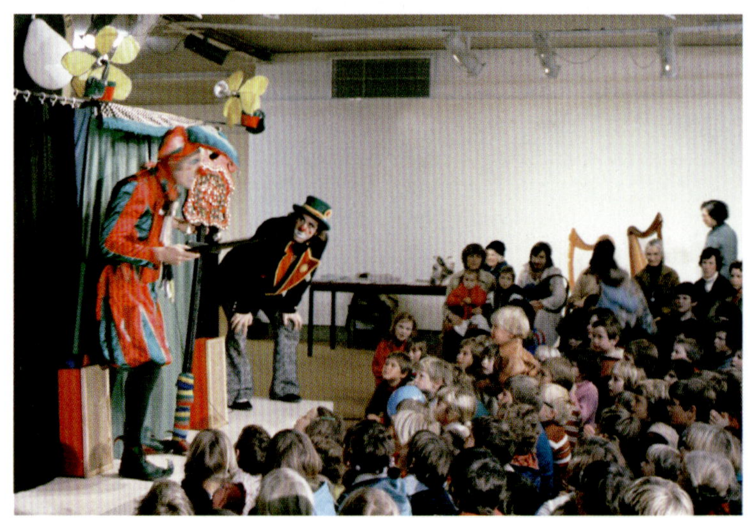

David Sheridan grew up on the rough side of London. His grandfather was a rag and bone man, an early recycler, and from the age of seven David would wag school and help him drive his cart, drawn by Katie, the Shetland pony. He moved to Auckland from the UK in the early 1970s and became involved in the road show/house truck world through the Kipling Avenue communal house. (The Kipling community, which ran the main food stall at Nambassa, had a demolition yard in Auckland to pay the rent, and had a strong connection with the Coromandel communities.) He was taught to eat fire by the Trunk Fools, mastered it and entered the world of performance. He created a very powerful silent fire show (to the music of 'Meddle' by Pink Floyd), and became a valuable part of the Nambassa Winter Show as well as on the main stages of both Nambassa and Sweetwater festivals.[38]

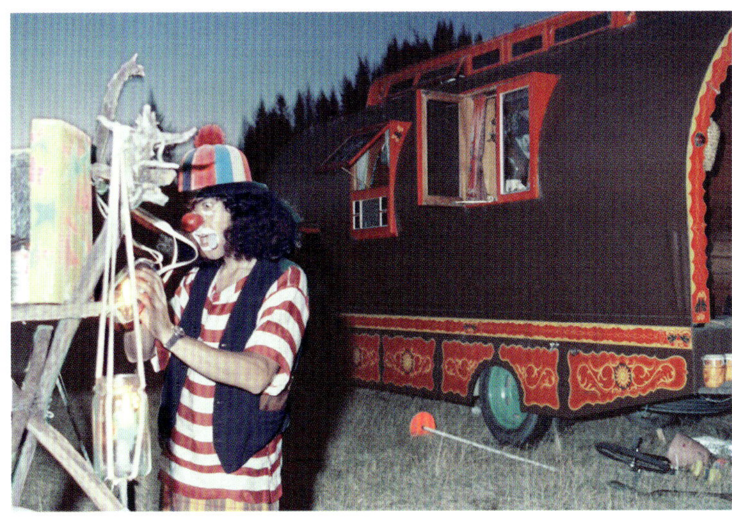

The house truck movement and the Truck & Mobile Shelter Convention, Aratiatia, 1984

In the early 1980s there was a steady increase in homemade house trucks. Apart from the professional entertainers and circus troupes who aspired to live on the road in permanent transience, there were many others with land or a house and disposable income, who made the trucks and travelled around the country on a part-time basis, helping at festivals and attending events. House truckers would converge on house-truck conventions and grassroots festivals of all themes that were held throughout New Zealand. They came not only for the event but also for the opportunity to connect and share information with other house truckers from across the nation.[39] Dom Ferry recollects:

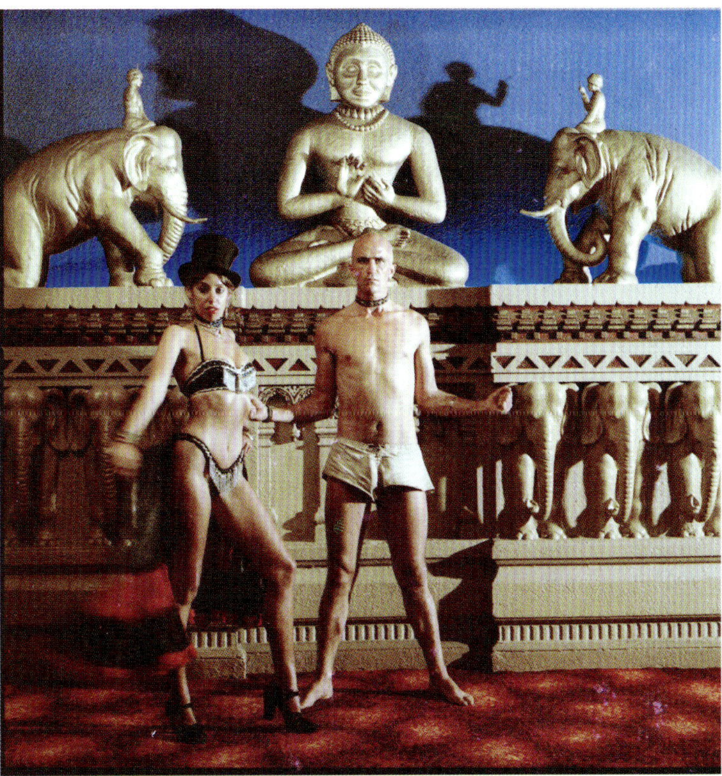

> As the years rolled on Jane and I realised that although there were other house trucks around, they weren't on the roads. Whereas we were. We got a lot of media attention, featuring in every local newspaper from Coromandel to Wellington. We also got a lot of police attention. We followed the traditional gypsy survival methods of fruit-picking for three months each year as the main income. We cooked daily on an open fire regardless of the weather and even in big cities (we didn't have a gas ring cooker inside the truck until 1983). We developed a constantly evolving roadmap of free-camping highway/back road parking spots between the south of Auckland and Wellington that governed the directions of our roamings. Only on the festival sites did we see others living like us. And meanwhile, the house truck movement slowly grew in numbers. One house truck parked up is cute, romantic, non-threatening. Ten and suddenly the authorities are out in force moving you on. It took us years to understand that without a strongly stated purpose (circus, road show, mobile library, travelling barn, whatever) sedentary communities freak out when the convoy rolls in. As we started to band together more and more, we began to realise that we could utilise the attention we generated to our advantage. And so the seeds for the People's Circus and the Gypsy Fairs were born as well as the possibility that performing could put fuel in the tank and food on the table.[40]

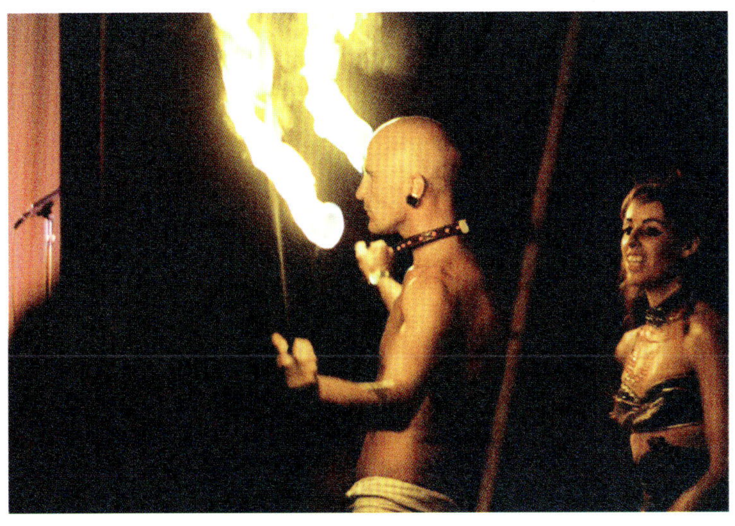

In 1980, Terry St George published *Road Lore*, a handbook for mobile house owners. The book outlined typical problems encountered by mobile home owners confronting the law. It explained basic traffic regulations and local council bylaws to make owners aware of their responsibilities and to help them avoid potential problems. Terry wrote to 39 councils in New Zealand to ascertain their attitudes to mobile caravans and to find out what individual council regulations were, and published their replies. His guide also gave useful suggestions for mobile house owners wanting to set themselves up as a group to establish their alternative lifestyle.[41]

The last event Paul Gilbert photographed for his *Road People* project was the House Truck & Mobile Shelter Convention held at Aratiatia near Taupō on 18–21 January 1984. This gathering was organised by David Sheridan, Dom Ferry, Jane Mitchell and Rob Tee, while they were all wintering at Moller's barn at Oratia. Dom Ferry recollects:

> Aratiatia was purely a gathering point for the ever-growing mobile tribe. Effortless to assemble as there was no need for security/fences/stalls/entertainment contracts etc. Everyone who came had their act together. We weren't running around nursing 60,000 city kids. It was all about cementing ties and learning about each other. New Zealand Broadcasting Corporation came in with a film crew which resulted in a five-minute documentary that got frequent national TV airplay. Rob Tee and I were the only ones who dared talk to the cameras. It felt right at the time as it wasn't just a few of us driving around NZ anymore. There wouldn't have been more than two hundred people there at any one time. And honestly, we were rapt with that as, with only 'word of mouth' for advertising, we thought it was going to be twenty of us around the fire. It was the baptism of the community that we'd become. That we'd grown big enough not to rely on the big festivals as rally points any more. It ran for four days and the following weekend about ten mobile units hit Taupo and did a 'Gypsy Fayre' in the heart of town.[42]

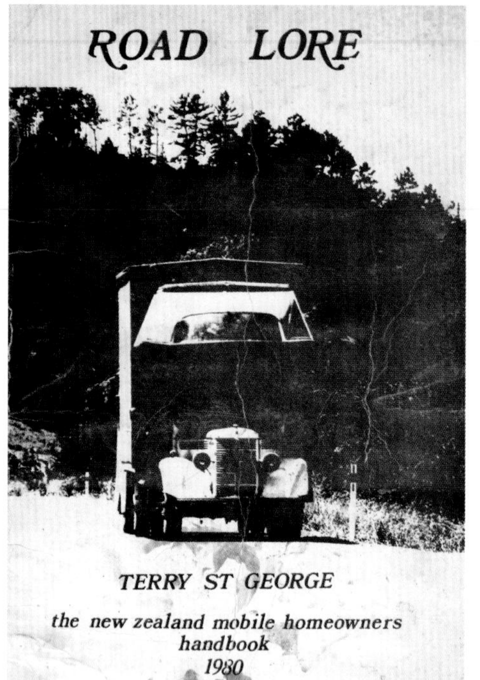

Road Lore, which featured Paul Gilbert's photos and research on the legal status of house trucks parked up on land in the 1980s in New Zealand.

Letterhead of the Flyer/ Invitation for the Aratiatia gathering.

Right:Dom and Jane's truck parked alongside David Sheridan's People's Circus bus, Aratiatia, 1984.

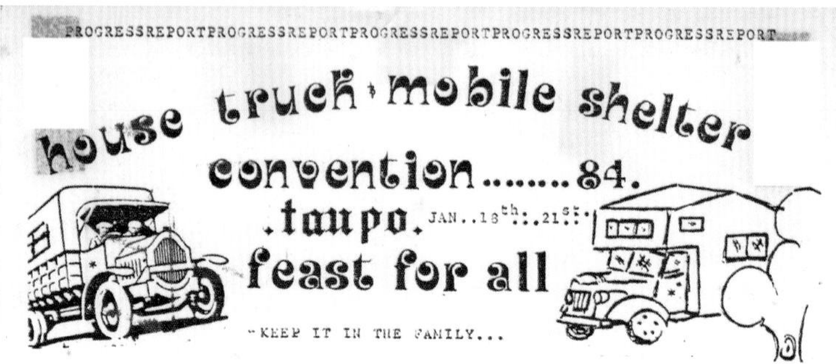

Dom and Jane sold their house truck in early 1985 and headed overseas to pursue their performing career, touring the cities and festivals of the world for many years as international street theatre performers. What they started as Gypsy Fayre in Taupō continues to this day in another form, as The Original Gypsy Fair,[43] a convoy of travelling house trucks offering townships throughout the country a choice of fairs, a merry-go-round, crafts and trinkets. It was officially registered as a business in 1990. More recently, The Extravaganza Fair[44] incorporated in 2015, continues the tradition of road people, by offering circus and entertainment troupe in house trucks, travelling across the country.

Many of the Mahana members went off to pursue their own careers in music. Cadzow Cossar has continued to play music professionally, and today he plays lead guitar with Moana and the Tribe who, prior to the Covid pandemic, toured extensively internationally. John Tucker also played around Auckland as solo performer Johnny Spacific, and has enjoyed a long career providing transport solutions for New Zealand's touring music groups, festivals and the film industry. In the early 1980s, Mahia Blackmore began an all-women band, Meg and the Fones, and one of the songs she wrote and recorded with the band, Little Tui, won the 1986 APRA Silver Scroll Best Song award. Mahia was also part of Ladies Sing the Blues, and provided backing vocals for Dame Kiri Te Kanawa on her *Māori Songs* album. Corben Simpson lives in Tauranga and plays at local venues and events with a variety of musicians, including Maurice Greer (Human Instinct), Billy TK (Human Instinct, Powerhouse) and Ara Mete (Powerhouse, Mahana). In 2019 Corben was invited to close the APRA Silver Scroll award ceremony with a performance of 'Dance All Around the World', the song he wrote for Blerta in 1972 with Geoff Murphy, that was inspired by a Margaret Mahy children's story. Stephen Kereama works as a session musician in Queensland, Australia.[45] Jonathon Acorn went on an international tour of street theatre and busking in many cities and festivals in Europe, as well as starring in film and television in New Zealand. David Sheridan returned to London in 1985 and spent a year there developing a chain-escape street performance with the help of Captain Keano and other Covent Garden performers. He returned to New Zealand in late 1986 to perform a regular busking act in downtown Auckland, to great response. From the street limelight, he went on to feature in many films, television shows and commercials as well as paid appearances at festivals, fairs and cabarets. Michael Colonna travelled with a number of circuses, toured the country 14 times performing in schools. He created and sold balloon sculptures at the Great NZ Craft Show circuit. He is now the editor of Variety Artists Club newsletter, which features many New Zealand entertainers.

> The concept of the Road Show was a mobile community of entertainers. Later the idea was picked up by a circus that got rid of their animals and instead just had human acts. For me, it was all about community, interrelationships and New Age Consciousness. All pretty trippy stuff now. Life, love, women and children came along and ruined everything (lol). Our kids are all chasing the material now; a house, car and television set! Good luck to them, I say.
>
> — Michael Colonna[46]

When I first met Paul Gilbert in the late 1990s at Elam, he was living primarily in his beautiful wooden boat. A dedicated and skilled mariner, he had sailed and photographed many parts of the New Zealand coastline, and when he was working at Elam, he moored in various harbours around Auckland. He remained a free spirit all of his life and lived pretty much true to the ethos of the road people he documented all those years ago. His memorial in August 2019 – appropriately, according to his wish – was held on board the rigged sailing scow *Jane Gifford* on Mahurangi River in Warkworth, and many of the road people turned up to pay tribute to their much loved old friend.

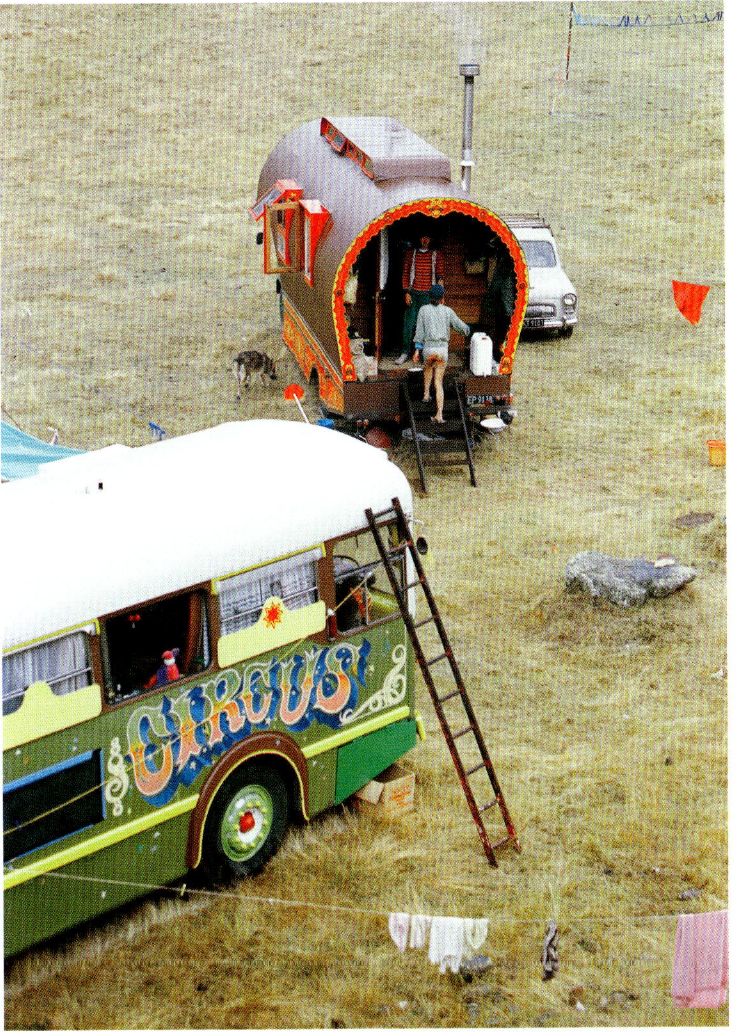

Notes:

1 This was about the same time as John Britten (1950–1995)'s conversion of the 1926 International truck in the South Island. The iconic Christchurch motorcycle engineer, restored and built a Romany inspired cabin on the chassis of 1926 International truck from 1972 to 1975, and travelled with it around the West Coast of the South Island, studying seabird flights. His truck is on view at the Geraldine Vintage Car & Machinery Museum.

2 Robyn Harris-Iles, email correspondence with H Sameshima, 28 August 2020.

3 'Alice's' was a large house on Tinakori Road in Wellington, named after the Arlo Guthrie song 'Alice's Restaurant'. For three years from 1974 to 1977, there were twenty or more people living there communally at any one time . . . 'Mt P' was a series of three neighbouring five-bedroom houses on Mount Pleasant Road at the end of the Aro Valley in Wellington, from 1975 to 1978. The houses and several others nearby on Holloway Road were communally linked: the occupants were kindred spirits who shared garden produce and music. Dom Ferry, email correspondence with H Sameshima, 5 March 2021.

4 Dom Ferry, email correspondence with H Sameshima, 28 August 2020.

5 Andrew Schmidt, *AudioCulture*, 15 March 2014, www.audioculture.co.nz/scenes/the-nambassa-festivals-and-the-counterculture-movement

6 Intentional communities is a community formed on the basis of shared belief (religious, political, social or environmental) by people from more than one family who come together to live cooperatively. Caren Wilton, 'Communes and communities', Te Ara - the Encyclopedia of New Zealand, http://www.TeAra.govt.nz/en/communes-and-communities/print (accessed 25 January 2021)

7 Tim Jones and Ian Baker, *A Hard-Won Freedom: Alternative communities in New Zealand*, Hodder & Stoughton, Auckland, 1975.

8 10,000 copies of *Nambassa Newsletter 1* were printed in 1976 by Goldfields Press in Paeroa and distributed free around New Zealand. Michael Colonna was handed one when he came out of a concert at the Auckland Town Hall around October 1977. He quit his job, and went to Golden Valley to help set up the festival.

9 Peter Terry, quoted in Robert Jenkin, 'Endless connections: New Zealand secular intentional rural communities founded in the 1970s', MA thesis, Massey University, 2012, p. 105.

10 Colin Broadley and Judith Jones (eds), *Nambassa: A new direction*, Reed Books, 1979, p. 117.

11 Michael Colonna, email correspondence with H Sameshima, 10 March 2021.

12 Blerta (Bruno Lawrence's Electric Revelation and Travelling Apparition) was a musical and theatrical co-operative active from 1971 until 1975: https://en.wikipedia.org/wiki/Blerta

13 Terry St George, email correspondence with H Sameshima, 25 September 2020.

14 Colin Broadley and Judith Jones (eds), *Nambassa: A new direction*, Reed Books, 1979, p. 88.

15 Ibid., p. 89.

16 Three slots for three days were offered; 28, 29, 30 January, 7:30–9:00am for acts for children, 10:30–6pm and 8:00–10:00pm. According to the invitation letter from Aerial Railway Music Co-op to John Tucker.

17 The band was initially named Palmerston North Theatre Troupe in the application to Nambassa.

18 Robyn Harris-Iles and John Tucker, email correspondences with H Sameshima, 20–28 August 2020.

19 *Nambassa Winter Show* programme, 1978, private collection.

20 Wayne Munro, *8 o'clock* (newspaper), 20 May 1978, clipping from John Tucker's scrapbook.

21 Michael Colonna, 'House Truckers (Gentle Revolution)', unpublished manuscript, private collection.

22 Wayne Munro in *8 o'clock*, 20 May 1978, newspaper clipping from John Tucker's scrapbook.

23 Michael Colonna, 'House Truckers (Gentle Revolution)', unpublished manuscript. Private collection.

24 Jonathon Acorn, in Alan Clay, *Angels Can Fly: A modern clown user guide*, Artmedia Publishing, 2005, p. 122.

25 Terry Bell, 'Trying Hard to become a Fool', *Women's Weekly* clipping, n.d., private collection.

26 Michael Colonna, 'The Original Travelling Road Show: Pine smoke. Feb 1979, unpublished manuscript from February 1979, private collection.

27 John Tucker, 'Open letter to all those people who have, at some stage or another, been associated with The Original Travelling Roadshow', 1978, private collection.

28 Robyn Harris-Iles, correspondence via Flickr comments section, 20 September 2020.

29 Letter from the Ministry of Recreation and Sport, 16 October 1978, private collection.

30 *Northern Advocate*, 'Novel twist to school work', 24 November 1978, p. 12.

31 https://en.wikipedia.org/wiki/Nambassa_Winter_Show_with_Mahana

32 'The Nambassa Festivals and the counterculture movement', Andrew Schmidt, *AudioCulture Iwi Waiata,* 25 March 2014

33 Dale Whitcombe, Flickr comments section. 5 October 2020

34 Dom Ferry, email correspondence with H Sameshima. 28 August 2020.

35 Michael Colonna, 'Jonathon Acorn: How many puppeteers does it take to change a light bulb?' n.d., unpublished manuscript in the author's private collection.

36 Brian Tracey, email correspondence with H Sameshima, 22 March 2021.

37 Dom Ferry, email correspondence with H Sameshima, 21 August 2020.

38 Dom Ferry, email correspondence with H Sameshima, 25 March 2021.

39 https://en.wikipedia.org/wiki/Housetrucker

40 Dom Ferry, email correspondence with H Sameshima, 28 August 2020.

41 *Hauraki Herald*, Saturday 9 February 1980, p. 35.

42 Dom Ferry, email correspondence with H Sameshima, 28 August 2020.

43 https://www.gypsyfair.nz/

44 https://extravaganzafair.co.nz/

45 Robyn Harris-Iles and John Tucker, email correspondence with H Sameshima, 22–24 March 2021.

46 Michael Colonna, email correspondence with H Sameshima, 23 February 2021.

Paul Gilbert remembered

– John B Turner

Reflecting on what Paul Gilbert achieved in his tragically short 64-year lifespan and how he did it, I was reminded of the teaching philosophy of Bill Jay, the dynamic photo historian: that the purpose of life is 'to become actually, what we are potentially'. Paul was an 18-year-old teenager when I first met him around 1972, and I was immediately struck by his passion for life, his open curiosity and independent spirit. I sensed that his character, his sensitivity and his dark humour had all been tempered by personal hardships. And as a voracious reader who couldn't afford university study, he was exactly the kind of person the 10-day photography workshop I was organising for the University of Auckland's Continuing Education and Fine Arts departments in January 1973 was intended for.

Jump to 2019, when he was facing imminent death from cancer with great equanimity, he wrote that only two things matter about photography:

1. It's where you stand.

2. It's when you trip the shutter. Practice refining this for as long as possible.[1]

Paul wasn't just thinking of photography – he was articulating a working philosophy based on his experience and the decisions he made about what kind of life he wanted for himself and his children.

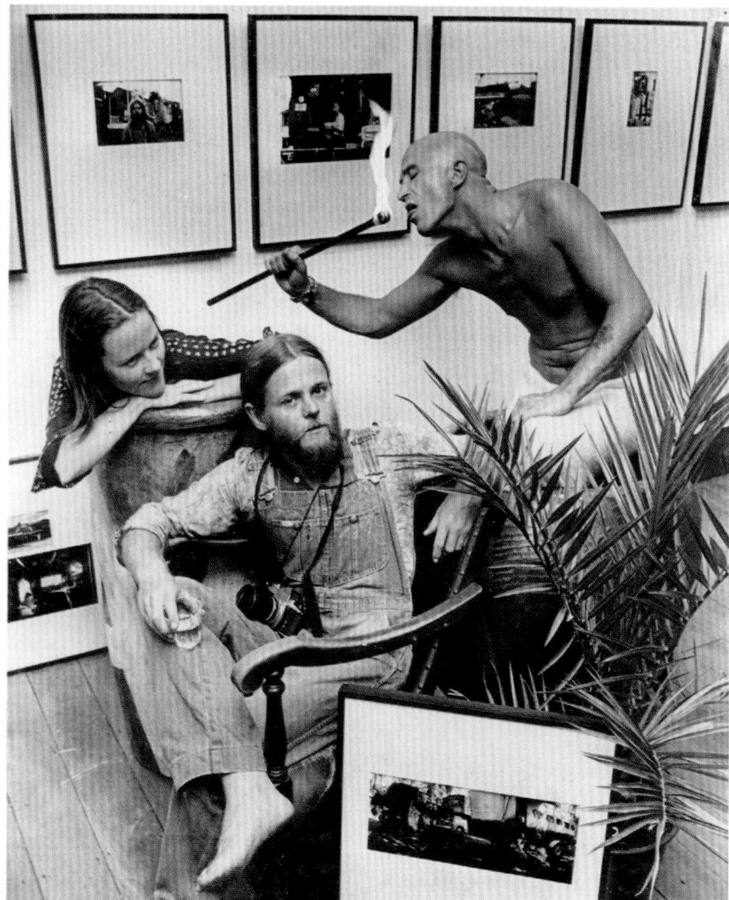

Paul Gilbert, with Monica Simmonds and Zingaro (David Sheridan) at the exhibition *Road People of Aotearoa* at Real Pictures Gallery, Auckland. Photo: Murray Job, Auckland Star, 20 December 1979.

It was the potential to shape one's life through personal enthusiasms, critical thinking, and a practical mix of idealism and perfection that he strove to live up to, both in his art and in daily life. It was seldom an easy road that he took, but it was one in which he became a remarkably skilled photographer, a devoted sailor, an outstanding teacher and a better person. His mother was an amateur artist with a passion for photography – an interest they shared. 'I have always been grateful that Mum passed on her interest in photography, which, at age seven, along with a boat trip, cemented two combined lifelong passions,' he wrote. 'I have lived aboard yachts, photographed them, and enjoyed that wonderful lifestyle that our sparkling waters provide.'[2] Paul's sense of perfection made him a valued teacher at the Elam School of Fine Arts, although the standards he set were not always appreciated by his students.

Paul's parents separated in 1968, and after the trauma of his mother's early death in 1974, Paul took on the responsibilities of being a substitute parent to his school-aged siblings, while earning his living as a trainee photographer at the Department of Scientific and Industrial Research in Auckland. He was exhausted by his life of parenting and frugal independence, and when the opportunity for a break for some rest and recreation presented itself in 1975, with the financial support of family and friends paving the way, he took the huge step of joining a Canadian couple on a three-month road trip from Mexico to Canada via the United States.

The timing was perfect. In what turned out to be a boost to his flagging energy and an introduction to a potential new direction in his work and lifestyle, the road trip exposed him to exciting aspects of the international peace movement, and to an alternative, mobile lifestyle – something that had all but disappeared in New Zealand since the transitory seasonal lifestyles that were common to Māori tāngata whenua and nineteenth-century pioneering colonists.

In March 1976, soon after he returned from overseas, Paul became the first photographer appointed to the Auckland City Art Gallery, where he worked with a congenial band of like-minded artistic types. His duties included the documentation of art conservation work and taking photographs for ACAG publicity and publications. In his spare time he pursued his own photography and maritime interests. In August 1977 he had a fortuitous meeting with John Tucker, who was in the process of founding and seeking government and private support for the artistic group that became The Original Travelling Road Show and included the musical group Mahana, comprised of Māori and Pākehā performers.

Stimulated by what he had glimpsed of such adventurous alternative 'hippie' lifestyles overseas, Paul was delighted to photograph some of the budding troupe's earliest performances – so much so, that he decided to leave his job at the art gallery in March 1978 to embrace the uncertain life of a freelance photographer. He was already committed to a long-term project documenting New Zealand's relatively neglected maritime history, and he could see the value in recording the unfolding story of the road people that he was becoming very much a part of, with his own converted bus to live in.

He successfully applied for a Queen Elizabeth II Arts Council Grant to continue his *Road People of Aotearoa* photo essay, which he presented in an exhibition at the newly founded Real Pictures Gallery in His Majesty's Arcade on Queen Street, Auckland, in December 1979.

The gallery became a natural offshoot of the pioneering colour-processing photography laboratory founded by the photographer and photo editor Ian Macdonald and the renowned gallerist, art publisher and auctioneer Peter Webb. Under Macdonald's visionary leadership Real Pictures became a photographic hub: while the gallery showcased the work of many talented young photographers, the laboratory produced exhibition-quality prints and provided jobs and facilities for photographers. It was Macdonald who encouraged Paul to use colour film for his *Road People* project.

Megan Jenkinson, who had just completed her final year as a student at the Elam School of Fine Arts at the time, recalls seeing Paul's exhibition: 'The world he portrayed in those photographs was exotic, but not distantly so, for it was on the fringe of our own world, not over some distant horizon.' She noted that Paul kept photographing the road people as an 'embedded witness' (what anthropologists call a 'participant observer') who, no matter how objective they try to be, cannot but alter the outcomes of their documentation due to their presence at each event. Documenting from the 'inside' was crucial for the series' success and 'the intimacy of these photographs is their strength'.[3]

Paul knew intuitively, I think, what the photography historian Beaumont Newhall eloquently explained about the subjective facet of documentary photography: that while it has a documentary *intention*, it draws on artistic expression (visual sophistication) to capture the attention of the audience to encourage an emotional as well as an intellectual response to the subject matter. Thus it is not surprising that Jenkinson described *Road People of Aotearoa* as an early example of a sustained colour photo essay that presents itself as both contemporary and relevant to today's art photography audience.[4]

Today it is hard to imagine how effectively the story of such a colourful journey could have been told wholly in a palette of black, grey and white. As it transpires, it is rare, worldwide, to find such an extensive personal photo essay *in colour* from the 1979–84 period – on the cusp of colour printing and reproduction becoming simpler and more commonplace due to the digital revolution. There were some outstanding and memorable photo essays published in colour before Paul made his first and favourite essay, of course: Larry Burrows' harrowing images of the Vietnam War (c.1966), and Brian Brake's travel essays from places such as Kashmir, China – and India, the source of his epic, and perhaps his most personal 1960s photo essay, *Monsoon*. Before them, Ernst Haas had come to prominence for his

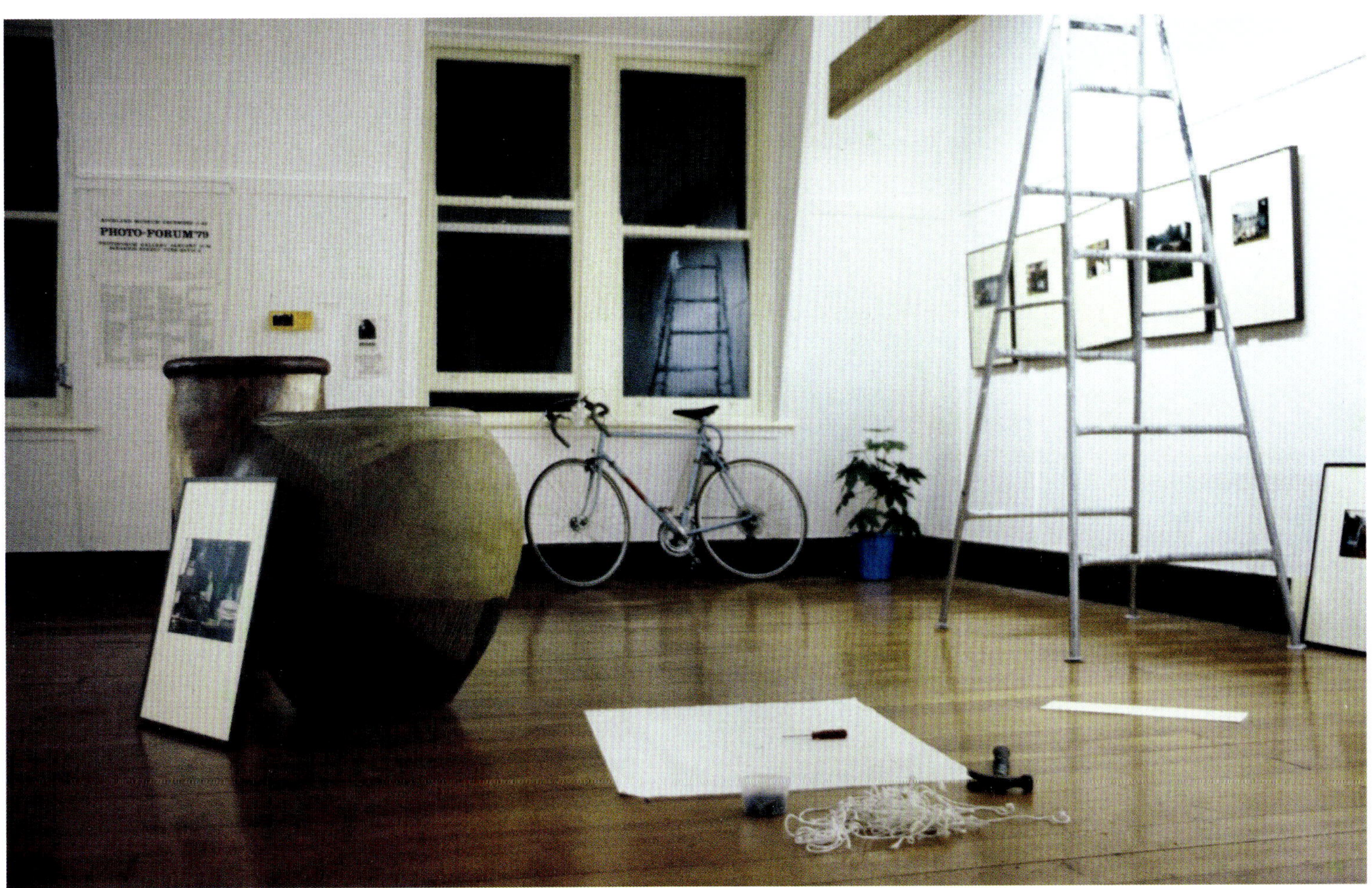

Road People of Aotearoa exhibition being set-up at Real Pictures Gallery, 1979.

pioneering exploratory colour essays of the early 1950s (New York, a bullfight, etc.). Their publication coincided with major improvements in the manufacture of colour film and in photolithography – the means by which photomechanical printing reproduced colour images in magazines and books – and advertisers poured their money into full-colour advertisements.

Needless to say, there was a personal element in their choice of subject matter and approach, as well as the challenge of mastering colour in those early days, but their essays were commissioned and presented through the filter of external editors and art directors. Overwhelmingly, though, the personal, self-assigned essays by some of the most influential photographers of the post World War II era, such as Bruce Davidson, Robert Frank, Danny Lyon, Mary Ellen Mark, Dennis Stock, and most others were all presented in black and white.

In New Zealand, Robin Morrison's magnum opus, *The South Island of New Zealand from the Road*, published by Alister Taylor in 1981, was a major exception. So, too, was the conservation-focused work of Craig Potton from the late 1970s, Fiona Clark's *Kai Moana* essay (1979/81), Bruce Foster's *Stockman Country* (1979), Glenn Jowitt's *Polynesia Here and There* (1981), Mark Adams' *Tatau* (1978–2000), and John Miller's exceptional, and largely unpublished colour documentation of the anti-apartheid protests during the Springbok rugby tour in 1981.

Unlike photographers such as George Silk, Tom Hutchins and Brian Brake, for whom the mass picture magazines such as *Life* and *Picture Post* provided the highest goalposts and the most generous rewards, Paul Gilbert's approach was firmly in the more modest New Zealand tradition. The curator and art historian Athol McCredie has recently defined this tradition as the 'New Photography' – part of an international response to the decline of mass photojournalism when the advertising industry switched from print media to television in the late 1960s. He notes that the techniques of photojournalism were 'decommercialised and repurposed away from the mass market to more personal ends. It was personal in that it wasn't made for a mass audience, like photojournalism, nor for an employer, or to win a prize in a camera club competition. But neither was it quite as personal as a family snapshot, to be shared with only an intimate circle and to record a family event. It was outward facing, recording a mostly public world, but driven from within. And if it depicted close friends or family it was in a way that was more universal than particular to those people. This was a significant departure from most photography that had gone before.'[5]

Paul Gilbert was following in the path of the 1960s practitioners who had turned away from commercial photography. Like them, he sought alternative ways to sustain the enjoyment and technical challenges he got from the medium. Hence his trajectory from scientific photographer to art gallery specialist, to freelancer and university teacher, and to freelancer again. It was a journey shaped by uncertainty, risks and hard work, with its own rewards in an art movement that was almost underground, McCredie concludes; and one that 'ultimately had more lasting impact' on the direction of New Zealand photography than the generation who gained an international reputation and some glamour through the pages of mass publications like *Life*.

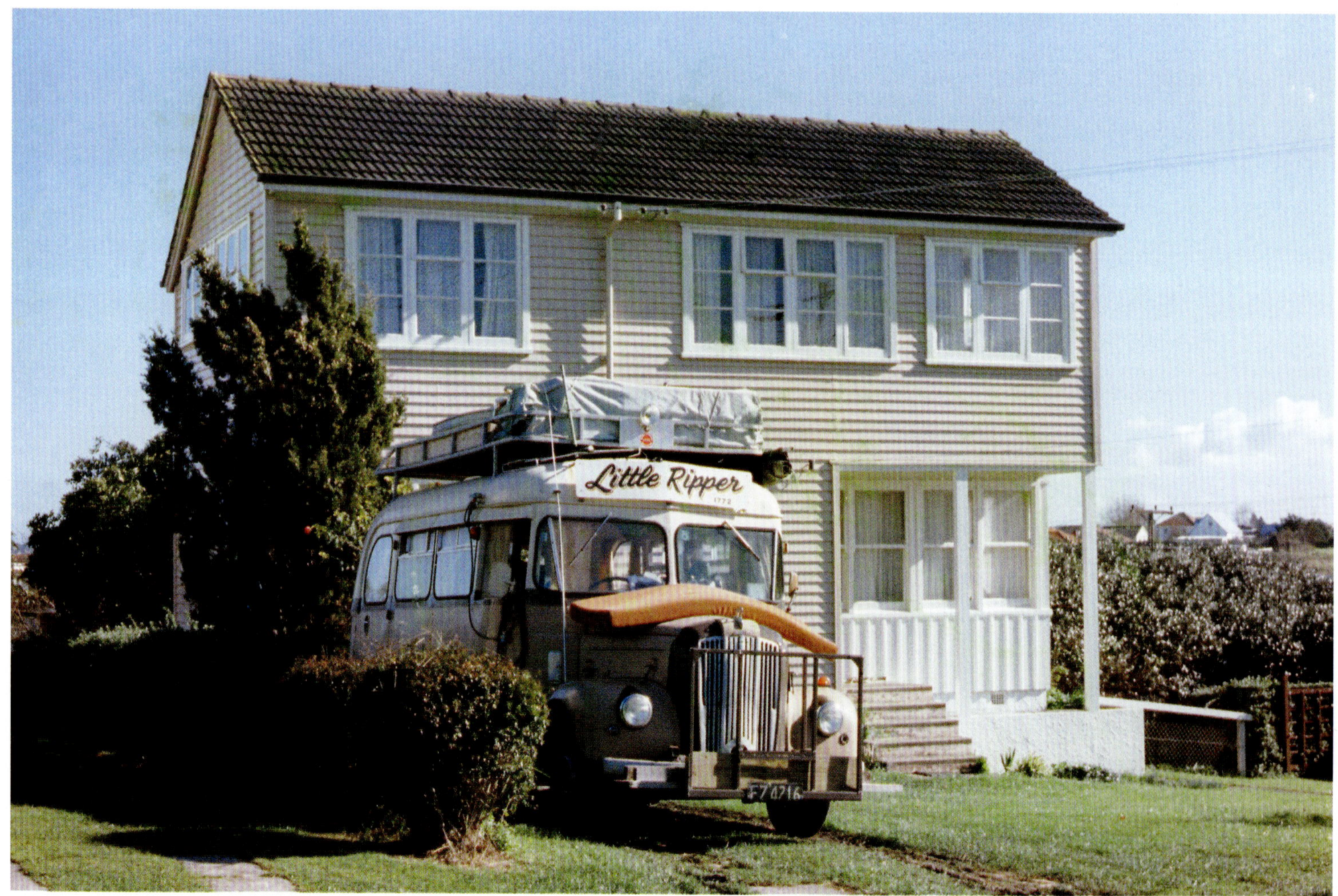

Little Ripper - unknown location, n.d.

As so many photographers imagine, and too often fail to realise, Paul always intended to make a book about his roadie experience – something for the historical record and to share. His original mock-up book, like the 1979 Real Pictures exhibition that I missed because I was visiting the US, now comes across as rather more romantic than the complete body of work. It emphasised his sensual admiration for the harmonious integration of the earnestly handcrafted mobile homes themselves and the relatively beautiful rural farmland they encamped on in places such as the Coromandel Peninsula, that still retained some of the flavour of the natural, pre-colonial landscape.

The emphasis of this book, with Paul's approval, includes more of his later images and gives a fuller picture of the work it took to create, and literally keep the show on the road. It records the daily life and growing confidence of an alternative community in the making, by artists who were dedicating themselves to the ideals of sustainable subsistence living as a peace-and-fun-loving musical and theatrical troupe. Theirs was a community for whom the *Whole Earth Catalog* was an essential bridge to the international social movement, focused on racial equality and alternative values, in opposition to the crass consumerist society promoted by the new growth of rampant capitalism.

Paul delighted in the juxtaposition of found incongruities, such as the sight of a handcrafted roadie's home parked outside a prim two-storey suburban property; or a colourful, artistic house on wheels drawing all of the attention away from the drab main street of a typical town. The road people stood out – intentionally, of course, because their spontaneous public antics and rehearsed performances were their chosen method for sharing their ideals and their ideas on the purpose of life, for the common good. It is evident from Paul's photographs that theirs was a cluster of artists and idealists. What they didn't know they studied and found out from others: the history and philosophy of gypsies; the art and craft of boat and house building. They had expert – and some perhaps fanatical – craftspeople among them. They learned basic truck maintenance, and how to share and live harmoniously in small spaces through making do, frugal living, and exploring and sharing communal values.

Today, during the life-threatening Covid-19 pandemic, we seem blessed by the digital revolution, which appears to be helping us cope and think again about adopting alternative, mobile and green living styles, in opposition to the financial burden of home ownership and working from a company or institution's office or workplace. I'm not sure that the current generation, those priced out of the housing market, have considered the possibilities of communal living, but I have observed that retirees, having downsized their fixed real estate, are opting for the caravan or motorhome for budget living and travel to new and favourite locations.

One lesson from Paul's experience, is that, despite one's best intentions, photographers need to pay more attention to editing and cataloguing their work and heritage as they create it. It makes no sense to leave these tasks to somebody else, when the link to the specific meaning and value to its author has been cut off. Life gets in the way, of course, and art tends to thrive on ambiguity, but leaving cherished images without identification is an invitation to overlook them. Our most active public collecting institutions are starting to run out of space for physical storage, and are no longer eagerly taking on this responsibility, no matter how significant the author or the content may be. It's a job that gets more difficult as time goes on but if we want our work to be valued as visual evidence and art, the first step is to ensure that we photographers take personal responsibility for properly documenting our own work to demonstrate its use value and thereby encourage its preservation for future generations.

Typical of so many prolific photographers, Paul never got around to clearly cataloguing all of his photographs as to subject matter and dates. His road people images were partially organised – clumped together as to their intended documentary purpose – but less individually than his maritime pictures, for which he created his own picture library. Like most of the independent photographers I know (including myself) who make pictures for themselves to scratch an artistic itch, rather than for commercial use, I'm sure Paul had the images for this book precisely catalogued in his imagination, before his time ran out. Nevertheless, it is perfectly clear that, alongside his maritime photography, he considered *Road People of Aotearoa* to be his most significant body of work.

John B Turner
Beijing, China, March 2021

Notes:

1 Paul Gilbert, quoted in Ilan Wittenberg, *Bare Truth – Paul Gilbert*, https://ilanwittenberg.com/paul-gilbert

2 Ibid.

3 Megan Jenkinson, pers. comm., 2020.

4 Ibid.

5 'Ten Questions with Athol McCredie', Museum of New Zealand Te Papa Tongarewa, https://www.tepapa.govt.nz/about/te-papa-press/our-authors/athol-mccredie-biography-and-interview-0

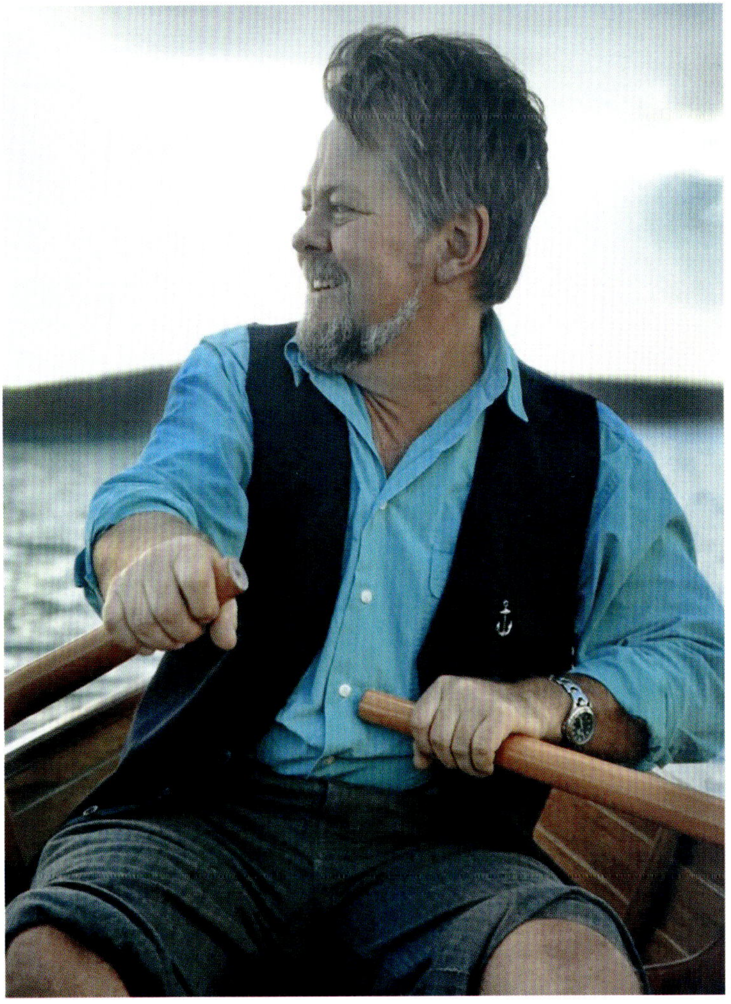

Paul C Gilbert, c.2009. Photo: Rachael Feather.

Select Bibliography

The Electric Kool-Aid Acid Test, Tom Wolfe, Farrar Straus Giroux, New York, 1968

Whole Earth Catalog, Stewart Brand, Portola Institute, California, 1968–1972

The Brutus Festival, Tim Shadbolt, Vanya McWilliams, photographs by Simon Buis, Brutus Festival, Auckland, 1969

The First New Zealand Whole Earth Catalogue Summer '72, Owen Wilkes, Jim Chapple, Dennis List, Tim Shadbolt, Alan Admore and John Pettigrew (eds), Alister Taylor, Wellington, 1972

The English Gypsy Caravan, CH Ward-Jackson & Denis E Harvey, Newton Abbot, 1972

Shelter, Lloyd Kahn (ed), Shelter Publications, California, 1973

Handmade Houses: A guide to the woodbutcher's art, Art Boericke and photographs by Barry Shapiro, A & W Visual Library, California, 1973

Mushroom: a magazine on alternative living in New Zealand, Alan Admore, Christchurch and Waitati, 1974–1985

Spiritual Midwifery, Ina May Gaskin, Book Publishing Company, Tennessee, 1975

A Hard-Won Freedom: Alternative communities in New Zealand, Tim Jones, photographs by Ian Baker, Hodder & Stoughton, Auckland, 1975

The 2nd New Zealand Whole Earth Catalogue, Dennis List and Alister Taylor (eds), Alister Taylor, Martinborough, 1975

Clowns, John H Towsen, Hawthorn Books, New York, 1976

Nambassa Festival Newsletter 1, Peter Terry, Lorraine Ward and Bernard Woods (eds), printed by Gold Field Press, Paeroa, 1976–77

The Third New Zealand Whole Earth Catalogue, Alister Taylor and Deborah McCormack (eds), Alister Taylor, Wellington, 1977

The Farm Vegetarian Cookbook, Louise Hagler, Book Publishing Company, Tennessee, 1978

'Red Mole', Brian McNeill, *Art New Zealand* 10, Winter 1978

The Nambassa Sun and *Nambassa Waves,* quarterly newspapers, 1978–81

Nambassa: A new direction, Colin Broadley and Judith Jones (eds), AH & AW Reed, 1979

Rolling Homes: Handmade Houses on Wheels, Jane Lidz, A & W Visual Library, 1979

'Road People of Aotearoa: Paul Gilbert's photographic series', Andrew Martin, *Art New Zealand* 15, Autumn 1980

Road Lore: The New Zealand mobile homeowners handbook, Terry St George, The Original Travelling Road Show Society., Auckland, 1980

Home Free: Housetrucking in New Zealand, Fiona Cunningham, photographys by Chris Hoult, Random House, 1994

Angels Can Fly: A modern clown user guide, Alan Clay, Artmedia Publishing, 2005

Monday Night Class, Stephen Gaskin, Book Publishing Company (originally published 1969), 2005

Tim Shadbolt: a Mayor of Two Cities, Tim Shadbolt, Hachette New Zealand, Auckland, 2008

Dirty Bloody Hippies, Dan Salmon (director), William Grieve (producer), Big Pictures, 2009
www.nzonscreen.com/title/dirty-bloody-hippies-2009/overview

The Elephant Man: A pictorial autobiography of the Whirling Bros Circus, Tony Ratcliffe, Op.cit Ltd, 2010

'Endless connections: New Zealand secular intentional rural communities founded in the 1970s', Robert Jenkin, MA thesis, Massey University, 2012

'The Nambassa Festivals and the counterculture movement', Andrew Schmidt, *AudioCulture Iwi Waiata,* 25 March 2014,
www.audioculture.co.nz/scenes/the-nambassa-festivals-and-the-counterculture-movement

'Mahia Blackmore Profile', John Dix, *AudioCulture Iwi Waiata,* 23 September 2014, www.audioculture.co.nz/people/mahia-blackmore

'The Corben Simpson story, part 1', Keith Newman, *AudioCulture Iwi Waiata,* 18 April 2016,
www.audioculture.co.nz/people/corben-simpson/stories/the-corben-simpson-story-part-1

Living in Utopia: New Zealand's intentional communities, Lucy Sargisson and Lyman Tower Sargent, Routledge, 2016 (orig. published by Ashgate, 2004)

"Inside Mahana: How a New Zealand Commune Turned to Rot, In a remote community in the forests of the Coromandel, a utopian dream has turned sour', Tess McClure, Vice, 2017, www.vice.com/en/article/7xzn34/inside-mahana-when-a-commune-turns-to-rot

First published 2021

Rim Books

www.rimbooks.com
info@rimbooks.com

ISBN 978-0-9951184-6-1

A catalogue record for this book is available from the National Library of New Zealand

Printed in China by Everbest Printing Investment Limited

Publisher's Acknowledgements

Paul's family for their support towards this book project:
Eamonn Honeyman-Gilbert, Freya Parmenter, Linda Gilbert, Bruce Gilbert, Rachael Feather

The road people who shared their stories:
Johnny (John) Tucker, Sioux (Sue) Honeyman, Michael Colonna, Dom Ferry, Robyn Harris-Iles, Jonathon Acorn, Marie Ward, Helen Driver, Ewan Morris, Brian Tracey, Dale Whitcombe, David Rees, John M Cameron, Terry St George, Wakatau Puki Kimura, Wilma Pucci

The book production team:
John B Turner, Megan Jenkinson
Text editor, Gillian Tewsley
Proofreader, Jan Young
Readers, Ron Brownson, Gregory O'Brien
Support letter, David Benge

Rim Books acknowledges the generous support of Creative New Zealand for the production of this book, and the Queen Elizabeth II Arts Council for facilitating Paul Gilbert's project in 1979

Captions for Paul's photographs were posthumously created by the publisher based on the recollections of the 40-year-old events supplied by the road people. If there are inaccuracies in names, dates and places as well as attribution of the makers and owners of the trucks, please notify the publisher so we can correct the records

Cover: Road Show camp, 309 Road, Coromandel, after Nambassa, 1979
Endpapers: House Truck & Mobile Shelter Convention, Aratiatia, near Taupō, 1984
Half-title: Dale Whitcomb and Trevor & Rangi McGlinchey's trucks, 309 Road, Coromandel, 1979
Frontispiece: Portrait of Paul Gilbert at Taupaki orchard, from the Road People of Aotearoa archive, c.1983

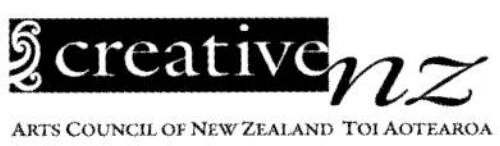

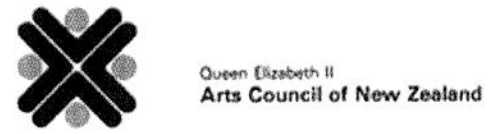

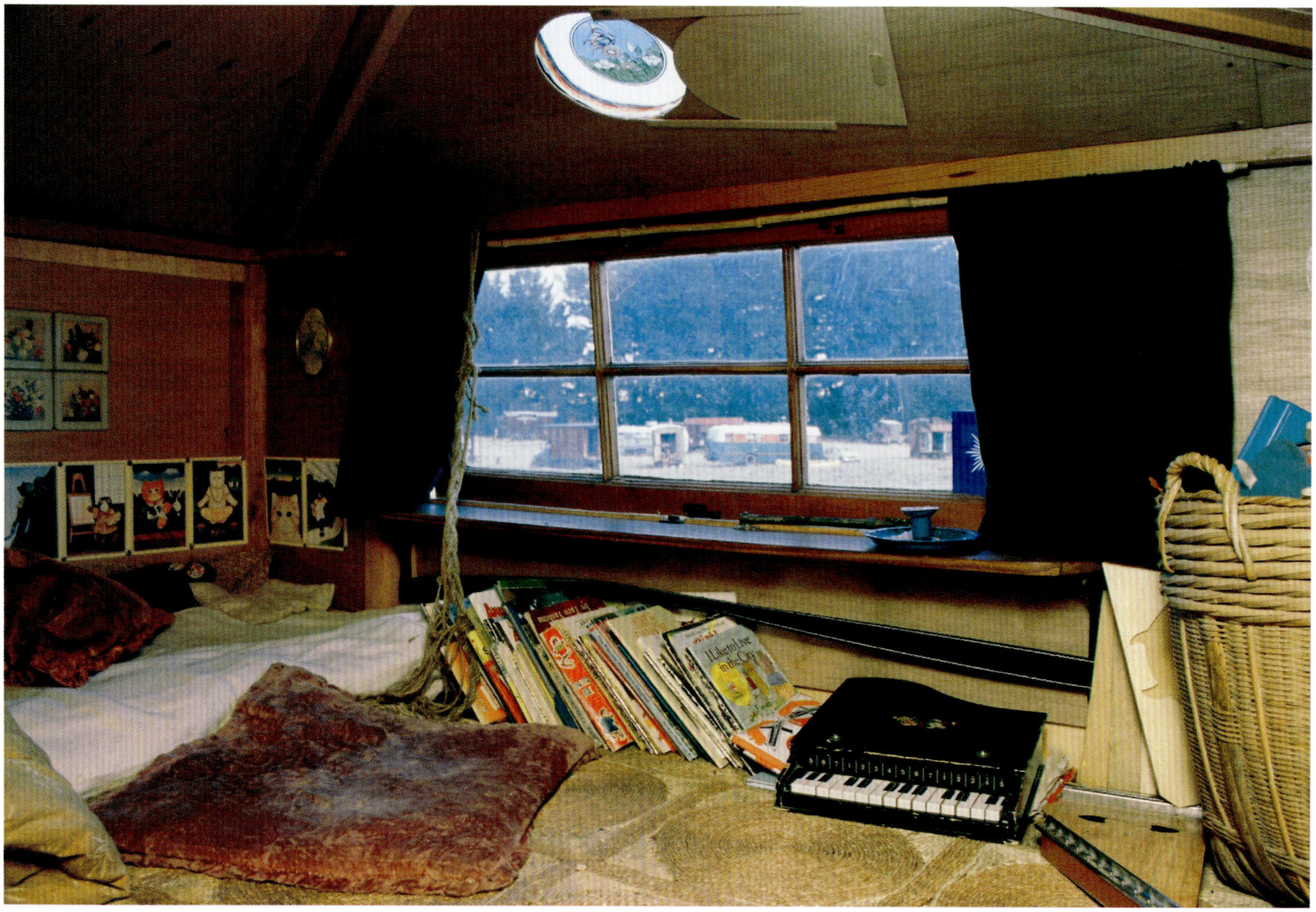

Aya's bedroom in Ewan and Helen's truck, Aratiatia, 1984

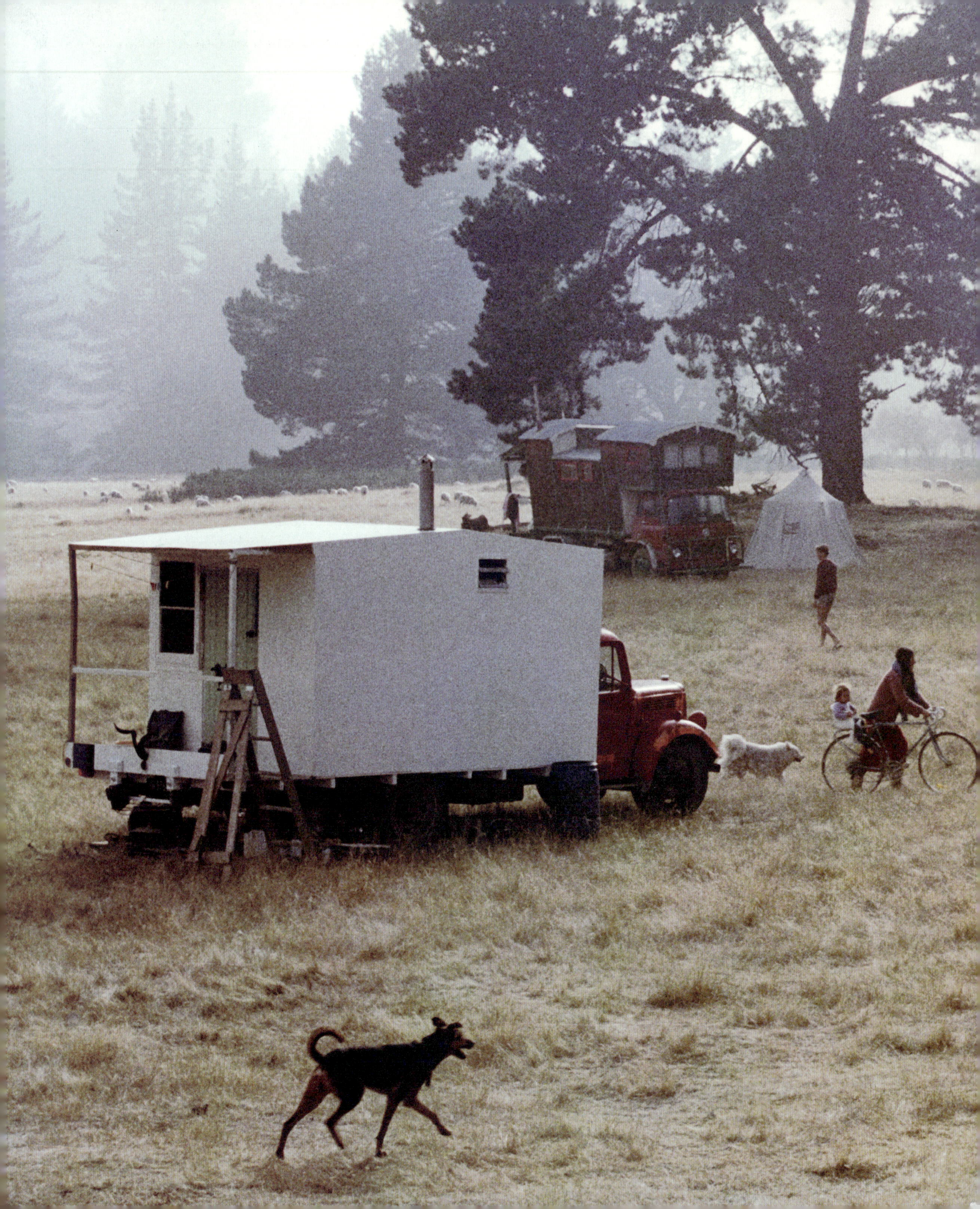